Flavour and Consumer Perception of Food Proteins

Food Chemistry, Function and Analysis

Series editors:
Gary Williamson, *Monash University, Australia*
Alejandro G. Marangoni, *University of Guelph, Canada*
Graham A. Bonwick, *AgriFoodX Limited, UK*
Catherine S. Birch, *AgriFoodX Limited, UK*

Titles in the Series:
1: Food Biosensors
2: Sensing Techniques for Food Safety and Quality Control
3: Edible Oil Structuring: Concepts, Methods and Applications
4: Food Irradiation Technologies: Concepts, Applications and Outcomes
5: Non-extractable Polyphenols and Carotenoids: Importance in Human Nutrition and Health
6: Cereal Grain-based Functional Foods: Carbohydrate and Phytochemical Components
7: Steviol Glycosides: Cultivation, Processing, Analysis and Applications in Food
8: Legumes: Nutritional Quality, Processing and Potential Health Benefits
9: Tomato Chemistry, Industrial Processing and Product Development
10: Food Contact Materials Analysis: Mass Spectrometry Techniques
11: Vitamin E: Chemistry and Nutritional Benefits
12: Anthocyanins from Natural Sources: Exploiting Targeted Delivery for Improved Health
13: Carotenoid Esters in Foods: Physical, Chemical and Biological Properties
14: Eggs as Functional Foods and Nutraceuticals for Human Health
15: Rapid Antibody-based Technologies in Food Analysis
16: DNA Techniques to Verify Food Authenticity: Applications in Food Fraud
17: Advanced Gas Chromatography in Food Analysis
18: Handbook of Food Structure Development
19: Mitigating Contamination from Food Processing
20: Biogenic Amines in Food: Analysis, Occurrence and Toxicity
21: Nutrition and Cancer Prevention: From Molecular Mechanisms to Dietary Recommendations
22: Health Claims and Food Labelling
23: Nutraceuticals and Human Health: The Food-to-supplement Paradigm
24: Nutritional Signalling Pathway Activities in Obesity and Diabetes
25: The Chemistry and Bioactive Components of Turmeric
26: Foodomics
27: Food Proteins and Peptides: Emerging Biofunctions, Food and Biomaterial Applications
28: Handbook of Antioxidant Methodology: Approaches to Activity Determination
29: Fats and Associated Compounds: Consumption and Human Health

30: Oral Processing and Consumer Perception: Biophysics, Food Microstructures and Health
31: Development of Trans-free Lipid Systems and their Use in Food Products
32: Advanced Spectroscopic Techniques for Food Quality
33: Berries and Berry Bioactive Compounds in Promoting Health
34: Metabolism of Nutrients by Gut Microbiota
35: Bio- and Nano-sensing Technologies for Food Processing and Packaging
36: Edible Fungi: Chemical Composition, Nutrition and Health Effects
37: Chemistry and Nutritional Effects of Capsicum
38: Pigmented Cereals and Millets: Bioactive Profile and Food Applications
39: Dietary Supplements with Antioxidant Activity: Understanding Mechanisms and Potential Health Benefits
40: Handbook of Cheese Chemistry
41: Flavour and Consumer Perception of Food Proteins

How to obtain future titles on publication:
A standing order plan is available for this series. A standing order will bring delivery of each new volume immediately on publication.

For further information please contact:
Book Sales Department, Royal Society of Chemistry, Thomas Graham House, Science Park, Milton Road, Cambridge, CB4 0WF, UK
Telephone: +44 (0)1223 420066, Fax: +44 (0)1223 420247
Email: booksales@rsc.org

Visit our website at books.rsc.org

Flavour and Consumer Perception of Food Proteins

Edited by

Jing Zhao
San Diego State University, USA
Email: jzhao2@sdsu.edu

and

Changqi Liu
San Diego State University, USA
Email: changqi.liu@sdsu.edu

ROYAL SOCIETY
OF **CHEMISTRY**

Food Chemistry, Function and Analysis No. 41

Print ISBN: 978-1-78801-758-9
PDF ISBN: 978-1-83916-504-7
EPUB ISBN: 978-1-83916-505-4
Print ISSN: 2398-0656
Electronic ISSN: 2398-0664

A catalogue record for this book is available from the British Library

The Royal Society of Chemistry is a charity, registered in England and Wales, Number 207890, and a company incorporated in England by Royal Charter (Registered No. RC000524), registered office: Burlington House, Piccadilly, London W1J 0BA, UK, Telephone: +44 (0) 20 7437 8656.

Visit our website at books.rsc.org

Preface

The United Nations predicted that the global population would reach 9.7 billion by 2050 and 10.4 billion by 2100. With this growth, there will be a heightened demand for high-quality proteins, as they are an essential nutrient and important food ingredient. Moreover, the rise in consumer awareness for health and fitness, increasing disposable income, growing demand from the millennial population for supplements, and the rise in popularity of alternative proteins as substitutes for meat and animal products are driving the growth of the protein ingredient market.

Many studies have investigated the functional properties of protein ingredients, such as their solubility, foaming, emulsification, gelation, water holding, and oil absorption properties. However, there is limited research on their flavour characteristics and protein–flavour interactions in different processing conditions and formulations. Certain proteins have distinct odours, hindering their widespread applications. For example, soy and pea proteins are often described as having a green, beany off-flavour by consumers. Profiling flavour compounds and removing or masking the off flavours are crucial for product development and quality control. This book provides consolidated information on the flavour of food protein ingredients and highlights areas for further research, serving as a resource for the food industry and researchers studying food proteins and flavours.

In this book, we use terms such as traditional proteins (from animal sources like meat, seafood, dairy, and eggs), plant/vegetable proteins (from plants like cereals, pseudocereals, and legumes), and novel proteins (from bacteria, fungi, algae, edible insects, and cultured meat) to describe food proteins from various sources. Leaf and oilseed proteins are grouped under novel proteins as they are not widely used. Chapter 1 provides an overview of the global protein market, sensory and flavour properties, problems associated with food protein ingredients, and the use of proteins and peptides

Food Chemistry, Function and Analysis No. 41
Flavour and Consumer Perception of Food Proteins
Edited by Jing Zhao and Changqi Liu

Published by the Royal Society of Chemistry, www.rsc.org

as precursors to generate flavours through the Maillard reaction. Chapter 2 discusses the current knowledge on consumer acceptance of food protein ingredients, *e.g.*, protein concentrates, isolates, and hydrolysates, and protein-based and enhanced food products. Consumer acceptance from both survey studies and consumer sensory studies is included. Chapters 3 and 4 cover protein–flavour interactions and the influencing factors. Specifically, Chapter 3 introduces the development of analytical methods for measuring both reversible and irreversible binding between proteins and flavour compounds. Chapter 4 discusses the various factors that could affect the formation of reversible or irreversible bonds between flavourings and various proteins, as well as their impact on flavour perception. Chapters 5–10 summarize the volatile and sensory characterizations of specific food proteins and protein-based products, including dairy (Chapter 5), fish (Chapter 6), soy (Chapter 7), pulse (Chapter 8), cereal and pseudocereal (Chapter 9), and novel (Chapter 10) proteins. The topics covered in these chapters include volatile profiles, volatile formation, factors influencing volatile formation during processing and storage, and off-flavour removal.

Jing Zhao and Changqi Liu

Contents

Food Chemistry, Function and Analysis No. 41
Flavour and Consumer Perception of Food Proteins
Edited by Jing Zhao and Changqi Liu

Published by the Royal Society of Chemistry, www.rsc.org

CHAPTER 1

Overview of Protein Flavours

JING ZHAO*, CHANGQI LIU AND ALI RAZA

5500 Campanile Drive, San Diego State University, San Diego, CA 92182-7251, USA
*E-mail: jzhao2@sdsu.edu

1.1 Introduction

The United Nations has predicted that the global population will reach 9.7 billion in 2050 and 10.4 billion in 2100.[1] Protein is a critical nutrient and an important food ingredient. Due to population growth, there is an increasing demand for high-quality proteins. Moreover, the increase in consumers' awareness of health and fitness, alarming rise in obesity rates, increase in disposable income, and increased demand from the millennial population for supplements are driving the growth of the protein ingredient market.

With the increasing demand and popularity of replacing meat and seafood with alternative proteins, high-quality protein ingredients with good manufacturing functionalities and flavours are highly sought after. A good number of studies have been done to investigate the functionalities of protein ingredients (*e.g.*, solubility, foaming, emulsifying, gelation, water holding, and oil absorption). However, the flavour and protein–flavour interaction of protein ingredients under various processing conditions and in different formulations have not been well investigated. It is known that the characteristic aroma of plant proteins and the allergenicity of some plant proteins restrict their wide applications. For example, the green and beany odours from soy and pea proteins are described as off-flavour by consumers. Profiling the flavour compounds and removing or masking

Food Chemistry, Function and Analysis No. 41
Flavour and Consumer Perception of Food Proteins
Edited by Jing Zhao and Changqi Liu

Published by the Royal Society of Chemistry, www.rsc.org

the odorants responsible for the off-flavours would be critical in product development and product quality control. This book aims to consolidate much-needed information on the flavour of food protein ingredients.

1.2 Protein Market Overview

The global protein market has grown strongly in the past decade and is projected to reach over US$114 billion by 2030 (Table 1.1).[2] Among the protein types, the demand for plant proteins has been steadily increasing due to religious reasons, rising health awareness among consumers, and environmental concerns. The global plant protein market value was estimated to be US$12.2 billion in 2022.[3,4] It has been forecast to achieve US$17.4 billion by 2027 by Markets and Markets,[4] US$17.8 billion by 2027 by Statista,[3] and US$22.7 billion by 2031 by Transparency Market Research.[5] Especially, plant-based protein substitutes are a fast-growing section.[6] Plant-based meat had a market value of US$6.67 billion in 2020 and is expected to reach US$16.7 billion in 2026.[6] Globally, the number

Table 1.1 Summary of the global protein market.

Product	Current market	Projected market	CAGR	Ref.
Total protein	US$72 billion in 2021	US$114 billion by 2030		2
Plant protein	US$12.2 billion in 2022	US$17.4 billion by 2027	7.3%	3,4
Plant protein	US$13.1 billion in 2021	US$22.7 billion by 2031	5.7%	5
Plant-based meat	US$6.67 billion in 2020	US$16.7 billion in 2026		6
Protein supplements	US$6.26 billion in 2021		8%	7
Whey protein	US$9.7 billion in 2021	US$16.4 billion in 2028		8
Soy protein	US$9.19 billion in 2021	US$15.65 billion in 2027		9
Pea protein	US$213.1 million in 2020		12.7%	10
Rice protein		US$218.48 million by 2028		11
Mung bean protein	US$226.6 million in 2022		3%	12
Fava bean protein	US$56.7 million in 2022		4.2%	13
Edible insects		US$9.60 billion by 2030	28.3%	14
Algae protein	US$700 million in 2021	US$1.05 billion by 2028	6%	15

CAGR: compound annual growth rate.

of companies producing plant-based dairy, meat, and egg alternatives has increased from less than 40 in 2010 to more than 160 by 2022.[16] In the United States, sales of plant-based alternatives amounted to US$5.8 billion, with about 43% (US$2.5 billion) coming from milk alternatives, 24% (US$1.4 billion) from meat alternatives, and 0.46% (US$27 million) from egg alternatives.[16]

Protein supplements are a major form of proteinaceous products. The global protein supplements market was valued at US$6.26 billion in 2021 and is expected to grow at a compound annual growth rate (CAGR) of 8.0% from 2021 to 2028.[7] In the United States, the market for protein supplements was US$1.9 billion in 2019. It is expected to reach US$3.6 billion by 2028, at a CAGR of 7.7%.[17] Protein supplements are available in the form of powder, ready-to-drink protein shakes, and protein bars. The most common protein supplements on the market are whey, soy, casein, and egg protein powders. Besides being used as protein supplements, proteins have been used as an ingredient in various products to enhance the protein content and as fat or egg replacers.[18–23]

1.2.1 Whey Protein Market

The dairy protein market comprises milk protein, milk powder, whey protein, and caseinate. Whey protein, a by-product of cheese production, is the main dairy protein product available on the market. It has been used as a protein supplement in the form of protein isolate and concentrate. Whey protein also pioneered the market for protein bars, protein shakes, *etc.* According to Vantage Market Research, the whey protein market was estimated to be worth approximately US$9.7 billion in 2021 and is expected to reach $16.4 billion by 2028. The continued growth is driven by the demand for whey protein in nutritional supplements, personal care, and the food and beverage industry.[8]

1.2.2 Plant Protein Market

Among plant proteins, soy protein has the largest market share at 64.4% and has gained a steady increase in the past decade.[24] The global soy protein market reached nearly US$9.19 billion in 2021 and is estimated to reach US$15.65 billion by 2027.[9] The main demand for soy protein has been observed for bakery and functional foods. Other uses include pharmaceuticals, meat additives, and feed, among others. Other major plant proteins include wheat and pea proteins. The pea protein market size was estimated at US$213.1 million in 2020 and is projected to grow at a CAGR of 12.7% between 2021 and 2028.[10] The major applications of pea protein are in dietary supplements, bakery products, meat substitutes, and beverages. Besides, protein powders of rice, almond, fava bean, broad bean, mung bean, hemp, *etc.* have been available in the market. Rice protein is

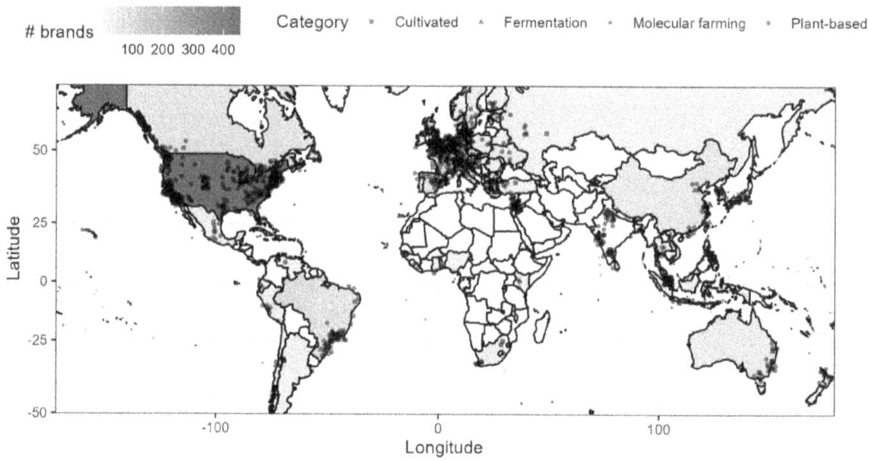

Figure 1.1 Companies producing alternative protein end-products. The figure is created based on data available on the GFI database.[25]

predicted to reach a global market size of US$218.48 million by 2028.[11] The drive for using rice protein is the rising need for dietary foods that are gluten-free, non-genetically modified organisms (GMOs), and hypoallergenic products. The mung bean protein market was estimated to be US$226.6 million in 2022.[12] Fava bean protein had a global market size of US$56.7 million in 2022.[13] The Good Food Institute has curated an extensive list that features more than 1400 companies specializing in alternative protein production.[25] These companies cover a range of products, including plant-based, cultivated, fermented, and molecular-farmed options. As shown in Figure 1.1, the majority of these innovative companies focus on plant-based protein alternatives.

1.2.3 Novel Protein Market

Novel proteins, such as edible insect proteins and algal proteins, are on the rise due to their sustainable production and environmental benefits. According to Research and Markets, the edible insect market is expected to reach US$9.60 billion by 2030, at a CAGR of 28.3% during the forecast period of 2022–2030.[14] The volume of the market is predicted to reach 3.14 million tons by 2030. The algae protein market size exceeded US$700 million in 2021 and is projected to grow at over 6% CAGR between 2022 and 2028.[15] The market growth is driven by the increasing product adoption in dietary supplements, food, and beverages.

1.3 Sensory Impacts and Flavour of Food Proteins

Flavour is a synthesis of three chemical senses: taste, smell, and chemesthesis. It is one of the most important sensory properties of foods, along with colour and texture. Protein ingredients play an important role in food texture, which is related to their gelling, foaming, emulsifying, water-holding, oil-absorption, and dough-formation properties. Protein ingredients could affect the colour of food products through their natural colours. For example, whey and most plant proteins are white to slightly yellow in colour while, hemp protein is dark brown. Also, proteins could undergo a Maillard reaction during the processing and/or storage of the products to form dark pigment melanoidins.[26]

Despite protein's effect on other sensory properties, the undesirable flavour of protein ingredients has been a significant concern for product quality and a limiting factor for the use of protein ingredients in foods. For example, the 'gluey' flavour of stored casein was studied in the 1960s.[27] The removal of beany flavour from soy and pulse proteins has been studied for decades and is still an active research topic today.[28–30] The following discussion will focus on the flavour, especially the taste and odour, of protein ingredients.

1.3.1 Taste of Amino Acids and Protein Hydrolysates

Sweet, salty, sour, bitter, and umami are the five basic tastes. In general, the tastes of food proteins are contributed by peptides (bitter, sweet, and umami) and amino acids (bitter, umami, sweet, salty, and sour) derived from protein hydrolysis.[31] Among the amino acids, L-glutamate and L-aspartate are known for having an umami sensation. In general, L-glycine, L-alanine, L-glutamine, L-proline, L-serine, and L-threonine have a sweet taste, while L-histidine has a bitter taste.[32]

Bitterness is a noticeable problem in protein hydrolysates or peptides. Enzymatic hydrolysis is a popular method of generating bioactive peptides and protein hydrolysates with improved functionalities.[33–35] Besides, hydrolysis can effectively lower the allergenicity of many food allergens, making it a very useful technique to produce hypoallergenic products and infant foods.[36–38] However, bitterness in the peptides is a hurdle to consumer acceptance. The taste of hydrolysates varies depending on the protein source, hydrolysis conditions, and proteolytic enzyme used.[39] It has been reported that several factors could determine the bitterness of peptides (*e.g.*, peptide chain length, hydrophobicity, and Q-parameter). The intermediate peptides with no more than 8 amino acid residuals have shown the strongest bitterness. Neither amino acids nor long-chain peptides are the major causes of bitterness. More importantly, the hydrophobicity of the peptides has a strong correlation with their bitterness. Another factor is the Q-parameter. It is defined as the average amount of free energy needed to transfer amino acid chains from ethanol to water and expressed in cal mol^{-1}. The Q is correlated to the hydrophobicity of the composing amino acids.[39] In an effort to reduce the bitterness in protein hydrolysates, 24 bitter taste inhibitors were documented for whey protein hydrolysate.[40] Many of the bitter taste inhibitors are sugars, salts, or nucleotides.

Besides the bitter taste from peptides and amino acids, some phytochemicals (*e.g.*, saponin, phenolic compounds, and alkaloids) and bitter lipids remaining in protein products could also be the source of the bitter taste (*e.g.*, many studies have reported saponin as a source of bitterness in pea protein).[41–43] Also, bitter-tasting lipids were identified from commercial pea protein isolates in a recent study using ultra-high-performance liquid chromatography–differential ion mobility spectrometry (DMS)–tandem mass spectrometry (MS).[44]

1.3.2 Taste of Proteins

Isolated protein ingredients from meat, dairy, eggs, and plant sources are generally considered bland in taste. A salty and sour taste could be noticeable in proteins due to the charged amino acid side chains.[45] But in general, salty and sour tastes are not significant in proteins or peptides. Some proteins derived from tropical plants have an intensely sweet taste (*e.g.*, thaumatin,[46] monellin,[47] and brazzein[48]). Besides, miraculin is a taste-modifying protein that makes acids taste sweet when the taste buds are exposed to it.[49] However, miraculin itself is sensory inactive. It activates the sweet taste receptor only at acidic pHs when a sour solution or compound co-exists.

1.3.3 Odours of Proteins

1.3.3.1 Odour Characteristics of Food Proteins

In protein isolates, mild yet significant odours exist due to the oxidation of residual lipids and amino acids, degradation of amino acids, and Maillard reaction of proteins with reducing sugars. Compared with tastes, odour sensations are much more complicated and difficult to describe. The odour of whey protein was described as sweet (desirable), cardboard (undesirable), brothy, cucumber, *etc.*[50–55] Stored casein was found to have a gluey off-odour.[56] Soy and pulse proteins have a strong beany odour.[28,30,57–59] Soy odours were described as beany, corn meal, musty, and toasted.[60] Most vegetable proteins also have a green or grassy off-odour.[61] Su *et al.* reviewed the odour lexicons used for dairy ingredients and the volatiles associated with the odours.[62] The sensory lexicon of whey and soy proteins was developed by Russell *et al.*, and that of pulse proteins was reviewed by Chigwedere *et al.*[63]

The describers people use to describe odours typically correspond to a mixture of compounds. For example, the cardboard flavour that was reported in whey protein is a combination of pentanal, heptanal, nonanal, 1-octen-3-one, and dimethyl trisulphide.[64] Even though compounds of similar chemical structures tend to have similar odours, chemicals from the same group could give off different odours.[62] In general, the major groups of volatile compounds found in protein isolates include aldehydes (*e.g.*, hexanal that has a green and grassy aroma), ketones (*e.g.*, 2-nonanone with a fresh, green, weedy, and earthy smell), alcohols (*e.g.*, hexanol with a herbaceous, woody, and green aroma), and sulphur-containing compounds (*e.g.*, dimethyl disulphide with a garlic-like odour).[65–67] A comprehensive description of the odour characteristics of various flavour ingredients can be found in *Fenaroli's Handbook of Flavor Ingredients*.[68] In this section, we only aim to provide a description of the odour characteristics of various food proteins. The volatile compounds associated with odours and off-odours will be discussed in later chapters.

1.3.3.2 Influencing Factors of Odour Formation

The odour of protein ingredients is complicated by the numerous factors that could influence odour formation in protein ingredients. The major factors include the protein source (inherent odours), preparation (extraction and drying) methods, and storage conditions. Protein isolates derived from different foods (*e.g.*, whey, egg, soy, pea, and mung bean) have a volatile profile that is unique to the food of origin. Soy and pea proteins have a noticeable beany flavour, while rice proteins have a high content of sulphur-containing volatiles.[57,65] Besides, proteins derived from the same food could have varied flavour properties due to the different starting materials. Smith *et al.* examined the effects of the source of whey (whey derived from cheddar, mozzarella, cottage cheese, and rennet casein) on the volatile profiles of whey protein isolates.[69] Their study demonstrated the changes in volatile compounds and their abundance due to the different sources of whey.

Preparation methods play a significant role in flavour formation. They often determine the physical (*e.g.*, particle size and thermal treatment history) and chemical (*e.g.*, residual lipid content and amount and activity of lipoxygenase) properties of the end protein product, therefore influencing its flavour profile. For example, several studies examined the drying methods and conditions on the flavour of derived protein products.[70–74] In whey protein concentrates, a low feed concentration of 10% led to increased overall aroma, intensified cabbage and cardboard off-odours, and higher concentrations of certain aldehydes (pentanal, hexanal, heptanal, decanal, (*E*)2-decenal, and 2,4-decadienal) and sulphur-containing volatiles (dimethyl trisulphide and dimethyl disulphide) as compared to products spray dried at 25% feed solids. The drying temperature is another key parameter. Product spray dried at a lower inlet temperature showed increased cardboard flavour compared to product spray dried at a higher inlet temperature.[70] Different processing techniques are also utilized to inhibit the formation of off-odours in protein isolates. These methods can be grouped as physical, chemical, and biological techniques. Physical methods include grinding, heating, and exposure to microwaves, magnetic fields, electric fields, irradiation, *etc.* Novel techniques in this area are constantly emerging, which provide new tools for protein and flavour modifications. The most commonly used chemical modifications utilize pH adjustment, organic solvents, reductants, supercritical carbon dioxide extraction,[75] *etc.* Biological methods include a wide range of approaches such as fermentation, germination, enzymatic treatment, and genetic engineering. The applications of some of these techniques to improve the flavour of pulse and soy protein isolates have been reviewed.[30,76,77]

The storage conditions, *e.g.*, temperature and water activity, also affect the flavour formation and stability of protein ingredients. Storage under elevated temperatures and high water activity conditions has been reported to increase the concentrations of volatile compounds in whey[78] and pea proteins.[79] For example, Mehle *et al.* reported that pea protein isolates stored at 37 °C and 0.501 water activity conditions had a greater degree of aroma change compared to samples stored at refrigerated temperatures and lower water activity conditions. The aroma changes were related to the generation of 1-octen-3-ol and 2-4-decadienal, which were products of lipid oxidation.[79]

As protein ingredients are widely used in various food products, especially plant-based meat alternatives, a better understanding of the flavour of the protein ingredients and their impact on the final products is much needed. The flavour of various food proteins and how they are affected by processing and storage conditions are covered in detail in Chapters 5 to 10.

1.3.3.3 Factors Influencing Odour Detection

Besides the inherent and environmental factors that influence flavour formation in food proteins, analytical and sample preparation methods could also affect the odours detected. The commonly used tools to study food aroma include sensory evaluation, gas chromatography (GC)-MS, GC-olfactometry (O), GC-O-MS, GC-flame-ionization detection (FID), electronic nose, *etc.* In GC-O-MS, the volatile compounds coming out of the GC column are simultaneously detected by MS and human participants at the olfactory detection port (ODP). This technique enables the identification of aroma-active compounds and the association of aroma characteristics with volatile compounds. The applications of GC-O-MS in flavour analysis were reviewed by Song and Liu.[80] In recent years, GC-ion mobility spectrometry (IMS) has been used in food flavour analysis to detect food adulteration, spoilage, and off-flavours.[81]

The most direct way to get samples for volatile analysis is equilibrium/static headspace (HS).[65,82,83] This method enables the detection of an unaltered volatile profile from a sample headspace. However, its sensitivity is very low.[84] Methods to concentrate HS volatiles include dynamic

HS (DHS, also known as purge and trap),[85,86] solid-phase microextraction (SPME),[87–89] in-tube extraction (ITEX), sorptive extraction,[90] *etc.* The analytical methods for detecting volatile compounds and flavour–protein interactions are discussed in detail in Chapter 2. In general, the sampling method and conditions, *e.g.*, extraction time, extraction temperature, extraction medium (pH and ionic strength), fibre, agitation, preheating time, adsorption time, adsorption temperature, desorption time, desorption temperature, injection depth, water content, and sample weight in the SPME method, could influence the analytical results. Singh *et al.* reported that increasing pH from 3 to 11 decreased recovery of volatiles while adding 1% NaCl increased the recovery.[61] Therefore, it is critical to understand the differences in the sampling techniques to properly choose the method for the targeted samples and interpret the results.

1.4 Protein and Peptide as Flavour Precursors

1.4.1 Protein and Peptide as Precursors of Meat Flavours

Raw meat is often described as having an unpalatable, smelly, and salty flavour.[91] However, when meat is cooked, it develops a pleasant aroma due to various thermal reactions. Non-volatile compounds, such as free amino acids and peptides, play a vital role in creating the desirable volatile compounds responsible for the meaty aroma.[92,93] Understanding the mechanism of meat flavour formation can aid in the development of meat-like aromas using non-animal proteins. Proteins, peptides, amino acids, lipids, sugars, and thiamine are all important components involved in the formation of meat flavours. Myofibrillar and sarcoplasmic proteins can bind to these flavour precursors and influence the release of flavour compounds.[94] Postmortem ageing has a positive impact on meat flavour precursors by increasing the breakdown of lipids and the release of small peptides.[95]

When meat is heated, these precursors undergo a complex array of chemical reactions. These reactions include lipid oxidation, sugar degradation, thiamine degradation, thermal degradation of amino acids and peptides, the Maillard reaction, and interactions between lipid oxidation and the Maillard reaction.[93] The proteinaceous precursors are specifically involved in the latter three reactions. When amino acids and peptides are heated above 125 °C, they can undergo decarboxylation and deamination reactions, resulting in the degradation of these compounds into amines, aldehydes, hydrogen sulphide, and aromatic compounds. Cysteine and cystine have been identified as key contributors to the formation of meat-like volatile compounds, particularly thiazoles.[96,97] Pyrazines, which significantly contribute to the meat flavour, can be produced from serine and threonine degradation.[98,99] During the Maillard reaction, the free amino groups of amino acids, amines, peptides, and proteins react with reducing sugars, leading to the formation of various Maillard reaction products. These products include alcohols, aldehydes, furans, ketones, pyrazines, pyridines, pyrroles, sulphides, and thiophenes, which collectively contribute to the meat flavour.[100] The interaction between lipid oxidation and Maillard reaction compounds can give rise to additional volatile compounds, such as oxazoles, pyrazines, pyridines, thiazoles, and thiophenes with alkyl side chains. These compounds contribute to the more pronounced aroma of cooked meat.[101]

1.4.2 Preparation of Maillard Reaction Products

Peptides obtained through enzymatic protein hydrolysis can be utilized to generate Maillard reaction products with unique flavour profiles. Peptides possess a wide range of taste characteristics, such as sweet, bitter, umami, sour, and salty tastes.[45] Furthermore, peptides can contribute to a taste-enhancing phenomenon known as kokumi, which is often described as imparting thickness, mouthfulness, and richness to the taste experience.[102,103] A variety of animal and plant proteins, as well as proteases, can be utilized to produce these peptides.[104] The use of exopeptidases has demonstrated a successful reduction of bitterness in peptides derived from animal and plant proteins.[105]

Maillard reaction products can be generated through the reaction between peptides and reducing sugars in various systems, including dry, wet, and organic solvent systems. Among these, the wet system is the most commonly employed method due to its ease of operation, low cost, and sustainability.[94] The flavour outcome of the Maillard reaction products can be influenced by factors such as peptide length, amino acid composition and sequence, and the susceptibility of peptide bonds to hydrolysis.[106] Furthermore, the flavour profiles can be manipulated by adjusting variables such as the type of reducing sugars and other additives, pH levels, temperatures, and heating durations. These factors allow for the creation of diverse flavour profiles in the Maillard reaction products.[94]

1.4.3 Taste of Maillard Reaction Products

The Maillard reaction products derived from peptides can exhibit a flavour-enhancing effect. For instance, the reaction between xylose and soybean peptides within the molecular weight range of 1–5 kDa has been found to enhance umami taste and kokumi sensation, indicating the significant potential of Maillard reaction products as taste enhancers.[107,108] Peptides within this specific molecular weight range have been shown to produce Maillard reaction products with reduced bitterness.[102,109] Similarly, the Maillard reaction between chicken peptides and xylose at 80 °C and with an extended heating duration resulted in the formation of products with low bitterness and a rich broth-like taste.[102] During the Maillard reaction between soybean peptides and xylose, peptide degradation and cross-linking occur simultaneously, with the cross-linking products contributing to the kokumi effect.[109] Enzymatic cross-linking of soybean peptides using transglutaminase can also generate Maillard reaction products that enhance mouthfulness and reduce bitterness.[110] Furthermore, Maillard reaction products prepared from soybean peptides and various carbonyl compounds have been found to influence human perception of salty taste.[111] Additionally, taste-active cyclic dipeptides, known as 2,5-diketopiperazines, generated from the Maillard reaction have been identified to elicit astringent, salty, metallic, or bitter tastes, and have been detected in different processed foods.[112]

1.4.4 Aroma of Maillard Reaction Products

The Maillard reaction products of peptides possess a strong aroma that is commonly associated with meat. In comparison with amino acids, peptides exhibit a greater ability to generate specific volatile compounds, including pyrazines, pyrazinones, thiazoles, and thiophenes.[113] Peptides participate in various pathways of the Maillard reaction, such as bond cleavage, cyclization, and glycation.[102,114] The formation of specific volatile compounds during the Maillard reaction heavily relies on the molecular weight distribution and the composition and configuration of peptides.[110] For instance, pyrazines responsible for the nutty and roasted meat aroma are more effectively produced by low molecular weight peptides (<500 Da) due to the high reactivity of amides in the reaction.[102] The structural characteristics of the N-terminal amino acid in peptides also influence the production of pyrazines.[115] Additionally, pyrazinones, which possess a distinct toasted aroma, are unique Maillard reaction products derived from the reaction between dicarbonyls and glycine dipeptide. They cannot be formed in a system containing free amino acids alone.[116,117] Apart from pyrazines and pyrazinones, sulphur-containing heterocyclic compounds, such as thiophenes and thiazoles, formed during the Maillard reaction or Strecker degradation, play a crucial role in producing the roasted and meaty aroma.[118] The release of hydrogen sulphide from sulphur-containing amino acids during the Maillard reaction leads to the formation of various sulphur-containing compounds with low odour thresholds when they react with Maillard reaction products.[113] However, peptide-bound cysteines generate fewer thiazoles compared to free cysteines during the Maillard reaction. This is attributed to the inability of peptides to undergo typical Strecker degradation due to the absence of a free carboxyl group at the α-carbon in relation to the free amino group, which hinders decarboxylation.[100,113,119]

1.4.5 Applications of Maillard Reaction Products

The Maillard reaction offers a valuable approach for the industrial production of meat flavourings, which are considered natural and green alternatives to traditional synthetic meat flavourings.[120,121] Protein hydrolysates contain a large number of peptides, which can result in a wide range of flavours that closely resemble the flavours of beef, chicken, pork, mutton, shellfish, salmon, cod, squid, and cocoa. They can also function as flavour-enhancing condiments.[94] These flavourings find applications in both meat products and plant-based meat analogues.

In meat products, the selection of flavourings depends on whether they are intended for low-temperature or sterilized meat products. For low-temperature meat products, it is crucial to choose flavourings with strong aromas and high volatility at low temperatures.[122,123] On the other hand, sterilized meat products undergo high-temperature treatment, which can lead to the loss of original flavours and the development of off-flavours. Therefore, it is essential to select flavourings with good heat stability or those generated through high-temperature reactions to enhance the flavour of sterilized meat products.[122,123]

When it comes to plant-based meat analogues, flavours can be incorporated by either adding flavour precursors[124] or directly incorporating the generated meat flavourings.[125] This enables the plant-based products to acquire meat-like flavours, thus enhancing their appeal to consumers.

1.5 Conclusions

Over the past decades, protein markets have been rapidly expanding due to the increased world population and consumer needs for nutritious foods. Proteins are important for the flavour of food products due to their unique inherent taste and odour, their capability to bind with other flavouring compounds, and their role as flavour precursors. Characterizing and improving the flavour characteristics of protein ingredients and understanding the effects of protein ingredients on the flavour of the final products are critical for product formulation and the creation of high-quality food products.

Abbreviations

CAGR	Compound annual growth rate
DHS	Dynamic headspace
DMS	Differential ion mobility spectrometry
FID	Flame-ionization detection
GC	Gas chromatography

GMO	Genetically modified organism
HS	Headspace
IMS	Ion mobility spectrometry
ITEX	In-tube extraction
MS	Mass spectrometry
O	Olfactometry
ODP	Olfactory detection port
SPME	Solid-phase microextraction

References

1. United Nations Department of Economic and Social Affairs, Population Division. World Population Prospects 2022: Summary of Results. UN DESA/POP/2022/TR/NO. 3, 2022.
2. M. Shahbandeh, Topic: Protein market, Statista, 2022, Available from: https://www.statista.com/topics/4232/protein-market/#topicHeader__wrapper. cited 30 Nov 2022.
3. M. Shahbandeh, Global plant protein market value 2022-2027, Statista, 2022, Available from: https://www.statista.com/statistics/658619/global-plant-protein-market-value/. Cited 30 Nov 2022.
4. Plant-based protein market, Markets and Markets, 2022, Available from: https://www.marketsandmarkets.com/Market-Reports/plant-based-protein-market-14715651.html. Cited 30 Nov 2022.
5. Plant-based Protein Market Forecast, 2021-2031, Transparency Market Research, 2022, Available from: https://www.transparencymarketresearch.com/plantbased-protein-market.html. Cited 30 Nov 2022.
6. Statista Research Department, Global: Meat substitutes market revenue 2016-2026, Statista, 2022, Available from: https://www.statista.com/forecasts/877369/global-meat-substitutes-market-value. Cited 30 Nov 2022.
7. Protein Supplements Market Size, Share &Trends Report, 2030, Grand View Research, 2020, Available from: https://www.grandviewresearch.com/industry-analysis/protein-supplements-market. Cited 30 Nov 2022.
8. Whey Protein Market Size, Share & Trends Analysis Report by Form (Powder, Liquid), by Nature (Organic, Conventional), by Grade (Food Grade, Pharma Grade, Feed Grade), by End Use (Infant Formula and Baby Foods, Dietary Supplements, Food Processing, Sports Nutrition), by Region (North America, Europe, Asia Pacific, Latin America) - Global Industry Assessment (2016 - 2021) & Forecast (2022 - 2028), Available from: https://www.vantagemarketresearch.com/industry-report/whey-protein-market-1550. Accessed November 2022.

9. Global soy protein market outlook, Expert Market Research, Available from: https://www.expertmarketresearch.com/reports/soy-protein-market #:~: text=The% 20global% 20soy% 20protein% 20market , to% 20reach%20USD%2015.65%20billion. Accessed November 2022.
10. Pea Protein Market Size, Share & Trends Analysis Report By Product (Hydrolysate, Isolates, Concentrates, Textured), By Application (Meat Substitutes, Dietary Supplement, Bakery Goods), By Region, And Segment Forecasts, 2021 – 2028, Available from: https://www.grandviewresearch.com/ industry-analysis/ pea-protein-market #:~: text=The% 20pea % 20protein %20market %20size,USD %20240.0 %20million%20in% 202021.AccessedNovember2022.
11. Global Rice Protein Market - Industry Trends and Forecast to 2028, Data Bridge Market Research, Available from: https://www.databridgemarketresearch.com/reports/global-rice-protein-market. Accessed May 2023.
12. Mung Bean Protein Market – Global Industry Analysis and Forecast (2023-2029) by Type, Application and Region, Maximize Market Research, Available from: https://www.maximizemarketresearch.com/market-report/mung-bean-protein-market/145967/. Accessed May 2023.
13. Fava Bean Protein Market Outlook (2022-2032), Fact MR, Available from: https://www.factmr.com/report/4929/faba-bean-protein-market. Accessed May 2023.
14. Global Edible Insects Market Report 2022-2030, Research and Markets, Available from: https://www.globenewswire.com/en/news-release/2022/06/ 15/2462794/28124/ en/Global -Edible -Insects -Market-Report- 2022-2030-Environmental-Benefits-of-Edible-Insects-Consumption-Rising-Demand- for- Insect- Protein- in- the- Animal- Feed- Industry. html. AccessedMay2023.
15. Algae protein market, Global Market Insights, Available from: https://www. prnewswire .com /news- releases/ the- algae- protein- market -is - growing-at-6-cagr-by-2028-says-global-market-insights-inc-301516472.html.AccessedMay2023.
16. The Good Food Institute Asia Pacific, Alternative Protein Company Database, GFI Asia Pacific, 2022, Available from: https://gfi-apac.org/industry/alternative-protein-company-database/. Cited 30 Nov 2022.
17. A. Onkar and D. Roshan, U.S. Protein Supplement Market, Allied Market Research, 2021, Available from: https://www.alliedmarketresearch.com/us-protein-supplement-market-A11938#:~: text= The%20U.S. % 20protein %20supplement %20market ,conscious %20customers %-20around%20the%20globe.Cited30Nov2022.
18. T. Ö. Yilsay, L. Yilmaz and A. A. Bayizit, *Eur. Food Res. Technol.*, 2006, **222**, 171–175.

19. R. Jyotsna, R. Sai Manohar, D. Indrani and G. Venkateswara Rao, *Int. J. Food Prop.*, 2007, **10**, 599–606.
20. H. D. Hoang, PhD thesis, *Evaluation of pea protein and modified pea protein as egg replacers*, North Dakota State University, 2012.
21. J. Söderberg, *Functional properties of legume proteins compared to egg proteins and their potential as egg replacers in vegan food*, Swedish University of Agricultural Sciences, 2013.
22. I. C. Torres, J. M. Amigo, J. C. Knudsen, A. Tolkach, B. Ø. Mikkelsen and R. Ipsen, *Int. Dairy J.*, 2018, **81**, 62–71.
23. Y. Wang, J. Zhao, S. Zhang, X. Zhao, Y. Liu, J. Jiang and Y. L. Xiong, *Food Hydrocolloids*, 2022, **126**, 107485.
24. Plant-based protein market forecast, 2021-2031, Transpanrency Market Research, Available from: https://www.transparencymarketresearch.com/plantbased-protein-market.html. Accessed November 2022.
25. Alternative protein company database, Good Food Institute, Available from: https://gfi.org/resource/alternative-protein-company-database/. Accessed May 2023.
26. M. Murata, *Glycoconjugate J.*, 2021, **38**, 283–292.
27. E. H. Ramshaw and E. A. Dunstone, *J. Dairy Res.*, 1969, **36**, 203–213.
28. Z. Gao, P. Shen, Y. Lan, L. Cui, J.-B. Ohm, B. Chen and J. Rao, *Food Res. Int*, 2020, **131**, 109045.
29. E. Fischer, N. Cayot and R. Cachon, *J. Agric. Food Chem.*, 2022, **70**, 4493–4508.
30. B. Wang, Q. Zhang, N. Zhang, K. H. Bak, O. P. Soladoye, R. E. Aluko, Y. Fu and Y. Zhang, *Trends Food Sci. Technol.*, 2021, **112**, 336–347.
31. O. Selamassakul, N. Laohakunjit, O. Kerdchoechuen, L. Yang and C. S. Maier, *J. Food Sci.*, 2020, **85**, 707–717.
32. A. A. Bachmanov, N. P. Bosak, J. I. Glendinning, M. Inoue, X. Li, S. Manita, S. A. McCaughey, Y. Murata, D. R. Reed and M. G. Tordoff, *Adv. Nutr.*, 2016, **7**, 806S–822S.
33. V. G. Tacias-Pascacio, R. Morellon-Sterling, E.-H. Siar, O. Tavano, Á. Berenguer-Murcia and R. Fernandez-Lafuente, *Int. J. Biol. Macromol.*, 2020, **165**, 2143–2196.
34. S. K. Ulug, F. Jahandideh and J. Wu, *Trends Food Sci. Technol.*, 2021, **108**, 27–39.
35. D. Panyam and A. Kilara, *Trends Food Sci. Technol.*, 1996, **7**, 120–125.
36. B. Cabanillas, M. M. Pedrosa, J. Rodríguez, Á. González, M. Muzquiz, C. Cuadrado, J. F. Crespo and C. Burbano, *Mol. Nutr. Food Res.*, 2010, **54**, 1266–1272.
37. B. Cabanillas, M. M. Pedrosa, J. Rodríguez, M. Muzquiz, S. J. Maleki, C. Cuadrado, C. Burbano and J. F. Crespo, *Int. Arch. Allergy Immunol.*, 2012, **157**, 41–50.

38. R. Kasera, A. B. Singh, S. Lavasa, K. N. Prasad and N. Arora, *Food Chem. Toxicol.*, 2015, **76**, 54–60.
39. K. Maehashi and L. Huang, *Cell. Mol. Life Sci.*, 2009, **66**, 1661–1671.
40. P. Leksrisompong, P. Gerard, K. Lopetcharat and M. Drake, *J. Food Sci.*, 2012, **77**, S282–S287.
41. W. S. U. Roland, L. Pouvreau, J. Curran, F. van de Velde and P. M. T. de Kok, *Cereal Chem.*, 2017, **94**, 58–65.
42. L. Heng, J.-P. Vincken, G. van Koningsveld, A. Legger, H. Gruppen, T. van Boekel, J. Roozen and F. Voragen, *J. Sci. Food Agric.*, 2006, **86**, 1225–1231.
43. K. R. Price, N. M. Griffiths, C. L. Curl and G. R. Fenwick, *Food Chem.*, 1985, **17**, 105–115.
44. P. Gläser, V. K. Mittermeier-Kleßinger, A. Spaccasassi, T. Hofmann and C. Dawid, *J. Agric. Food Chem.*, 2021, **69**, 8768–8776.
45. P. A. Temussi, *J. Pept Sci.*, 2012, **18**, 73–82.
46. H. van der Wel and K. Loeve, *Eur. J. Biochem.*, 1972, **31**, 221–225.
47. J. A. Morris, R. Martenson, G. Deibler and R. H. Cagan, *J. Biol. Chem.*, 1973, **248**, 534–539.
48. D. Ming and G. Hellekant, *FEBS Lett.*, 1994, **355**, 106–108.
49. T. Misaka, *Semin. Cell Dev. Biol.*, 2013, **24**, 222–225.
50. B. J. Wright, S. E. Zevchak, J. M. Wright and M. A. Drake, *J. Food Sci.*, 2009, **74**, S17–S29.
51. M. C. Whetstine, A. E. Croissant and M. A. Drake, *J. Dairy Sci.*, 2005, **88**, 3826–3839.
52. J. M. Wright, M. E. Carunchia Whetstine, R. E. Miracle and M. Drake, *J. Food Sci.*, 2006, **71**, C86–C90.
53. J. Evans, J. Zulewska, M. Newbold, M. Drake and D. Barbano, *J. Dairy Sci.*, 2010, **93**, 1824–1843.
54. P. P. Leksrisompong, R. E. Miracle and M. Drake, *J. Agric. Food Chem.*, 2010, **58**, 6318–6327.
55. S. Mahajan, L. Goddik and M. Qian, *J. Dairy Sci.*, 2004, **87**, 4057–4063.
56. E. Ramshaw and E. Dunstone, *J. Dairy Res.*, 1969, **36**, 215–223.
57. W. L. Boatright and Q. Lei, *J. Food Sci.*, 1999, **64**, 667–670.
58. L. Xiang, W. Zhu, B. Jiang, J. Chen, L. Zhou and F. Zhong, *Food Chem.*, 2023, **402**, 134275.
59. L. Cui, J. Kimmel, L. Zhou, J. Rao and B. Chen, *Carbohydr. Polym.*, 2021, **245**, 116546.
60. J. Kalbrener, A. Eldridge, H. A. Moser and W. Wolf, *Cereal Chem.*, 1971, **48**, 595–600.
61. A. Singh, Y. Shi, P. Magreault, D. D. Kitts, M. Jarzębski, P. Siejak and A. Pratap-Singh, *Molecules*, 2021, **26**, 4104.

62. X. Su, M. Tortorice, S. Ryo, X. Li, K. Waterman, A. Hagen and Y. Yin, *Molecules*, 2020, **25**, 569.
63. C. M. Chigwedere, J. P. D. Wanasundara and P. J. Shand, *Compr. Rev. Food Sci. Food Saf.*, 2022, **21**, 999–1023.
64. M. Whitson, R. Miracle and M. Drake, *J. Sens. Stud.*, 2010, **25**, 616–636.
65. J. Zhao and W. L. Boatright, *Food Chem.*, 2017, **221**, 345–350.
66. A. Karolkowski, E. Guichard, L. Briand and C. Salles, *Foods*, 2021, **10**, 3140.
67. M. Solina, P. Baumgartner, R. L. Johnson and F. B. Whitfield, *Food Chem.*, 2005, **90**, 861–873.
68. G. A. Burdock, in *Fenaroli's Handbook of Flavor Ingredients*, CRC press, 2016.
69. T. J. Smith, E. A. Foegeding and M. A. Drake, *J. Food Sci.*, 2016, **81**, C849–C857.
70. C. W. Park, E. Bastian, B. Farkas and M. Drake, *J. Food Sci.*, 2014, **79**, C19–C24.
71. C. W. Park, MSc thesis, *The effect of acidification, feed solids concentration, and inlet temperature on the flavor of spray dried whey protein concentrate*, North Carolina State University, 2013.
72. Y. Lan, M. Xu, J.-B. Ohm, B. Chen and J. Rao, *Food Chem.*, 2019, **278**, 665–673.
73. J. Zhang, J. Cao, Z. Pei, P. Wei, D. Xiang, X. Cao, X. Shen and C. Li, *Food Res. Int.*, 2019, **123**, 217–225.
74. C. W. Park, M. A. Stout and M. Drake, *J. Dairy Sci.*, 2016, **99**, 9598–9610.
75. S. Vatansever, M. Xu, A. Magallanes-López, B. Chen and C. Hall, *Processes*, 2021, **9**, 489.
76. S. Fernando, *J. Sci. Food Agric.*, 2022, **102**, 892–897.
77. F. Zha, J. Rao and B. Chen, *Compr. Rev. Food Sci. Food Saf.*, 2021, **20**, 3036–3060.
78. Y. B. Lee, I. Laye, Y. D. Kim and C. V. Morr, *Int. Dairy J.*, 1996, **6**, 485–496.
79. H. Mehle, L. Paravisini and D. G. Peterson, *Food Funct.*, 2020, **11**, 8309–8319.
80. H. Song and J. Liu, *Food Res. Int.*, 2018, **114**, 187–198.
81. S. Wang, H. Chen and B. Sun, *Food Chem.*, 2020, **315**, 126158.
82. J. L. Chávez-Servín, A. I. Castellote and M. C. López-Sabater, *Food Chem.*, 2008, **107**, 558–569.
83. D. B. de Alencar, J. C. Diniz, S. A. S. Rocha, K. M. dos Santos Pires-Cavalcante, J. O. Freitas, C. S. Nagano, A. H. Sampaio and S. Saker-Sampaio, *J. Appl. Phycol.*, 2017, **29**, 1571–1576.

84. M. Fabre, V. Aubry and E. Guichard, *J. Agric. Food Chem.*, 2002, **50**, 1497–1501.
85. E. Ha, S. Lee, E. Jung, Y. Lee and C. Morr, *Prev. Nutr. Food Sci.*, 2002, **7**, 299–304.
86. S. Schindler, K. Zelena, U. Krings, J. Bez, P. Eisner and R. G. Berger, *Food Biotechnol.*, 2012, **26**, 58–74.
87. I. Javidipour and M. C. Qian, *Dairy Sci. Technol.*, 2008, **88**, 95–104.
88. E. Fischer, R. Cachon and N. Cayot, *Food Res. Int.*, 2021, **150**, 110760.
89. H. H. Jeleń, M. Majcher and M. Dziadas, *Anal. Chim. Acta*, 2012, **738**, 13–26.
90. C. C. Grimm, E. T. Champagne, S. W. Lloyd, M. Easson, B. Condon and A. McClung, *Cereal Chem.*, 2011, **88**, 271–277.
91. H. Van Ba, I. Hwang, D. Jeong and A. Touseef, *Latest Research into Quality Control*, 2012, **2**, 145–176.
92. A. Spanier, M. Flores, F. Toldrá, M. Aristoy, K. L. Bett, P. Bystricky and J. Bland, in *Quality of Fresh and Processed Foods*, ed. F. Shahidi, A. M. Spanier, C.-T. Ho and T. Braggins, Springer, 2004, pp. 33–49.
93. M. I. Khan, C. Jo and M. R. Tariq, *Meat Sci.*, 2015, **110**, 278–284.
94. A. Sun, W. Wu, O. P. Soladoye, R. E. Aluko, K. H. Bak, Y. Fu and Y. Zhang, *Food Res. Int.*, 2022, **151**, 110823.
95. Y. Luo, B. Wang, C. Liu, R. Su, Y. Hou, D. Yao, L. Zhao, L. Su and Y. Jin, *Food Sci. Nutr.*, 2019, **7**, 2796–2805.
96. C. Cao, J. Xie, L. Hou, J. Zhao, F. Chen, Q. Xiao, M. Zhao and M. Fan, *Food Res. Int.*, 2017, **99**, 444–453.
97. L. Hou, J. Xie, J. Zhao, M. Zhao, M. Fan, Q. Xiao, J. Liang and F. Chen, *Food Chem.*, 2017, **232**, 135–144.
98. X. Ma, M. Yu, Z. Liu, D. Deng, Y. Cui, Z. Tian and G. Wang, *J. Food Sci. Technol.*, 2020, **57**, 404–412.
99. R. Zhou, J. Grant, E. M. Goldberg, D. Ryland and M. Aliani, *J. Sci. Food Agric.*, 2019, **99**, 1728–1739.
100. M. A. J. S. van Boekel, *Biotechnol Adv.*, 2006, **24**, 230–233.
101. C. Diez-Simon, R. Mumm and R. D. Hall, *Metabolomics*, 2019, **15**, 41.
102. J. Liu, H. Song, Y. Liu, P. Li, J. Yao and J. Xiong, *J. Sci. Food Agric.*, 2015, **95**, 3183–3194.
103. T. Feng, Z. Zhang, H. Zhuang, J. Zhou and Z. Xu, *Mini-Rev. Org. Chem.*, 2016, **13**, 255–261.
104. C. C. Udenigwe and R. E. Aluko, *J. Food Sci.*, 2012, **77**, R11–R24.
105. L. K. Y. Cheung, R. E. Aluko, M. A. Cliff and E. C. Li-Chan, *J. Funct. Foods*, 2015, **13**, 262–275.
106. Y. Fu, Y. Zhang, O. P. Soladoye and R. E. Aluko, *Crit. Rev. Food Sci. Nutr.*, 2020, **60**, 3429–3442.

107. M. Ogasawara, T. Katsumata and M. Egi, *Food Chem.*, 2006, **99**, 600–604.
108. M.-R. Rhyu, A.-Y. Song, E.-Y. Kim, H.-J. Son, Y. Kim, S. Mummalaneni, J. Qian, J. R. Grider and V. Lyall, *Nutrients*, 2020, **12**.
109. X. Lan, P. Liu, S. Xia, C. Jia, D. Mukunzi, X. Zhang, W. Xia, H. Tian and Z. Xiao, *Food Chem.*, 2010, **120**, 967–972.
110. N. Song, C. Tan, M. Huang, P. Liu, K. Eric, X. Zhang, S. Xia and C. Jia, *Food Chem.*, 2013, **136**, 144–151.
111. T. Katsumata, H. Nakakuki, C. Tokunaga, N. Fujii, M. Egi, T.-H. T. Phan, S. Mummalaneni, J. A. DeSimone and V. Lyall, *Chem. Senses*, 2008, **33**, 665–680.
112. A. D. Borthwick and N. C. Da Costa, *Crit. Rev. Food Sci. Nutr.*, 2017, **57**, 718–742.
113. F. Van Lancker, A. Adams and N. De Kimpe, *Chem. Rev.*, 2011, **111**, 7876–7903.
114. C. Yang, R. Wang and H. Song, *Food Chem.*, 2012, **133**, 373–382.
115. F. Van Lancker, A. Adams and N. De Kimpe, *J. Agric. Food Chem.*, 2012, **60**, 4697–4708.
116. H. V. Izzo and C.-T. Ho, *Trends Food Sci. Technol.*, 1992, **3**, 253–257.
117. Y. C. Oh, C. K. Shu and C. T. Ho, *J. Agric. Food Chem.*, 1992, **40**, 118–121.
118. D. D. Jayasena, D. U. Ahn, K. C. Nam and C. Jo, *Asian-Australas. J. Anim. Sci.*, 2013, **26**, 732.
119. Y. Zhang and C. T. Ho, *J. Agric. Food Chem.*, 1991, **39**, 760–763.
120. H. Cui, C. Jia, K. Hayat, J. Yu, S. Deng, E. Karangwa, E. Duhoranimana, S. Xia and X. Zhang, *RSC Adv.*, 2017, **7**, 45442–45451.
121. F. B. Siewe, T. G. Kudre, B. K. Bettadaiah and B. Narayan, *Ultrason. Sonochem.*, 2020, **65**, 105055.
122. S. Ji, M. Yang, J. Zhang and G. Zhang, *Meat Ind.*, 2007, **3**, 35–38.
123. M. Font-I-Furnols and L. Guerrero, *Meat Sci.*, 2014, **98**, 361–371.
124. D. S. Mottram, *J. Sci. Food Agric.*, 1985, **36**, 377–382.
125. N. R. Rubio, N. Xiang and D. L. Kaplan, *Nat. Commun.*, 2020, **11**, 6276.

CHAPTER 2

Consumer Perceptions of Food Proteins and Protein-enriched Foods

JING ZHAO*[a], MINWEI XU[b], JESSE BAKER[a] AND SARAH KINSEY[a]

[a]5500 Campanile Drive, San Diego State University, San Diego, CA 92182-7251, USA; [b]Department of Plant Sciences, North Dakota State University, Fargo, North Dakota, USA
*E-mail: jzhao2@sdsu.edu

2.1 Introduction

Meat, seafood, dairy, and eggs are well-accepted traditional protein sources that are nutritious and palatable. Due to religious beliefs and increasing concerns about the environment and animal welfare, the vegetarian population has increased to 1.5 billion worldwide, according to Leahy, Lyons, and Tol.[1] The demand for and consumer interest in plant proteins are quickly increasing.[2] Besides plant proteins, novel proteins can now be derived from bacteria, fungi, algae, cultured meat, and edible insects for human food. Due to the considerations of sustainable food production, alternative proteins (non-animal protein-based animal food analogs) are gaining increasing interest and investment.[3] Despite the current trends and movements in the protein market, a good understanding of consumer perceptions of food proteins is critical for further product development and business decisions.

Food Chemistry, Function and Analysis No. 41
Flavour and Consumer Perception of Food Proteins
Edited by Jing Zhao and Changqi Liu

Published by the Royal Society of Chemistry, www.rsc.org

Proteins are available to consumers in the form of protein powders, protein shakes, protein beverages (ready-to-drink or ready-to-mix powders), protein bars, *etc*. Besides, protein ingredients are very commonly used in the formulation of various products to enhance the protein content (nutritional value) and product quality (*e.g.*, texture, color, emulsion stability, oxidative stability). In recent years, alternative meat products have been developed using plant-based, fermentation-enabled, and cultured proteins. This section will discuss consumer studies of protein ingredients (in the form of protein concentrate, protein isolate, and protein hydrolysate) and protein-based products. Besides, the sensory impacts of protein ingredients in various protein-enhanced or protein-based products are also included.

To understand consumer perceptions of food products, consumer studies are usually carried out using surveys (focus groups, online surveys, paper surveys, *etc.*) or consumer sensory evaluations. Survey studies can be performed during or before the development of a product to understand the market potential of the product or after the product is launched to understand the drivers of consumer purchases. The food industry often conducts consumer sensory evaluations during the product development process and before the launch of a new product. However, the sensory study information available to the public is limited. Among the available studies, most have investigated the sensory impact of protein ingredients in foods. Sensory studies, especially consumer studies, that focus on protein ingredients have been scarce. In this chapter, both sensory evaluation and survey-based consumer studies on protein ingredients and protein-based foods are covered. This chapter aims to focus on consumer studies (preference sensory tests) of protein ingredients. Descriptive sensory analysis and volatile characterizations of protein ingredients are discussed in Chapters 5–10.

2.2 Protein Concentrate, Isolate, and Hydrolysate

Many studies have investigated the sensory impact of protein ingredients in protein-enhanced foods. However, sensory studies, especially consumer studies, that focus on protein ingredients have been scarce. The consumer studies (preference sensory tests and survey studies) on protein ingredients that are available in the current literature are described below. Due to the limited studies available, the section also included online consumer reviews for some commercial protein products.

2.2.1 Whey Protein Concentrate, Isolate, and Hydrolysate

Numerous studies have investigated the flavor properties of whey protein ingredients using descriptive sensory tests and a trained panel.[4,5] The descriptive studies are critical in characterizing the odor profiles of foods and understanding the sensory impacts of volatile compounds. An early study characterized the flavor of whey protein concentrate (WPC) and whey protein isolate (WPI) following the rehydration of the powders. Sensory evaluations using a seven-member trained panel found that whey proteins exhibited sweet aromatic, cardboard/wet paper, animal/wet dog, soapy, brothy, cucumber, and cooked/milky flavors, along with the basic bitter taste and the feeling factor astringency.[4]

On the other hand, it is valuable to understand consumer preferences for these products. Evans *et al.* tested the consumer acceptance (n = 77) of 80% WPC in the form of an acidified 6% protein beverage.[6] Overall liking, appearance liking, flavor liking, and texture liking were rated using a 9-point hedonic scale. The overall liking of WPC ranged from 3.7 to 5.8. Compared to the sensory attributes rating from a descriptive sensory analysis by a trained panel, sweet aroma is positively related to consumer preference, while cardboard odor negatively affects consumer preference.

Lee *et al.* investigated the consumer acceptance of WPI coatings as an alternative to shellac in confectionery products. Consumer participants (n = 180) evaluated five different chocolate-coated almonds that had either a WPI coating or a shellac coating for overall liking, taste, texture, overall appearance, surface texture/appearance, and gloss using a 9-point hedonic scale. Results from the hedonic testing revealed that the WPI coating with lipids had higher hedonic scale ratings overall as compared to the WPI coatings without lipids and the shellac-coated sample. The WPI coating with the highest lipid content had the highest overall liking score (6.22/9).[7]

2.2.2 Online Consumer Reviews of Soy Protein Isolate Powders

Soy protein is a high-quality protein that has many health benefits. Its use in food products can be beneficial for both nutritional and functional properties. Many consumers consider protein powders a healthy and convenient way to add protein to their diet. In terms of soy protein isolate (SPI) powders, consumer acceptance appears to be satisfactory when reviewing various popular products in the market and their consumer reviews (Table 2.1). The products "NOW Foods SPI - Natural Unflavored," "Puritan's Pride - Soy Protein Isolate Powder Natural," and "Bob's Red Mill - Soy Protein Powder" were assessed for consumer acceptability based on the sensory-related comments found in online consumer reviews. In general, consumers appreciated these products because they were unflavored and provided good textural properties and blendability. However, some consumers did not enjoy the texture of these products and noted a chalky and grainy texture. Some consumers noted earthy and beany aromas as well. Although reviews were mixed, most consumers had good acceptability toward these popular products in the market. Evaluating consumer reviews does have limitations of bias, varying preparation methods for consumption, and a lack of a controlled environment. Future studies within this domain would be needed to further analyze SPIs for general consumer acceptability and preference.

2.2.3 Rice Protein Concentrate, Isolate, and Hydrolysate

Kaewka *et al.* examined the sensory aroma characteristics of acid-hydrolyzed rice bran protein concentrate (RBPC) and the effects of drying methods.[8] A descriptive sensory analysis was performed by a trained panel of 10 members. The overall aroma liking was evaluated by 30 untrained participants using the 9-point hedonic scale (1 = dislike extremely, 9 = like extremely). The results showed that spray-dried hydrolysate (acid hydrolyzed for 12 h) received the highest overall aroma-liking score (6.2/9) compared to other RBPC hydrolysates that were hydrolyzed for 36 h or were freeze-dried (4.8–5.6/9). According to the descriptive sensory analysis, the spray-dried 12 h hydrolysate exhibited the highest sweet- and cereal-like aromas and the lowest rice bran-like and earthy aromas.[8]

Arsa and Theerakulkait published several studies on the flavor of RBPC and its hydrolysates.[9–13] In one of their studies, the aroma characteristics of RBPC and hydrolysates (alcalase-hydrolyzed, sugar added) were studied using both instrumental and sensory evaluation methods.[11] The aroma characteristics were evaluated in a descriptive sensory analysis by 16 trained panelists. The overall aroma-liking score of the samples was evaluated using a 9-point hedonic scale with 30 untrained panelists (1 = dislike extremely, 9 = like extremely). The results from the descriptive sensory analysis revealed that when adding various sugars (sucrose, glucose, and fructose) to the hydrolysates before spray drying, the sweet, cocoa-like, and milk powder-like aroma was the most intense in fructose-added rice bran protein hydrolysate (RBPH) ($P \leq 0.05$). Even though glucose-added hydrolysate showed the lowest rice bran-like aroma, the fructose-added hydrolysate received the highest overall aroma-liking score (7.18/9) in the preference test compared with other RBPH (5.76–6.46/9) and the concentrate (5.29/9). The study showed that spray drying, hydrolysis, and sugar addition could improve the sensory aroma characteristics of RBPC.[11]

One group performed a preference test of untreated and glutaminase-deamidated rice protein powders.[14] The overall taste liking was evaluated on a "7-point scale" ranging from 0 (absence) to 7 (extremely intense) by 15 participants. Even though the number of participants in this study was very small, it found an average overall taste liking of 4.3 for untreated rice protein on a 7-point scale. The deamidation treatment improved the overall taste-liking score to 4.9, probably due to reduced grittiness and lumpiness and improved suspendability of the protein powder. A comparison of the consumer scores from the above-mentioned studies is summarized in Table 2.2.

Table 2.1 Sampled online consumer reviews on selected SPI powders.

Name of soy protein powder	Sensory-related comments (no. of comments on the indicated property)	Summary	Link
NOW Foods SPI - Natural Unflavored	• Powdery taste (1) • Good texture/blendability (10) • Poor texture/blendability (4) • Grainy/gritty texture (2) • CChalky taste (1) • Flavorless (5) • Enjoys unsweetened flavor (8) • Good taste (9) • Poor taste (4) • Poor aroma (1) • Creamy texture (1) • Earthy soy flavor (2) • Strong soy aftertaste (1)	Consumers tend to enjoy the unsweetened/unflavored aspect of this product as they can add their own flavors, fruits, or vegetables to a smoothie without having an overpowering taste. Overall good texture and taste.	[a,b]
Puritan's Pride - SPI Powder Natural	• Good taste (10) • Good texture/blendability (11) • Subtle creamy flavor (1) • Neutral taste - positive (6) • Natural favor - positive (5) • Good unsweetened taste/flavor (12) • Poor taste (5) • Poor texture/blendability (6) • Decent taste (2) • Bland/flavorless - negative (2) • Thick texture - negative (1) • Chalky texture - negative (3) • Gritty texture - negative (1)	Consumers tend to think this product has a good taste and texture. The unflavored/unsweetened taste of this product is also generally favored by consumers. There are still some mixed reviews in which some consumers did not enjoy the taste, flavor, and texture.	[c]

(continued)

Table 2.1 (*continued*)

Name of soy protein powder	Sensory-related comments (no. of comments on the indicated property)	Summary	Link
Bob's Red Mill - Soy Protein Powder	• Good taste (13) • Good texture/blendability (19) • Good unflavored taste (14) • Mild taste (2) • Poor taste (7) • Poor texture/blendability (7) • Good aroma (2) • Chalky texture (2) • No chalky aftertaste (1) • Decent taste (4) • Bland taste - negative (1) • Creamy texture (1) • Good soy taste (1) • Bitter taste (1)	Overall, this product seems to have good texture and blendability as compared to other soy protein powders. The taste also tends to appeal to consumers, as many enjoy the unflavored/unsweetened taste.	[d]

[a]https://www.swansonvitamins.com/p/now-foods-soy-protein-isolate-unflavored-2-lbs-pwdr?SourceCode = INTL4071&showPopup = f&DFA = 1&UTM_Medium = Shopping&UTM_Source = GOOGLE&UTM_Campaign=+&UTM_Content=&SourceCode = INTL4071&gclsrc = aw.ds&gclid = Cj0KCQjwk7ugBhDIARIsAGuvgPYVGK6j243rLyeNRPFoVkiqKIVinuYh2AzSPo5KkiTzPlWpsjWoRgIaAhPAEALw_wcB

[b]https://www.iherb.com/r/now-foods-sports-soy-protein-isolate-pure-unflavored-2-lbs-907-g/786?lang = en-us&sort = 6&isshowtranslated = true&*P* = 2

[c]https://www.puritan.com/puritans-pride-brand-0102/soy-protein-isolate-powder-natural-002660?scid = 45182&utm_source = pla&utm_medium = google_ads&utm_campaign = G_PLA_Pmax_1st_Party_HIGH&utm_term=&utm_content = pmax&tf_adid=&tf_asid=&tf_cid = 15504397666&gclid = Cj0KCQjwk7ugBhDIARIsAGuvgPbFcmGdgTKiMPXz9-mUfVKc6hoC_alhwo59nOvzZ3HaoHDReDdT_KAaAhQKEALw_wcB&gclsrc = aw.ds

[d]https://www.iherb.com/pr/bob-s-red-mill-soy-protein-powder-gluten-free-14-oz-397-g/36163?gclid = Cj0KCQjwk7ugBhDIARIsAGuvg-PaiKFDp4LfDbzIsE8s5ein7sY2RTOSa1rf6g-EmNSjF3HgBUhSTIuAaAleTEALw_wcB

Table 2.2 Consumer preference rating of rice protein ingredients.[8,11,14]

Product evaluated	Sensory panel	Scale	Measurement	Score	Ref.
Rice bran protein concentrate	30 untrained panelists	9-point hedonic scale	Overall aroma liking	5.29	11
Rice bran protein hydrolysate (alcalase)	30 untrained panelists	9-point hedonic scale	Overall aroma liking	5.76–7.18	11
Rice bran protein acid hydrolysates	30 untrained panelists	9-point hedonic scale	Overall aroma liking	4.8–6.2	8
Rice protein powder	15 participants	7-point scale	Overall taste liking	4.3	14
Deamidated rice protein powder	15 participants	7-point scale	Overall taste liking	4.9	14

2.3 Consumer Acceptance of Protein-based Foods

2.3.1 Protein Shakes and Beverages

Protein shakes and beverages are used by athletes and consumers for weight management, muscle building, and nutrition purposes. Even though various flavors are available on the market, vanilla and chocolate are the dominant and most well-accepted flavors. Liu *et al.* performed descriptive analysis on 33 vanilla flavor ready-to-drink protein beverages and consumer acceptance testing on 10 representative beverages.[15] The protein beverages tested included 3 pea-based, 1 soy-based, and 29 dairy protein-based products. The positive flavor properties that relate to consumer liking were identified as "cooked/milky and caramelized flavors, sweet taste, and viscosity," while the negative properties included "sweet aromatic, green pea, and cereal/grainy flavors and astringent mouthfeel." The study also concluded that the ideal vanilla ready-to-drink beverages would be sweet, viscous, with a milky and caramelized flavor, high in protein, and at a reasonable price.

2.3.2 Protein Bars

Childs *et al.* conducted an online survey to assess consumers' perceptions of whey and soy protein in meal replacement products. Consumers ranked nutritional content, price, and flavor as the three most important attributes for purchasing. The study suggested that most consumers chose protein products for their health benefits. However, most consumers were dissatisfied with the flavor of meal replacement products they had tried in the past. Among all consumers, bars were the most popular product type.[16]

Jones *et al.* conducted a study that surveyed the consumer perception of soy- and dairy-based protein bars among the US (n = 147) and New Zealand (n = 126) consumers. The survey study investigated cross-cultural differences regarding the health benefits and product claims associated with soy and dairy products. The study showed that New Zealand consumers tended to favor dairy products, while US consumers did not clearly prefer a specific protein source over the other.[17]

2.3.3 Animal Food Alternatives

2.3.3.1 Sensory Evaluation Studies of Plant-based Meat Alternatives

Due to environmental concerns, plant-based animal food products have attracted a lot of interest in recent years. Short *et al.* reviewed 12 sensory evaluation studies of plant-based cheeses that aimed to improve the sensory properties of the products.[18] Lactic acid bacteria fermentation combined with other strategies (*e.g.*, co-fermented with another culture, glucono-delta-lactone coagulation, or enzymatic hydrolysis) improved the sensory properties and consumer liking of soy-based spreadable and semi-hard cheese products.[19–21] Blending other plant-based milk with soymilk was found to reduce the beany flavor[22] and improve the flavor and appearance[23,24] of plant-based cheese made from soymilk only.

The high moisture meat analogs (HMMAs) made by pulse proteins are a hot topic in the modern food industry. The HMMAs can be employed to produce vegetable hamburger patties. Kim and coworkers produced HMMAs with pea, lentil, and fava bean protein concentrates using a twin-screw extruder.[25] And vegetable hamburger patties with HMMAs were produced with the addition of spices and binders. The sensory evaluation of HMMAs and vegetable hamburger patties was conducted by 9 trained panelists and 80 untrained consumers, respectively. Compared to the HMMAs with soy protein concentrates, pulse proteins scored higher for flavor attributes that were bean-like, salty, sweet, umami, musty-earthy, and malt-like. In contrast, these samples were lower for soy and cardboardy. The protein source did not affect the color of raw or cooked patties for lightness and cooked patties for yellowness compared to the control sample. However, the protein source significantly affected the redness and yellowness of raw patties and the redness of cooked patties. Overall, pulse protein can be an alternate source of soy to produce HMMAs. as consumers scored a similar liking for vegetable proteins containing different pulse proteins.[25]

Plant-based meat analogs are a fast-growing market. Even though there is an increasing interest in innovative food products made from plant proteins, and the majority of people, especially male and highly educated populations, are willing to try plant-based meat analogs, studies have also indicated that most non-vegetarians are still unwilling to remove meat from their diet.[26] One of the major obstacles to adopting a plant-based diet is the sensory characteristics of plant-based analogs, including color, texture, taste, and aroma. Research is needed in this area to further improve the quality and consumer acceptance of alternative protein-based food products.

2.3.3.2 Sensory Evaluation Studies of Plant-based Dairy Alternatives

Soymilk is a popular plant-based milk substitute. Lawrence *et al.* evaluated 12 different soymilk products through a consumer sensory study with 235 regular US consumers of original-flavored soymilk products. The study found that sweetened shelf-stable and refrigerated soymilks received the highest overall flavor, sweetness, and aftertaste-liking scores as compared to other samples. Ethnicity influenced consumer liking scores, while age did not. Asian consumer scores were mainly in the middle of the scale, while Caucasian and African American consumers had a wider range of scores. Universal drivers for consumer acceptance were soymilks with a vanilla flavor, a sweet taste, and a high viscosity.[27]

Soybean has been used to make cheese analogs. A study by Oyeyinka *et al.* looked at the consumer acceptability of cheese analogs in which combinations of soybean and cashew were used. Cashew milk was substituted for soymilk at proportions of 20, 30, 40, 50, 60, and up to 100%. Participants (n = 30) evaluated samples in terms of flavor, taste, color, mouthfeel, and overall acceptance using a 9-point hedonic scale. The sample with the highest overall acceptability ratings contained 60% soymilk and 40% cashew milk, while the least favorable sample contained 100% cashew milk. The sample with the second-best overall acceptance rating contained 100% soymilk. This study suggests that the most optimal ratio for consumer acceptance of cheese analogs containing soy and cashew is 60% soymilk and 40% cashew milk, followed by 100% soymilk.[28]

Lopes *et al.* developed pulse-based milk from chickpea and lupin proteins.[29] The process includes seed soaking, cooking, milling, sieving, and pasteurizing. The sensory evaluation was carried out to observe its sensory acceptance. Twenty-nine untrained panelists were employed to evaluate the samples in terms of color, appearance, taste, flavor, consistency, and overall appreciation. Comparing lupin- and chickpea-based beverages, lupin-based beverages had better color and appearance, while chickpea beverages had better flavor, taste, and consistency. Chickpea- and lupin-based beverages still had a beany flavor before the cooking process. Thereafter, the cooking process significantly decreased the beany flavor. The cooking step was observed to be essential for the reduction of the unpleasant beany flavor.

2.3.3.3 Survey-based Consumer Studies on Plant-based Alternative Products

Consumers' interest in alternative proteins is on the rise. Numerous survey studies have been done to understand the drivers and barriers to consumer acceptance of these alternative protein products. These studies have been summarized in several review articles.[30,31] Environment sustainability, animal welfare, culture, neophobia, human nutrition, palatability, appearance, flavor, texture, and price are the key factors examined in various studies.[31] Through an assessment of the demographic characteristics across various studies, it shows that young and highly educated consumers have a higher acceptance of plant-based proteins.[30,32–34] The major driving forces of alternative meat consumption include environmental concerns about traditional meat, health concerns, and animal welfare.[31]

On the other hand, by systematically reviewing 38 articles, Hartmann and Siegrist concluded in 2017 that consumers' willingness to change meat consumption behavior in terms of reducing or substituting meat (*e.g.*, by eating insects or meat substitutes) is low.[26] Similarly, several studies also indicated that even though there is a gradual increase in the acceptance of plant-based meat alternatives, the majority of consumers are unwilling to replace meat with plant-based diets.[31,34,35]

Cereals, grains, vegetables, and fruits are familiar components of human diets. Among alternative protein products, plant protein-based foods are reported to have the highest consumer acceptance.[30,36] A survey study of 495 Denmark consumers suggested that indicating the source of the protein used led to a favorable perception among consumers who are familiar with plant-based proteins.[37] A study among Italian consumers identified plant-based foods and edible insects as the most viable alternative protein sources, while plant-based foods have the highest acceptance among all alternative protein products.[36] The study suggested that providing information on all novel protein alternatives would be the best strategy to increase the acceptance of these products among Italian consumers.

Based on an online survey of 872 respondents, Faber *et al.* reported that the respondents showed an overall interest in plant protein-based foods regardless of their current frequency of plant-based food consumption or if they are vegetarians.[38] Interestingly, the study showed that the respondents were in favor of plant-based products that do not mimic the taste and texture of their meat counterparts. On the other hand, other studies have indicated that consumers demonstrate stronger preferences for alternative proteins when the product imitates the taste and texture of traditional meat.[30,34]

2.4 Protein-enriched Foods

Besides being used as the major component of a product, proteins have been added to various products, *e.g.*, meat, baked foods, snacks, beverages, and wine, for enhanced nutritional value and improved product quality. Protein ingredients used in foods include proteins from fish,[39,40] dairy,[41–44] soy,[19,45–49] pulses (*e.g.*, pea,[50–52] lupin,[53,54] fava bean), rice,[55,56] *etc.* Moreover, proteins from pulses and rice are hypoallergenic and gluten-free, making these ingredients suitable for a large variety of food products, such as infant formula, baked food, plant-based meat, and beverages. The section includes examples of consumer acceptance studies performed for various types of foods made with or enhanced with food proteins.

2.4.1 Meat Products

In meat products, non-meat proteins have been used as a binder. Zamuz *et al.* examined the sensory impacts of adding different alternative proteins, including soybean, pea, lentil, bean, *Chlorella*, and *spirulina* proteins, to pork liver pâté, turkey patties, cooked turkey breast, fresh pork sausages, and marinated turkey breast.[57] The sensory characteristics and preferences for the alternative protein-added meat products were evaluated by a semi-trained panel of 16 panelists. Despite the small number of participants for the preference sensory test, the study found that products made with soybean and bean proteins were most preferred, followed by products made with dairy and pea proteins. The addition of *Chlorella* and *spirulina* proteins negatively affected consumer preference.[57]

2.4.2 Dairy Products

A study by Drake and Gerard looked at consumer acceptance of four different yogurts. These yogurts were either strawberry- or lemon-flavored, contained no soy, or were fortified with 2.5% soy protein. Participants for this study were from either a college campus (n = 388) or a grocery store (n = 388), and they rated yogurts on texture, flavor, and overall acceptability using a 9-point hedonic scale. College campus participants evaluated all four yogurts, while grocery store participants only evaluated strawberry-flavored yogurts with and without soy and also completed a purchase intent questionnaire. Half of all participants were given information on soy protein's health benefits prior to completing this study. Results from the sensory evaluation indicated that the flavor of the yogurt had no impact on texture, flavor, or acceptability scores at the college campus. This indicates that the grocery store participants' evaluation of just strawberry yogurt was adequate for evaluation in conjunction with college campus responses. For both groups of participants, flavor and texture were drivers of acceptability (r = 0.98). The yogurts without soy had higher overall acceptance scores as compared to yogurts with added soy (6.41–7.42 and 4.86–5.41, respectively). Flavor and texture acceptability scores were also higher for yogurts without added soy. Although hedonic scores were lower for soy-fortified yogurts, results from the consumer purchase intent survey indicated a strong willingness to purchase soy-fortified yogurts. When soy protein health benefits were disclosed to participants, this increased the acceptability scores as well. Results from this study show that consumers value products that provide health benefits, and marketing soy-fortified yogurts in this way can have a positive effect on consumer acceptance.[58]

2.4.3 Baked Foods

Baked foods are a major category of foods that incorporate protein ingredients. Especially non-gluten proteins are used in gluten-free products. Miñarro *et al.*[59] compared the consumer acceptance of four gluten-free breads made with chickpea flour, pea isolate, carob germ flour, and soy flour. A consumer test (n = 66) was performed using a 9-point hedonic scale. No difference in consumer preference was found in taste, while bread with pea protein isolate was rated lower for its odor. Garcia-Segovia *et al.*[60] performed a consumer study (n = 106) on wheat flour breads partially substituted with 5% or 10% pea protein isolate, mealworm, or buffalo. Compared to the control wheat flour bread, consumer acceptance of pea protein and buffalo worm breads was lower, while bread with 10% mealworm showed increased overall liking by the consumers.

A study by Ivanovski *et al.* looked at consumer acceptance of soy protein-enhanced bread as compared to wheat bread. Although the soy flour-containing bread had a notable beany flavor, its acceptance was comparable to that of the wheat bread during consumer sensory studies using a 9-point hedonic scale. The SPI-enhanced bread used in this study had the lowest acceptance ratings, which may be attributed to poor sensory attributes of an astringent and bitter flavor as analyzed by a trained panel.[61]

Siddiq *et al.*[62] studied the sensory attributes of cookies with the ingredients navy and pinto bean flour. The cookie samples were presented in random order, and 107 panelists were asked to rate their assessment of appearance, aroma, flavor, texture, and overall acceptability. Cookies with pinto bean flour have a significant ($P < 0.05$) higher flavor, texture, and overall acceptability than those with navy bean flour. Interestingly, over seven-eighths (94/107) of the panelists could not detect beany flavor in their cookies, which could be attributed to the process of extrusion cooking or steaming.[62]

Potato protein isolate and concentrate are by-products of the potato starch industry. Waglay and Karboune investigated the incorporation of potato protein isolate and concentrate into rice and/or wheat flour cookies. Consumer (n = 70) sensory results revealed that adding 12.5% potato protein isolate or more significantly lowered consumer acceptance, while the addition of 5% or less did not affect consumer preference.[63]

2.4.4 Snacks

Extruded snacks are a popular form of high-protein food. Saint-Eve *et al.* produced and evaluated the liking of eight extruded high-protein snacks that were made of 50–80% pea flour and 20–50% wheat or rice flour.[64] The study found that products made with peas and wheat were preferred over those made with peas and rice. Also, for both wheat- and rice-containing products, the increase in pea flour content negatively affected the consumer's liking score. The trend was consistent even though the decrease in consumer liking score for pea and wheat products was not statistically significant. The study showed that both pea and rice ingredients have lower consumer acceptance compared to wheat.

Saldanha do Carmo *et al.* produce expanded snacks entirely based on pea and oat ingredients using extrusion technology.[65] Ingredients with a lower moisture level (ML), 11.2% dry basis, and higher ML, 16.8% dry basis, were investigated, respectively. Ten assessors were trained following ISO 8586:2012 and employed for the sensory evaluation. They concluded that lower ML was associated with positive sensory attributes such as crispy, juicy, sweet, umami, and sour. On the other hand, samples processed at high ML were described as hard, chewy, granular in texture, and with pea flavor, cloying, and old/stuffy flavors. Therefore, samples processed with extrusion at lower ML were considered sensorially and texturally better than at higher ML for consumption.[65]

Shaviklo *et al.* studied the sensory properties and consumer acceptance of corn snacks fortified with 3%, 5%, 7%, and 9% fish protein powder made from saithe (*Pollachius virens*) surimi. Products with different fish protein levels were evaluated by an expert panel of four, and it was concluded that adding 9% fish protein powder significantly lowered the liking for odor, texture, flavor, and overall acceptability compared to the other three prototypes. A product with 7% fish protein was further tested along with commercial corn snack products (with no fish protein) in a 7-point hedonic test by ninety 6- to 16-year-old children from Iran and Iceland, respectively. The fish protein fortified corn snack was rated higher than the commercial corn snack by Iranian kids, while it was less preferred by Icelandic kids.[40]

Kreger *et al.* evaluated the consumer acceptability of a corn meal-based snack with varying levels of whey and soy protein (23–48%) added in 20 different formulations. One hundred consumers utilized a 9-point hedonic scale during sensory testing, and scores for most samples were fairly neural at ~5/9; however, the sample with 50% soy protein had the highest rating overall. Samples were also more favored at lower protein concentrations. Further formulation of a corn meal-based snack with added protein would be favorable in order to draw general conclusions around consumer acceptability.[66]

De Farias Silva *et al.* examined the sensory properties (flavor, aroma, color, smell, and appearance) of various fruit pulps (acerola, cashew, guava, soursop, passion fruit, and mandarin) fortified with 1–5% agglomerated concentrated rice protein powder.[56] The study found that the addition of rice protein affected the lightness, redness, and yellowness of the fruit pulps, increased product viscosity, and decreased consumer acceptance. The consumer acceptance of all fruit pulps decreased with an increasing amount of rice protein inclusion. Except for cashew and soursop pulps that maintained an acceptance between 70% and 80% with 5% rice protein inclusion, consumer acceptance of other pulps tested dropped from around 90% to below 70% when 5% rice protein was added. One limitation of the study was that it only had 30 tasters for a hedonic test.[56]

2.4.5 Dessert

Walker *et al.* investigated the consumer acceptability and purchase intent of four low-fat, sugar-free SPI-containing sherbets. All sherbets had an orange flavor and varied in levels of SPI (6.0, 6.5, 7.5, and 7.9 g per serving). Untrained consumers (n = 140) were asked to evaluate each sample for overall liking and acceptability of appearance/color, flavor, sweetness, sourness, and texture/mouthfeel, using a 9-point hedonic scale. Results from this study showed that all samples were rated highly in terms of flavor and sweetness but poorly for texture acceptability. The overall liking scores were highest for sherbets containing 6.0 g and 7.9 g of SPI per serving. The mean scores were 5.66 and 5.60, respectively. Acceptance ratings (65.0%) and purchase intent (55.7%) after being informed of soy's health benefits were highest for the sherbet containing 6.0 g of SPI per serving. Through predictive discriminant analysis and logistic regression analysis, purchase intent and overall acceptability were determined to be most affected by texture and overall liking. To further improve consumer acceptance of these sherbets, looking at these critical attributes would be most beneficial.[67]

2.4.6 Wine and Beverages

It is known that the content of protein components in glutinous rice significantly affects the quality of Chinese rice wine. Chen *et al.* examined the individual and combined effects of added rice glutelin and albumin on the quality of the wine.[55] The volatile compounds were characterized, and a descriptive sensory panel of ten judges was selected and trained to evaluate the sensory attributes of rice wine flavor. It was found that adding glutelin enhanced the fruity and honey aromas and umami and bitterness sensations while reducing the caramel-like, herb, smoky, Qu aroma, and sweetness in the resulting wines. An increase in the albumin component lowered the caramel-like, fruity, smoky, Qu aroma, and acidity and enhanced the herb odor, sweetness, and bitterness.

Pulse proteins are considered a potential supplement to beverages such as apple juice and orange juice.[68] Apple juice and orange juice were supplemented with 1–4% pulse ingredients (pea protein, chickpea flour, lentil flour, and pea fiber). Considering the lower solubility of pulse protein at pH 3–6, the turbidity increased as a result of supplementation. The L, a, and b values of juice added with pea protein increased, which indicated increased whiteness, redness, and yellowness, respectively. Sensory evaluation for flavor, mouthfeel, and overall acceptance was conducted by 25 untrained panelists. Overall, a 1% supplementation level of pulse ingredients showed a significantly higher score than the control sample (without supplementation) in terms of flavor, mouthfeel, and overall acceptance.

2.4.7 Other Products

Yeu *et al.* studied the consumer acceptability of soy-based breakfast cereals. Soy flour was included at 41%, 47%, 54%, and 60% (w/w) in the extruded product. The consumer participants (n = 120) evaluated the experimental breakfast cereals and five commercial counterparts on overall acceptability and the acceptance of overall aroma, texture, and taste using a 9-point hedonic scale. Among the 16 laboratory samples (4 soy flour levels with or without cinnamon flavor and consumed with or without milk), some were commercial cereals, while others were high-protein soy-based cereals. A subgroup of 26 participants received cost and nutrition information about the samples they assessed. The commercial cereals had an overall acceptance score of 3.4–7.5. When cost and nutrition information were disclosed to the participants, the acceptability of the soy-based, high-protein cereals improved. The study indicated that the sensory properties of soy-based, high-protein cereals need further improvement.[69]

Soybean was studied as a low-cost solution to help with malnutrition in Malawi by replacing maize. Ng'ong'ola-Manani *et al.* looked at Malawian consumer acceptance of pastes containing soybean and soybean–maize blends with 100%, 90%, and 75% soybean, which were either naturally fermented or lactic acid bacteria fermented. Consumers (n = 150) evaluated 6 soybean pastes using a 7-point hedonic scale on acceptance of taste, smell, color, smoothness, and overall acceptability. The results indicate that most consumers preferred the naturally fermented pastes as compared to the lactic acid-fermented pastes. The naturally fermented pastes had sensory attributes such as a strong yellow color, a higher pH (5.36–5.81), a fried-egg aroma, a sweet taste, and soft textural properties, which were drivers of consumer acceptance. The lactic acid fermented pastes differed in that they had a strong brown color, a lower pH, were bitter, and had a burnt aroma. Bitter tastes are often undesirable among consumers, which may have contributed to low consumer acceptance scores for lactic acid fermented pastes. The 100% and 90% soybean pastes that were naturally fermented were most favorable among consumers (~5.00–6.25 hedonic scores). Findings from this study suggest that naturally fermented soybean pastes are more favorable among Malawian consumers.[70]

A study by Detchewa *et al.* had consumers participate in sensory evaluation to analyze consumer acceptance of gluten-free rice spaghetti (GFRS) with the addition of various SPI concentrations (0%, 2.5%, 5.0%, 7.5%, and 10.0%). The overall acceptability for the GFRS samples was highest for the GFRS with 5.0% SPI, in which scores of ~6/9 using a 9-point hedonic scale were found for sensory attributes of appearance, color, flavor, texture, and overall acceptability. However, commercial wheat spaghetti received the highest score in comparison to all GFRS samples. The overall sensory properties of the GFRS were improved with the addition of SPI at a 5.0% concentration. However, the study concluded that further formulation and sensory testing are needed to better improve consumer acceptance of GFRS.[71]

2.5 Consumer Studies on Novel Proteins

In this book, we use the term novel proteins to indicate proteins derived from bacteria, fungi, algae, cultured meat, edible insects, and plants that are not commonly used for human foods. Many studies have shown that familiarity is a significant factor influencing consumer acceptance.[30,72,73] When it comes to novel protein ingredients, due to a lack of previous knowledge and experience with the product, consumer acceptance tends to be low. This section will review sensory and survey-based consumer studies on a few novel proteins.

2.5.1 Algae Proteins

Algae include two major categories, microalgae and macroalgae (seaweed). Microalgae is very high in protein and omega-3 fatty acids and contains antioxidants and phenolics. Spirulina (*Arthrospira platensis*) and chlorophyta (*Dunaliella salina*) are two of the most widely grown and consumed microalgae species in the world. Macroalgae, which are commonly called seaweed, mainly include green, red, and brown algae.

2.5.1.1 Survey Studies

Edible algae are known to have fresh seashore, seafood-like, cucumber green, and earthy aromas.[74] Even though seaweed has long been a dietary component in Pacific and Asian cultures,[75] consumer acceptance of using algae and algae proteins in Western societies is low. A study in Australia surveyed 521 consumers nationwide. It was reported that 75% of the respondents had experience consuming seaweed, while only 37% regularly consumed seaweed. The key barriers to seaweed consumption include lack of knowledge, familiarity, and cost. A consumer study ($n = 476$) performed in the United Kingdom investigated the consumer's attitude toward using algae as an alternative protein source.[76] The study found that taste/edibility and familiarity are significant predictors of consumer acceptance. Using hypothetical food products, the study also identified differences in acceptability among these foods, with seaweed-based sushi and burgers, being more acceptable and juice drinks and baby sugar kelp being the least acceptable. Another study in the United Kingdom conducted six focus groups with a total of 34 participants. Despite limited knowledge of edible algae, consumers expressed a willingness to try algae products.[77] Another study investigated the acceptance of spirulina ingredients among 1325 Belgian consumers. Health consciousness and willingness to compromise on taste were identified as the major drivers of adoption.[78] The studies consistently showed that flavor (taste and odor) had been an obstacle to the adoption of algae ingredients. Besides, due to the significant impact of food neophobia and consumers' unfamiliarity with algae-based foods, studies suggested using algae as an ingredient rather than by itself.[79,80]

2.5.1.2 *Sensory Studies*

2.5.1.2.1 Microalgae. In a 2022 study, 50 semi-trained panelists evaluated instant jasmine rice coated with spirulina and edible polymers. The final coating formulation tested was 20% maltodextrin, 1% carboxymethyl cellulose, and 2% spirulina. The panelists rated the overall acceptability of the pasta enriched with spirulina at 7.1 ± 1.1 on a 9-point hedonic scale (slightly higher than the uncoated rice at 6.6 ± 1.0). However, the panelists did not like the green color of the pasta. In addition, a higher concentration of spirulina (3–5%) was reported as having an unnatural flavor and unpleasant odor. Edible coating with spirulina may be a favorable way to improve the health benefits of functional foods if there is a way to overcome the unfavorable green sight and unpleasant flavor and odor.[81]

A 2019 study by Fradinho *et al.* evaluated gluten-free pasta with 2% added spirulina from two different sources. The tagliatelle pasta was made with a 50/50 ratio of rice flour and psyllium gel, then two different concentrations (from 1% to 15%) of two sources of spirulina (*A. platensis* F&M-C256 and *A. platensis* Ox Nature) were incorporated into the rice pasta dough. Samples with a higher concentration of spirulina were excluded from the sensory evaluation because there were issues with a fishy odor and high cooking loss. Untrained panelists (n = 31) evaluated the pasta on the basis of color, odor, flavor, extensibility, texture, and overall liking on a 5-point scale from "very unpleasant" to "very pleasant." The spirulina pasta had a similar texture to the control pasta but a less enjoyable flavor. Although the spirulina pastas rated similarly to each other in the sensory analysis, they resulted in different bioactive components and antioxidant capacities. Fifty-eight percent of consumers said they would probably buy the spirulina pasta.[82]

A study in India added spirulina biomass in increasing concentrations (2%, 5%, 7.5%, 10%, and 12.5%) to fusilli pasta prepared with durum wheat semolina, water, 1% carboxymethyl cellulose, and cheese spread.

Semi-trained panelists (n = 83) evaluated the appearance, color, aroma/flavor, taste, mouthfeel, and overall acceptability of the pasta using a 9-point hedonic scale. The pasta with 12.5% spirulina biomass was liked "very much." It was rated the second highest (7.56) in overall acceptability, following the control (7.91). The bright green color improved the appearance of the pasta. The 12.5% pasta was also rated high in terms of color, flavor, and appearance (7.81, 7.48, and 7.82, respectively). This study was one of the first to report consumer favor for spirulina with regard to flavor. The authors concluded that the pleasant flavor could be due to the addition of cheese to the pasta.[83]

A study in Turkey facilitated an untrained panel (n = 20) to examine the color, appearance, and flavor of cookies enriched with (1% and 2%) spirulina and Dunaliella biomass. The green color of the spirulina cookies led to a less favorable appearance rating than the yellow-orange Dunaliella cookies. The 1% Dunaliella-enriched cookies scored the highest in terms of flavor (score of 4.70 ± 0.02 on a 5-point hedonic scale) compared to the spirulina and control cookies. The Dunaliella cookies also scored favorably in color (4.40 ± 0.11) and appearance (4.10 ± 0.06). The spirulina cookies were characterized by their hardness and crispness, potentially due to the increased protein content and, therefore, a high water-holding capacity. Overall, the cookies with Dunaliella showed high consumer acceptance in their appearance, color, and flavor but had a less favorable texture.[84]

A 2021 study evaluated functional semolina flour couscous fortified with 6% algae biomass. The study used four *Chlorella vulgaris* biomasses (organic Chlorella and three heterotrophic *C. vulgaris*: smooth Chlorella, honey Chlorella, and white Chlorella) and one mixture of *C. vulgaris* with two macroalgae biomasses (called *AlgaEssence*, a mixture of *Ulva rigida* and *Fucus vesiculosus*). An untrained panel (n = 33) assessed the couscous on the following traits: color, odor, taste, texture, and overall liking on a 9-point hedonic scale. The participants favored the flavor and color of the honey and smooth couscous (yellow and light green, respectively). The flavor and odor of the organic couscous were less appreciated compared to the others. Twenty-three percent of the panelists "would usually buy" and "would always buy" the honey and smooth couscous, respectively, and 32% "would occasionally buy" the organic couscous, despite the reported fishy flavor. Overall, the sensory analysis shows that couscous enriched with algae biomass would be well accepted in the market.[85]

2.5.1.2.2 Macroalgae. A 2017 study explored the sensorial properties of commercially unavailable, wild-harvested Australian seaweeds prepared in soup and salad dishes. These seaweeds had a higher total lipid content than the commercially available seaweeds, and 9 out of the 13 had a favorable omega-6 to omega-3 ratio (≤1). Forty panelists between 18 and 70 evaluated foods on a 9-point hedonic scale. The Australian seaweed species scored equally high as the commercially available seaweed strains when prepared in the soup. The *Cystophora torulosa* was described as being "tender" and held its form like the commercially available hijiki. However, *Cystophora polycystidea* and *Pseudoterranova decipiens* had a bitter flavor and a tough, stringy mouthfeel. Australian wild-harvested seaweeds may be a sustainable option for human consumption.[86]

2.5.2 Cultured Alternative Products

Cultured meat and seafood are an emerging industry that produces meat and seafood from cell cultures using the *in vivo* method. Due to innovative technology and regulations, cultured meat, especially cultured chicken, is only sold in Singapore. In the United States, the FDA announced on November 16, 2022, that they have "evaluated the information UPSIDE Foods submitted to the agency and have no further questions at this time about the firm's safety conclusion."[87] The announcement indicates the FDA's support for the safety of cultured meat products and will accelerate progress toward bringing cultured meat to the market.

Numerous groups studied and reviewed the consumer acceptance of cultured meat.[30,88–108] Due to the limited products available to consumers, studies on consumer acceptance of cultured products are conducted on a predictive basis using surveys and focus groups.[108] Based on the survey studies, consumer acceptance of cultured meat is generally promising. Consumer acceptance varies among cultures and in different countries. Also, as the industry is evolving and growing very quickly, consumer perspectives on cultured meat are also changing as more information becomes available. Wilks and Phillips reported in 2017 that 65.3% of 673 respondents from the United States were willing to try cultured meat, and 31.5% of them were willing to eat it instead of farmed meat.[88] Similarly, Bryant *et al.* reported in 2019 that 66.4% of 1185 respondents in the United States would try cultured meat and 55.2% of them would eat it instead of conventional meat.[109] A study in Canada reported that 11% of 533 respondents would eat cultured meat over conventional and plant-based meat.[34] Three studies came out in 2020 in Germany. Weinrich *et al.* found that 57% of 713 German consumers showed willingness to try cultured meat, while less than 30% were willing to consume it regularly.[92] Dupont and Fiebelkorn surveyed 718 German children and adolescents and reported that 63.2% of boys and 53.2% of girls were willing to consume the cultured meat burger.[110] Bryant *et al.* compared German and French consumers and reported that 42.5% of German and 29.2% of French consumers preferred cultured meat over plant-based products, and 49% of German and 33.9% of French consumers preferred cultured meat over edible insects.[102] Mancini and Antonioli reported in 2019 that 54% of 525 Italian respondents would like to try cultured meat.[111]

2.6 Conclusion

Consumer acceptance is a key determinant of the success of food products. With the increasing demands for protein, novel protein ingredients have been developed from various sources, including but not limited to pulses, algae, bacteria, fungi, and edible insects. The creation of protein-based and protein-enhanced foods will continue to be an important trend in the modern food industry. However, consumer acceptance for protein ingredients, especially plant proteins and novel protein ingredients from algae, edible insects, and cultured sources, is limited. Improving the flavor characteristics of protein ingredients, especially proteins from plants, algae, and edible insects, is essential to boosting the use of these ingredients in human foods.

Abbreviations

GFRS	Gluten-free rice spaghetti
HMMAs	High moisture meat analogs
ML	Moisture level
RBPC	Rice bran protein concentrate
RBPH	Rice bran protein hydrolysate
SPI	Soy protein isolate
WPC	Whey protein concentrate
WPI	Whey protein isolate

References

1. E. Leahy, S. Lyons and R. S. Tol, *An estimate of the number of vegetarians in the world*, ESRI working paper, 2010.
2. M. Shahbandeh, Global plant protein market value 2022-2027, Statista, Available from: https://www.statista.com/statistics/658619/global-plant-protein-market-value/. Cited 30 Nov 2022.
3. The Good Food Institute Asia Pacific, Alternative Protein Company Database, GFI Asia Pacific, 2022, Available from: https://gfi-apac.org/industry/alternative-protein-company-database/. Cited 30 Nov 2022.
4. M. C. Whetstine, A. Croissant and M. Drake, *J. Dairy Sci.*, 2005, **88**, 3826–3839.
5. M. A. Mortenson, Z. M. Vickers and G. A. Reineccius, *Int. Dairy J.*, 2008, **18**, 649–657.
6. J. Evans, J. Zulewska, M. Newbold, M. Drake and D. Barbano, *J. Dairy Sci.*, 2009, **92**, 4773–4791.
7. S. Y. Lee, K. Dangaran, J. X. Guinard and J. Krochta, *J. Food Sci.*, 2002, **67**, 2764–2769.

8. K. Kaewka, C. Therakulkait and K. R. Cadwallader, *J. Cereal Sci.*, 2009, **50**, 56–60.
9. S. Arsa and C. Theerakulkait, *Effect of pH and fructose addition on aroma and color characteristics of papain hydrolyzed rice bran protein concentrate*, in 55. Kasetsart University Annual Conference, Bangkok (Thailand), 2017.
10. S. Arsa and Y. Puechkamutr, *J. Food Sci. Technol.*, 2022, **59**, 890–897.
11. S. Arsa and C. Theerakulkait, *J. Food Sci. Technol.*, 2015, **52**, 5285–5291.
12. S. Arsa, C. Theerakulkait and K. R. Cadwallader, *J. Agric. Food Chem.*, 2019, **67**, 8205–8211.
13. S. Arsa and C. Theerakulkait, *J. Sci. Food Agric.*, 2018, **98**, 4479–4487.
14. Y. Hu, D. Sun-Waterhouse, P. z. Liu, C. Cui and W. Wang, *Int. J. Food Sci. Technol.*, 2019, **54**, 2458–2467.
15. Y. Liu, R. S. D. Toro-Gipson and M. Drake, *J. Sens. Stud.*, 2021, **36**, e12704.
16. J. L. Childs, J. L. Thompson, J. S. Lillard, T. K. Berry and M. Drake, *J. Sens. Stud.*, 2008, **23**, 320–339.
17. V. Jones, M. Drake, R. Harding and B. Kuhn-Sherlock, *J. Sens. Stud.*, 2008, **23**, 65–79.
18. E. C. Short, A. J. Kinchla and A. A. Nolden, *Foods*, 2021, **10**, 725.
19. Q. Li, Y. Xia, L. Zhou and J. Xie, *Food Bioprod. Process.*, 2013, **91**, 429–439.
20. S. Chumchuere, D. B. MacDougall and R. K. Robinson, *Int. J. Food Sci. Technol.*, 2000, **35**, 577–581.
21. Y. Li, X. Zhang, J.-J. Yang, X.-Y. Ma, X.-D. Jia, P. Du and A.-I. Li, *J. Food Process. Preserv.*, 2020, **44**, e14823.
22. S. Khodke, M. Pardhi, M. Pramodini and A. Kakade, *IOSR J. Environ. Sci. Toxicol. Food Technol.*, 2014, **8**(9), 12–16.
23. M. Butool and S. Butool, *Int. J. Dev. Res.*, 2015, **5**, 4124–4130.
24. M. A. Balogun, S. A. Oyeyinka, F. L. Kolawole, J. K. Joseph and G. E. Olajobi, *Ceylon J. Sci.*, 2019, **48**, 353–358.
25. T. Kim, R. Miller, H. Laird and M. N. Riaz, *Food Sci. Nutr.*, 2021, **9**, 4048–4056.
26. C. Hartmann and M. Siegrist, *Trends Food Sci. Technol.*, 2017, **61**, 11–25.
27. S. E. Lawrence, K. Lopetcharat and M. A. Drake, *J. Food Sci.*, 2016, **81**, S463–S476.
28. A. T. Oyeyinka, J. O. Odukoya and Y. S. Adebayo, *J. Food Process. Preserv.*, 2019, **43**, e14285.
29. M. Lopes, C. Pierrepont, C. M. Duarte, A. Filipe, B. Medronho and I. Sousa, *Foods*, 2020, **9**, 1458.

30. M. C. Onwezen, E. P. Bouwman, M. J. Reinders and H. Dagevos, *Appetite*, 2021, **159**, 105058.
31. J. He, N. M. Evans, H. Liu and S. Shao, *Compr. Rev. Food Sci. Food Saf.*, 2020, **19**, 2639–2656.
32. G. T. Tonsor, J. L. Lusk and T. C. Schroeder, 2021, Available from: https://www.agmanager.info/sites/default/files/pdf/PlantBasedProteinAlternatives_FullReport.pdf.
33. E. J. Van Loo, V. Caputo and J. L. Lusk, *Food Policy*, 2020, **95**, 101931.
34. P. Slade, *Appetite*, 2018, **125**, 428–437.
35. C. Tobler, V. H. M. Visschers and M. Siegrist, *Appetite*, 2011, **57**, 674–682.
36. M. C. Mancini and F. Antonioli, *Meat Sci.*, 2022, **193**, 108942.
37. J. Aschemann-Witzel and A. O. Peschel, *Food Hydrocolloids*, 2019, **96**, 20–28.
38. I. Faber, K. Henn, M. Brugarolas and F. J. Perez-Cueto, *J. Sci. Food Agric.*, 2022, **102**, 5034–5043.
39. G. R. Shaviklo, S. Arason, G. Thorkelsson, K. Sveinsdottir and E. Martinsdottir, *J. Sens. Stud.*, 2010, **25**, 316–331.
40. G. R. Shaviklo, A. Olafsdottir, K. Sveinsdottir, G. Thorkelsson and F. Rafipour, *J. Food Sci. Technol.*, 2011, **48**, 668–676.
41. M. Díaz-Ramírez, G. Calderón-Domínguez, M. García-Garibay, J. Jiménez-Guzmán, A. Villanueva-Carvajal, M. de la Paz Salgado-Cruz, D. Arizmendi-Cotero and E. Del Moral-Ramírez, *Food Hydrocolloids*, 2016, **61**, 633–639.
42. R. Jyotsna, R. Sai Manohar, D. Indrani and G. Venkateswara Rao, *Int. J. Food Prop.*, 2007, **10**, 599–606.
43. I. C. Torres, J. M. Amigo, J. C. Knudsen, A. Tolkach, B. Ø. Mikkelsen and R. Ipsen, *Int. Dairy J.*, 2018, **81**, 62–71.
44. T. Ö. Yilsay, L. Yilmaz and A. A. Bayizit, *Eur. Food Res. Technol.*, 2006, **222**, 171–175.
45. S. Mazinani, A. Motamedzadegan, S. Nghizadeh Raeisi and M. Alimi, *J. Food Meas. Charact.*, 2021, **15**, 5515–5527.
46. M. Drake, X. Chen, S. Tamarapu and B. Leenanon, *J. Food Sci.*, 2000, **65**, 1244–1247.
47. M. Dervisoglu, F. Yazici and O. Aydemir, *Eur. Food Res. Technol.*, 2005, **221**, 466–470.
48. S. Adeyeye, A. Adebayo-Oyetoro and S. Omoniyi, *Cogent Food Agric.*, 2017, **3**, 1278827.
49. S. M. Mohsen, H. H. Fadel, M. Bekhit, A. E. Edris and M. Y. Ahmed, *Int. J. Food Sci. Technol.*, 2009, **44**, 1705–1712.
50. S. Mazinani, A. Motamedzadegan, S. Naghizadeh Raeisi and M. Alimi, *J. Food Process. Preserv.*, 2020, **44**, e14448.

51. N. Omrani Khiabanian, A. Motamedzadegan, S. Naghizadeh Raisi and M. Alimi, *J. Texture Stud.*, 2020, **51**, 488–500.
52. M. Wee, D. Loud, V. Tan and C. Forde, *Food Chem.*, 2019, **294**, 152–159.
53. M. Danowska-Oziewicz and L. Kurp, *Meat Sci.*, 2017, **131**, 158–165.
54. A. Mubarak, *Food/Nahrung*, 2001, **45**, 241–245.
55. T. Chen, F. Wu, J. Guo, M. Ye, H. Hu, J. Guo and X. Liu, *J. Sci. Food Agric.*, 2020, **100**, 3297–3307.
56. C. E. De Farias Silva, R. C. Vieira, I. C. C. da Silva, R. B. de Oliveira Cerqueira, N. P. Andrade, F. Claudino da Silva, F. P. de Andrade, A. K. de Souza Abud, K. Andreola and O. P. Taranto, *J. Food Sci. Technol.*, 2021, **58**, 4194–4204.
57. S. Zamuz, L. Purriños, F. Galvez, N. Zdolec, V. Muchenje, F. J. Barba and J. M. Lorenzo, *J. Food Process. Preserv.*, 2019, **43**, e13940.
58. M. Drake and P. Gerard, *J. Food Sci.*, 2003, **68**, 1118–1122.
59. B. Miñarro, E. Albanell, N. Aguilar, B. Guamis and M. Capellas, *J. Cereal Sci.*, 2012, **56**, 476–481.
60. P. García-Segovia, M. Igual and J. Martínez-Monzó, *Foods*, 2020, **9**, 933.
61. B. Ivanovski, K. Seetharaman and L. M. Duizer, *J. Food Sci.*, 2012, **77**, S71–S76.
62. M. Siddiq, S. Kelkar, J. B. Harte, K. D. Dolan and G. Nyombaire, *LWT-Food Sci. Technol.*, 2013, **50**, 215–219.
63. A. Waglay and S. Karboune, *Food Bioprocess Technol.*, 2020, **13**, 1645–1660.
64. A. Saint-Eve, P. Granda, G. Legay, G. Cuvelier and J. Delarue, *J. Sci. Food Agric.*, 2019, **99**, 3983–3991.
65. C. S. do Carmo, P. Varela, C. Poudroux, T. Dessev, K. Myhrer, A. Rieder, H. Zobel, S. Sahlstrøm and S. H. Knutsen, *LWT*, 2019, **112**, 108252.
66. J. W. Kreger, Y. Lee and S.-Y. Lee, *J. Food Sci.*, 2012, **77**, S161–S169.
67. J. Walker, C. Boeneke, S. Sriwattana, J. A. Herrera-corredor and W. Prinyawiwatkul, *J. Food Qual.*, 2010, **33**, 27–41.
68. F. Zare, V. Orsat and J. I. Boye, *J. Food Res.*, 2015, **4**, 143.
69. K. Yeu, Y. Lee and S.-Y. Lee, *J. Food Sci.*, 2008, **73**, S20–S25.
70. T. A. Ng'ong'ola-Manani, A. M. Mwangwela, R. B. Schüller, H. M. Østlie and T. Wicklund, *Food Sci. Nutr.*, 2014, **2**, 114–131.
71. P. Detchewa, M. Thongngam, J.-L. Jane and O. Naivikul, *J. Food Sci. Technol.*, 2016, **53**, 3485–3494.
72. S. A. Siddiqui, T. Alvi, A. Sameen, S. Khan, A. V. Blinov, A. A. Nagdalian, M. Mehdizadeh, D. N. Adli and M. Onwezen, *Sustainability*, 2022, **14**, 15370.
73. E. Woolf, Y. Zhu, K. Emory, J. Zhao and C. Liu, *LWT*, 2019, **102**, 100–105.

74. N. Francezon, A. Tremblay, J.-L. Mouget, P. Pasetto and L. Beaulieu, *J. Agric. Food Chem.*, 2021, **69**, 11753–11772.
75. L.-E. Rioux, L. Beaulieu and S. L. Turgeon, *Food Hydrocolloids*, 2017, **68**, 255–265.
76. R. Embling, L. Neilson, T. Randall, C. Mellor, M. D. Lee and L. L. Wilkinson, *Food Qual. Prefer.*, 2022, **100**, 104613.
77. C. Mellor, R. Embling, L. Neilson, T. Randall, C. Wakeham, M. D. Lee and L. L. Wilkinson, *Foods*, 2022, **11**.
78. I. Moons, C. Barbarossa and P. De Pelsmacker, *Ecological Economics*, 2018, **151**, 151–161.
79. A. S. Chapman, P. Stévant and W. E. Larssen, *Bot. Mar.*, 2015, **58**, 423–433.
80. D. Birch, K. Skallerud and N. A. Paul, *Br. Food J.*, 2018, **121**, 603–615.
81. N. Utama-ang, I. Kuatrakul, P. Walter, P. Rattanapitigorn and A. Kawee-Ai, *Sci. Rep.*, 2022, **12**, 7699.
82. P. Fradinho, A. Niccolai, R. Soares, L. Rodolfi, N. Biondi, M. R. Tredici, I. Sousa and A. Raymundo, *Algal Res.*, 2020, **45**, 101743.
83. D. K. Koli, S. G. Rudra, A. Bhowmik and S. Pabbi, *Foods*, 2022, **11**, 979.
84. O. I. Şahin, *J. Food Sci. Technol.*, 2020, **57**, 3639–3646.
85. S. Khemiri, M. C. Nunes, R. J. Bessa, S. P. Alves, I. Smaali and A. Raymundo, *Foods*, 2021, **10**, 3159.
86. V. M. Skrzypczyk, K. M. Hermon, F. Norambuena, G. M. Turchini, R. Keast and A. Bellgrove, *J. Appl. Phycol.*, 2019, **31**, 709–724.
87. U.S. Food and Drug Administration (FDA), 2022, Available from: https://www.fda.gov/food/cfsan-constituent-updates/fda-completes-first-pre-market-consultation-human-food-made-using-animal-cell-culture-technology.
88. M. Wilks and C. J. Phillips, *PLoS One*, 2017, **12**, e0171904.
89. C. Bryant and J. Barnett, *Meat Sci.*, 2018, **143**, 8–17.
90. C. Bryant and J. Barnett, *Appl. Sci.*, 2020, **10**, 5201.
91. W. Verbeke, P. Sans and E. J. Van Loo, *J. Integr. Agric.*, 2015, **14**, 285–294.
92. R. Weinrich, M. Strack and F. Neugebauer, *Meat Sci.*, 2020, **162**, 107924.
93. N. Palmieri, M. A. Perito and C. Lupi, *Br. Food J.*, 2020, **123**, 109–123.
94. M. Zhang, L. Li and J. Bai, *Food Control*, 2020, **118**, 107390.
95. A. Pakseresht, S. A. Kaliji and M. Canavari, *Appetite*, 2022, **170**, 105829.
96. N. C. M. Rolland, C. R. Markus and M. J. Post, *PLoS One*, 2020, **15**, e0231176.
97. M. Siegrist, B. Sütterlin and C. Hartmann, *Meat Sci.*, 2018, **139**, 213–219.

98. M. C. Mancini and F. Antonioli, *Animals*, 2020, **10**, 656.
99. C. Bryant and C. Dillard, *Front. Nutr.*, 2019, **6**, 103.
100. S. A. Siddiqui, N. A. Bahmid, I. Karim, T. Mehany, A. A. Gvozdenko, A. V. Blinov, A. A. Nagdalian, M. Arsyad and J. M. Lorenzo, *LWT*, 2022, **172**, 114192.
101. I. T. Kadim, O. Mahgoub, S. Baqir, B. Faye and R. Purchas, *J. Integr. Agric.*, 2015, **14**, 222–233.
102. C. Bryant, L. van Nek and N. C. Rolland, *Foods*, 2020, **9**, 1152.
103. S. A. Siddiqui, S. Khan, M. Q. U. Farooqi, P. Singh, I. Fernando and A. Nagdalian, *Appetite*, 2022, 106314.
104. W. Verbeke, A. Marcu, P. Rutsaert, R. Gaspar, B. Seibt, D. Fletcher and J. Barnett, *Meat Sci.*, 2015, **102**, 49–58.
105. M. Siegrist and C. Hartmann, *Appetite*, 2020, **155**, 104814.
106. L. Zhang, Y. Hu, I. H. Badar, X. Xia, B. Kong and Q. Chen, *Trends Food Sci. Technol.*, 2021, **116**, 434–444.
107. X. Guan, Q. Lei, Q. Yan, X. Li, J. Zhou, G. Du and J. Chen, *Future Foods*, 2021, **3**, 100032.
108. N. Treich, *Environ. Resour. Econ.*, 2021, **79**, 33–61.
109. C. J. Bryant, J. E. Anderson, K. E. Asher, C. Green and K. Gasteratos, *Meat Sci.*, 2019, **154**, 37–45.
110. J. Dupont and F. Fiebelkorn, *Food Qual. Prefer.*, 2020, **85**, 103983.
111. M. C. Mancini and F. Antonioli, *Meat sci.*, 2019, **150**, 101–110.

CHAPTER 3

Chemical and Instrumental Characterization of Protein–Flavor Interactions

VAIDHYANATHAN ANANTHARAMKRISHNAN*[a] AND GARY A. REINECCIUS*[b]

[a]Research and Development, Lux Flavors, Chennai 600114, India; [b]Department of Food Science and Nutrition, University of Minnesota, 1334, Eckles Avenue, St. Paul, Minnesota 55108, USA
*E-mail: anant040@umn.edu, greinecc@umn.edu

3.1 Introduction

As early as 1920, researchers reported that the sensory character and intensity of flavor compounds depended on the food components (*e.g.*, salt, sugar, fat, complex carbohydrates, and proteins) present in the mixture.[1] Salts and sugars typically decrease the sensory thresholds of added flavor compounds, while the addition of lipids, some complex carbohydrates, and proteins increases the sensory threshold levels. While the influence of fat was readily understood, the interactions between flavor compounds, carbohydrates, and proteins were more complicated. The interactions of flavor components with proteins were found to be very complex, involving both weak and transient bonds (*e.g.*, hydrophilic, ionic, or hydrophobic bonds) and very rigorous covalent bonds.[2] These interactions were considered to be responsible for undesirable changes in flavorings added to

Food Chemistry, Function and Analysis No. 41
Flavour and Consumer Perception of Food Proteins
Edited by Jing Zhao and Changqi Liu

Published by the Royal Society of Chemistry, www.rsc.org

Table 3.1 Methods used to measure reversible binding between proteins and flavor compounds. Reproduced from ref. 3 with permission from John Wiley & Sons, Copyright 2017.

Technical approaches	Sources of proteins	Flavor compounds	Ref.
Distillation/high vacuum/ extraction/ purging	Soy protein	Hexanol, hexanal	2,4
Static headspace-gas chromatography	Soy protein, board bean globulins, β-lactoglobulin	Aldehydes, ketones, alcohols	5,6,7,8,9,10,11
Equilibrium dialysis	Soy proteins, bovine serum albumin, β-lactoglobulin	Lactones, alkanones,	12,13,14,15,16
Solid-phase microextraction	Milk proteins, animal proteins	Aldehydes, ketones, esters, alcohols	17,18,19,20,21, 22
In-tube extraction with *Tenax* trap	Pea, canola, wheat	Aldehydes, ketones	23,24,25,26,27
High-performance liquid chromatography	Soy protein, Fababean proteins, β-lactoglobulin, sodium caseinate, bovine serum albumin, whey protein isolate	Vanillin	28,29,30,31
Radioisotopes	Soy protein, β-lactoglobulin	Benzyl alcohol, benzaldehyde	32,33

protein-containing food products initially and potentially during storage. Thus, the academic community embarked on a journey to develop ways to measure these interactions and find solutions to minimize their deleterious effects. A summary of the methods that have been developed to measure protein–flavor interactions is presented in Table 3.1 and discussed in the following sections.

3.2 Methods to Measure Reversible Protein–Flavor Interactions

3.2.1 Early Methods

Arai *et al.* (1970)[34] used three different methods to study the interaction of hexanol and hexanal (two volatiles identified to be present in soy) with soy protein. The first method, *Vacuum distillation* (30 °C under reduced pressure), was used to initially remove the water and entrained volatiles from an aqueous reaction system (protein + flavor compound), then a high vacuum treatment was applied for 10 h to produce a powder. The powder was dissolved in an alkaline solution (to dissolve the protein), neutralized with HCl, brought to the desired volume, and analyzed by gas chromatography (GC). The second method, *sweeping evaporation*, was used with an enzyme treatment to monitor how an enzyme might alter protein structure to release bound volatiles from soy protein. (The binding was considered to be non-covalent in nature, and thus, the protein structure was considered a key factor in determining protein–flavor binding.) It started with the same initial vacuum distillation (30 °C), followed by placing the concentrated sample in a flask and passing nitrogen gas over the sample. The exit gas was sampled and analyzed for volatiles by GC every 6 min. This provided a time release profile for the enzyme treatment. The third method, *countercurrent extraction*, was used to further study the effect of hydrolysis on volatile binding/release. This method measured the model volatile concentrations partitioning into hexane from the aqueous reaction system. As a summary, the authors noted that the model volatiles were "hardly extractable" from the aqueous protein solution with n-hexane. While the bonding was non-covalent, volatiles were considered to be strongly bonded to the protein.

Gremli (1974)[4] also worked with soy and used two methods in analysis. The first method was *traditional headspace analysis* (discussed later), and the second method was the *high vacuum transfer* method. Gremli studied the interaction of 14 volatiles (mostly aldehydes, with some ketones and alcohols) with soy protein isolate. The *high-vacuum transfer* method started with the model flavor compound being added to an aqueous slurry of soy protein and held for an hour at 22 °C to allow a complete reaction. The sample was shell frozen at −70 °C and connected to a second flask that was held at −20 °C, and the first flask was warmed to 10 °C. Over 6 h, the collected water (with associated volatiles) was transferred to the second flask, yielding an aqueous solution, which was then extracted with dichloromethane, concentrated, and analyzed by GC. Retention/binding were determined relative to the recovery of the same model volatiles from an aqueous system (no protein). Their summary was that alcohols have no reaction with soy protein, ketones are reversibly bound (reduced headspace concentrations but were completely recovered on high vacuum transfer), and aldehydes reduced headspace concentrations but were partially recovered by the high vacuum system (a portion was irreversibly bound).

3.2.2 Equilibrium Headspace Analysis

The methods of study initially focused on the equilibrium measurements of headspace above aqueous samples due to the simplicity of there being no requirement for the separation of volatiles from the model system.

The headspace gas above a system provided a simple sample for the measurement that reflected the system effects on the volatility of various flavor molecules. Materials that would make the model system less favorable for flavor molecules would force them into the headspace, and those that involved binding and a reduction in volatility would lessen the headspace concentration (Figure 3.1). Using this method, substantial data were obtained on the interactions between various food components and flavoring materials in solution (see Table 3.1).[5–7]

The limitation of equilibrium headspace methodology tends to be its sensitivity.[8,9] Headspace methods are not particularly suitable for measuring volatiles at very low concentrations, such as those typically found in foods. Schaefer noted that a volatile must be present in sampled headspace at 10^{-5} to 10^{-6} g L^{-1} to be detected by headspace methods.[10] Also, the methodology would work best for volatiles with high vapor pressures and not be adequately sensitive to those with very low vapor pressures. The

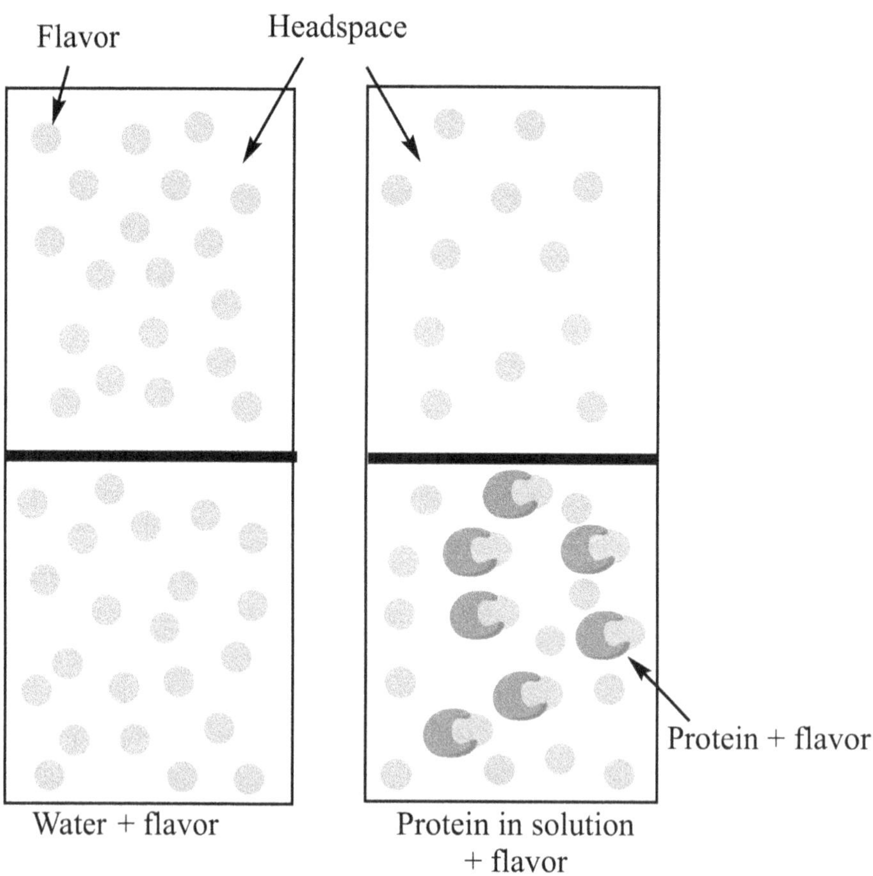

Figure 3.1 Illustration of headspace analysis.

low sensitivity across flavor compounds suggested a need for a different methodology or a modification to improve sensitivity.

Within these limitations, the method was very simple and provided substantial data.[11] The method involved preparing an aqueous solution of protein and adding the desired model flavor compound in a septum-capped vial to readily permit headspace sampling. The vial was held in a water bath to control temperature and sampled at selected time intervals (*e.g.*, every 5 min). Sampling was done by simply withdrawing an aliquot of headspace (*e.g.*, 0.5 mL) and injecting it into a GC. Equilibrium was generally reached in *ca.* 20 min.

3.2.3 Headspace Solid-phase Microextraction

Pawliszyn and Arthur[12] developed a method for headspace concentration that proved useful in studying equilibrium headspace protein–flavor interactions at much lower flavor concentrations. They termed this extraction method solid-phase microextraction (SPME). In this technique, an inert fiber is coated with an adsorbent that is placed in the headspace above a sample or in the sample itself for a given time period (see Figure 3.2).[13] This allows the fiber to adsorb/concentrate volatiles from the sample or sample headspace.[14] The loaded fiber is then thermally desorbed in the injection port of a GC or specialized thermal desorber-cryfocusing unit to release the volatiles for analysis. The system can be operated manually or automatically.

This technique has been applied to measure protein–flavor binding by four research groups (see Table 3.1). Fabre *et al.* (2002)[15] initiated the use of this technique (SPME headspace) by comparing binding results (selected volatiles in dairy protein emulsions) obtained using SPME headspace, static headspace, and dynamic headspace (aka Purge and Trap). They reported that the best results were obtained with the SPME headspace method. Gianelli *et al.* (2005)[16] initiated studies using SPME headspace on the binding of pork muscle components with various model volatiles. The primary objective was to determine the influence of curing agents on the binding between peptides/proteins and flavor compounds. Perez and Flores (2006, 2007, 2008)[17–19] built upon this work using similar methodology to study the influence of curing salts associated with pork processing on the binding of selected volatiles to pork proteins. These studies on the effect of various salts on flavor binding/release from pork proteins were partially motivated by a consumer's desire to reduce the salt (NaCl) in his/her diets by substituting other salts for NaCl. Kuhn *et al.* (2007)[20] used SPME headspace concentration to study the effect of thermal processing on volatile binding using whey protein isolates. Their work focused on the effect of protein denaturation on binding. Their work also studied the effect of high pressure processing on protein–flavor binding.

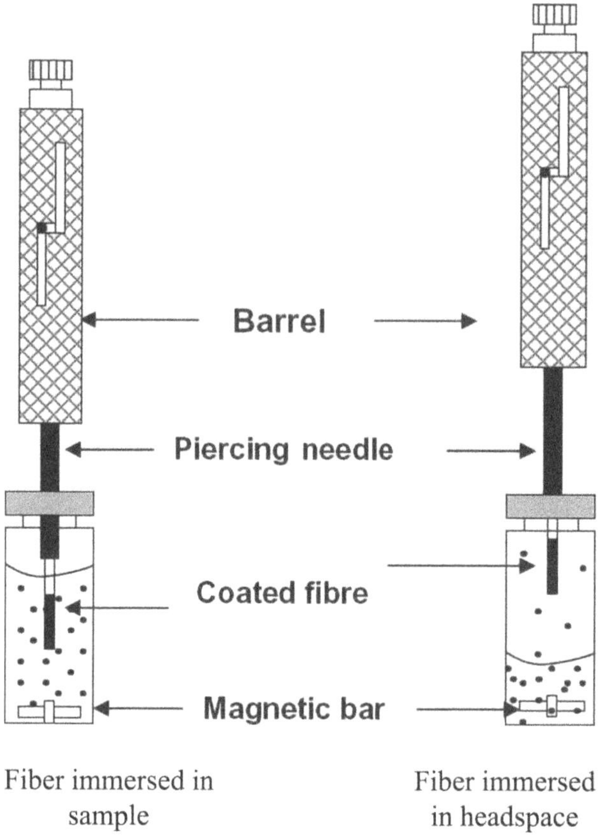

Figure 3.2 Use of SPME to concentrate headspace volatiles for GC analysis. Reproduced from ref. 13 with permission from John Wiley & Sons, Copyright 2010.

3.2.4 Headspace In-tube Extraction

In this system (see Figure 3.3), a sorbent-packed (or coated) needle is fitted to a gas tight syringe (with side port) surrounded by an external heater (Figure 3.3A). Sample vapors are drawn into the syringe through the adsorbent/coating. Pumping the syringe several times draws volatiles through the adsorbent, resulting in the adsorbent picking up/concentrating volatiles (Figure 3.3B). After the desired number of syringe pumps, the adsorbent is heated, and the adsorbed volatiles are thermally desorbed into the GC injector with desorption gas or aspirated carrier gas from the injector (Figure 3.3C). After the syringe is withdrawn from the injection port, the plunger is moved above the side port, and the heated trap is flushed with inert gas several times to avoid carryover (Figure 3.3D). Laaks *et al.* (2010)[21] compared this methodology for the volatile analysis

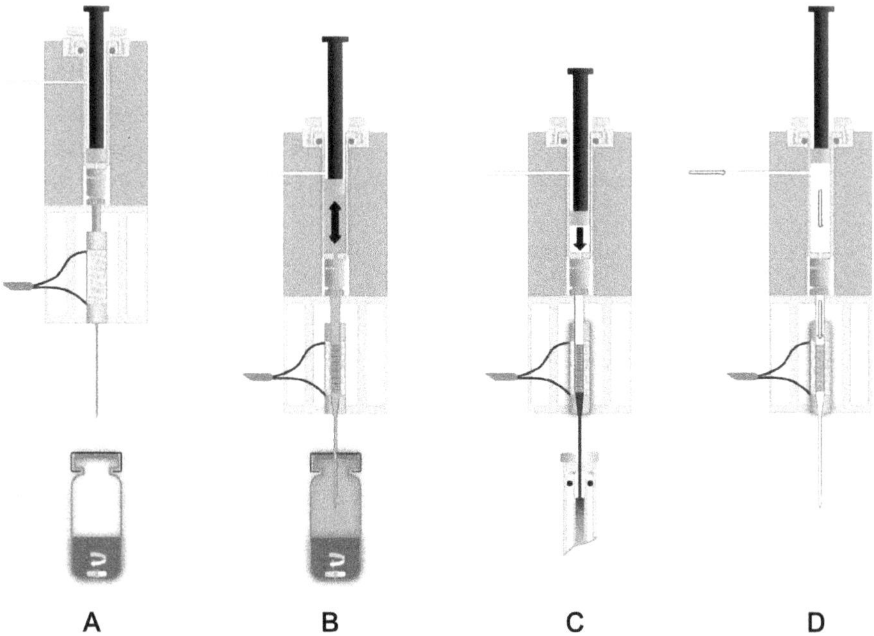

Figure 3.3 Headspace concentration using ITEX. (A) A sorbent-packed needle is fitted to a gas tight syringe with an external heater. (B) Headspace volatiles are drawn into the syringe through the adsorbent. (C) The adsorbed volatiles are thermally desorbed into the GC injector. (D) The plunger is moved, and the heated trap is flushed with inert gas to avoid cross contamination. Reproduced from ref. 21 with permission from American Chemical Society, Copyright 2010.

of water quality and concluded that the in-tube extraction (ITEX) system gives similar sensitivity to that obtained by purge and trap systems, but that it requires smaller sample volumes and is less prone to contamination. They also found the method to be much simpler, more flexible, and more affordable than purge and trap. The average relative standard deviations were below 10% for all analytes.

Laaks *et al.* (2010)[21] also offered comparisons with SPME and noted several weaknesses of SPME that were less problematic with the ITEX method:

- Fragility of the fused-silica fiber
- Mechanical damage to the coating due to scraping and needle bending during agitation

- Limited flexibility in fiber length and coating thickness, resulting in low amounts of stationary phase [about 0.6 μL on a 100 μm polydimethylsiloxane (PDMS) fiber]
- Ghost peaks from septum particles or fiber glue, a possible memory effect from incomplete sample desorption, and the limited lifetime.

Wang and Arntfield published four papers using this methodology to study flavor interactions with pea (2015, 2016a, b); salt-extracted pea and canola (2015); and canola, pea, and wheat proteins (2014).[22–26]

3.2.5 Equilibrium Membrane Dialysis

Equilibrium membrane dialysis was used as another means of studying protein–flavor interactions at lower concentrations and their work with less volatile flavor compounds. The process is simple in that volatiles added to one side of a dialysis chamber would pass through the membrane and come to equilibrium in the two chambers. If there is protein in solution on one side of the chamber and the flavor compound binds to the protein, the concentration of free flavor compound on both sides would decrease, reflecting the extent of the binding (see Figure 3.4).

By changing the experimental conditions, binding parameters such as binding constants and the number of binding sites can be determined. A weakness of this method is that, like the previously described headspace equilibrium method, it does not differentiate the type of reaction that occurs between the protein and the flavor, and the results are highly sensitive to solution temperature, type of dialysis, and volume shifts during dialysis.

One might question why it is necessary to use a dialysis system as opposed to simply adding the flavor to an aqueous protein solution, allowing the desired reaction time, and solvent extracting to quantify the free flavor. There are two reasons why this approach cannot be done: the first is that the dialysis membrane provides a protein-free, equilibrated sample that one can solvent extract without emulsion formation, *i.e.*, only water plus flavorants (Figure 3.4, lower right). Proteins tend to be very good emulsifiers, and simply adding a flavoring to the protein and attempting to solvent extract the flavoring would generally result in an inseparable emulsion. A second reason is that one might solvent-extract flavor molecules that are weakly bound to the protein. These weakly bound flavor molecules may not be sensed in food but can be extracted and measured by solvent extraction.

Damodaran and Kinsella (1981)[5] used this method to study the reaction of alkanones and one aldehyde with soy protein (native, partially denatured, or succinylated). They put the protein solution (1%) on one

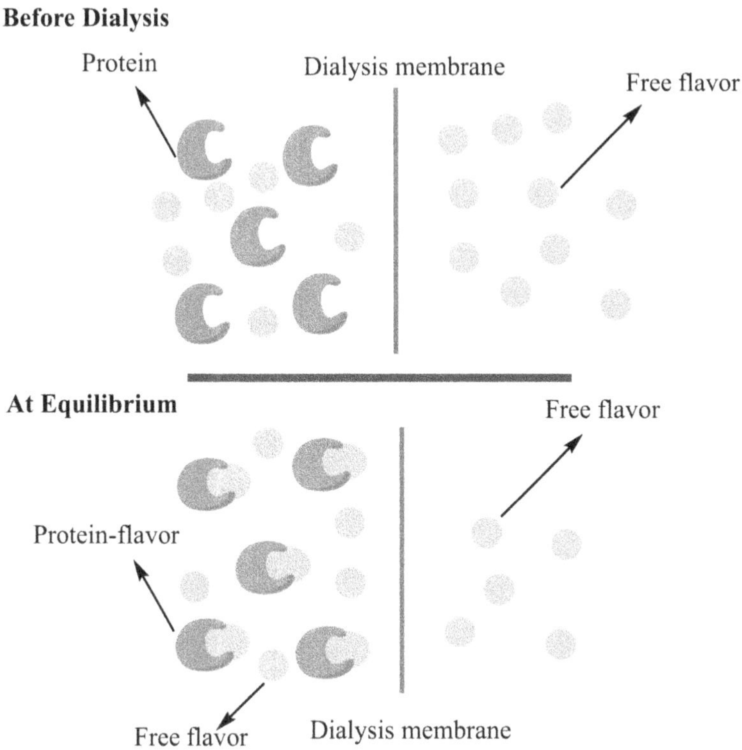

Figure 3.4 Example of equilibrium membrane dialysis.

side of the dialysis chamber and the buffer/flavor on the other side. The dialysis cells were shaken for at least 18 h to allow time for equilibrium to be reached. One millimeter of the aqueous buffer side of the dialysis apparatus was added to pure isooctane (1 mL), shaken to extract the model volatile, decanted, and analyzed by GC. Damodaran and Kinsella[6] used this same method to study how protein structure may influence the binding of 2-nonanone. O'Neill and Kinsella (1988)[7,8] used this method to study alkanone interactions with β-lactoglobulin (BLG), while Druaux *et al.* (1995)[9] studied γ-decalactone interactions with bovine serum albumin under conditions related to alcoholic beverages.

3.2.6 Inverse Gas Chromatography and Radioisotopes

While the two equilibrium methods (headspace and dialysis) discussed above provided the greatest amount of data to the literature (Table 3.1), two additional methods are also mentioned: inverse gas chromatography (IGC) and radioisotopes.

3.2.6.1 Inverse Gas Chromatography

In IGC, the column is packed with a protein of interest, and a flavor molecule is injected into the carrier gas. One can calculate the dead volume of the column and the rate of carrier gas flow to give an expected elution time, assuming there is no interaction between the protein and the flavor molecule. The degree to which a compound retards the elution of a compound reflects the binding between the flavor molecule and the protein.[27] The advantage of IGC is that it is capable of measuring the adsorption of flavor compounds to a protein at very low concentrations and at low moisture content/water activities.[28,29] All of the other methods of analysis for measuring protein–flavor interactions are conducted in solution: many foods exist in dry or semi-dry water activities. Similar to the headspace and equilibrium methods, one can measure adsorption but cannot differentiate the type of linkages (reversible or irreversible).

This method has been used to study the influence of flavor compound functional group and relative humidity on flavor interaction with soy protein isolates[27,29] and the binding of butter flavor components with wheat-based soda cracker systems.[28]

3.2.6.2 Radioisotopes

The use of radioisotopes to monitor protein–flavor binding offers some major advantages over the previously discussed methods. Issues of volatility, solubility, ease of analysis, and sensitivity tend to be non-issues when using radioisotopes to measure binding. King and Solms (1979)[30] were the first to use this method to measure the binding of benzyl alcohol to either heat-denatured bovine serum albumin or Soyamin 90 (denatured, dried soy protein). They started with a 1% aqueous solution of the protein being studied and then added the desired amount of labeled and unlabeled benzyl alcohol. They allowed a 16 h reaction time at 23 °C. They then withdrew a sample to count/determine the total concentration of radioactivity. Another sample was taken and centrifuged to precipitate the protein, and the liquid (non-protein phase) was counted again to determine the unbound benzyl alcohol. From these counts, one could determine the amount of C14-labeled benzyl alcohol that was bound to the protein.

Andriot *et al.* (1999)[31] used radioisotopes to study the kinetics of protein–flavor interactions. They added C14-labeled benzaldehyde to a BLG solution and then injected an aliquot of the solution for high-performance liquid chromatography (HPLC) every 30 min. The HPLC separated the free unreacted benzaldehyde from the reacted (labeled) BLG: the radioactivity was measured at the column exit using a Berthold LB 506-CI LC radioactivity detector. The increase in the size of the LC peak for bound benzaldehyde indicates the amount of bound radioactivity (see Figure 3.5).

These are the only two studies found in the literature to measure protein–flavor reactions using radioisotopes.

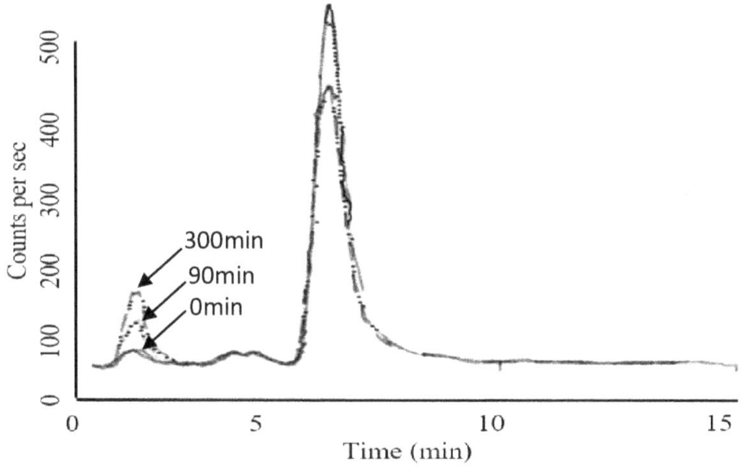

Figure 3.5 Kinetic effect in binding of radiolabeled benzaldehyde to BLG: chromatograms with radioactivity detection. (Peak at *ca.* 2.3 min – BLG bound benzaldehyde, 7.6 min = free benzaldehyde) (*r'x* time = 0 min, ... = 90 min, --...-- = 300 min). Reproduced from ref. 31 with permission from EDPSciences, Copyright 1999.

3.3 Irreversible Bonding

3.3.1 Default Methods

There have been several decades of studies focused on the reversible interactions occurring between food ingredients and flavoring materials. However, few studies have been done on irreversible reactions.[32] There are two main reasons for this lack of study on irreversible reactions (covalent). The first reason is that there are no adequate methods to accurately measure the covalent bonding of a flavor molecule to a protein. This capability has only recently come into being due to advances in mass spectrometry (MS). The second reason is that it was not considered necessary or valuable to differentiate between the mechanisms of flavor loss (reversible or irreversible) since the flavor was lost from sensory response under either circumstance. However, it is important to determine the mechanism by which flavor is lost. Reversible reactions are exactly that, reversible, and more importantly, they come to equilibrium that can potentially be managed through the formulation of flavoring. The irreversible reactions do not come to an equilibrium but continue until the flavor is completely consumed or all the reactive sites on the protein have reacted. It is important to be able to distinguish between the two mechanisms if one wishes to address the problem of flavor change during thermal processing and, ultimately, during shelf life.

The following sections will present methods used to measure the irreversible reactions that occur between proteins and flavor compounds.

3.3.2 Vacuum Stripping

The methods described in the reversible binding Section 3.2.1 of this chapter were methods that hypothetically would measure/differentiate both reversible and irreversible binding. The processes used a rigorous distillation or vacuum treatment, proposing that the volatiles that were removed by these treatments were reversibly bound and those that could not be recovered were irreversibly bound. Unfortunately, it was not possible at that time to accurately determine whether volatiles not recovered by the vacuum treatment were simply very tightly bound or whether they were truly irreversibly bound (*i.e.*, covalently bound). Thus, the effort to determine irreversibly bound volatiles may have been in error. However, as noted earlier, since the volatiles were bound tightly enough to no longer make a sensory contribution, researchers were not overly concerned about the inability to differentiate binding method.

3.3.3 Indirect Methods

3.3.3.1 Ellman's Test

The Ellman's test quantifies free protein thiols.[33,35] Ellman's reagent is 5,5′-dithio-bis-(2-nitrobenzoic acid), a compound used for measuring free sulfhydryl groups in solution. When the reagent reacts with a sulfhydryl group, a yellow-colored product is produced (see Figure 3.6). This product formation is quantified using a spectrophotometer based on its absorbance at 412 nm. This is a rapid and simple test, but it has limitations in its specificity.[36,37] It assumes that a decrease in free sulfhydryl groups is the result of a reaction with a flavor compound. In addition, it can be used only when sulfur chemistry is involved in the reaction between flavor and protein. This method could potentially be used to measure protein–flavor reactions, but no studies to date have been reported in the literature using this approach.

3.3.3.2 o-Phthaldialdehyde Test

The *o*-phthaldialdehyde (OPA) test measures free amino groups in a protein.[38–40] OPA, in conjunction with reduced sulfhydryl groups, reacts with primary amines to form fluorescent moieties (see Figure 3.7). Fluorescence is measured spectrophotometrically at 460 nm. There are several factors that influence the reactivity of OPA with proteins. The OPA method measures free amine groups in proteins, but in samples that have been

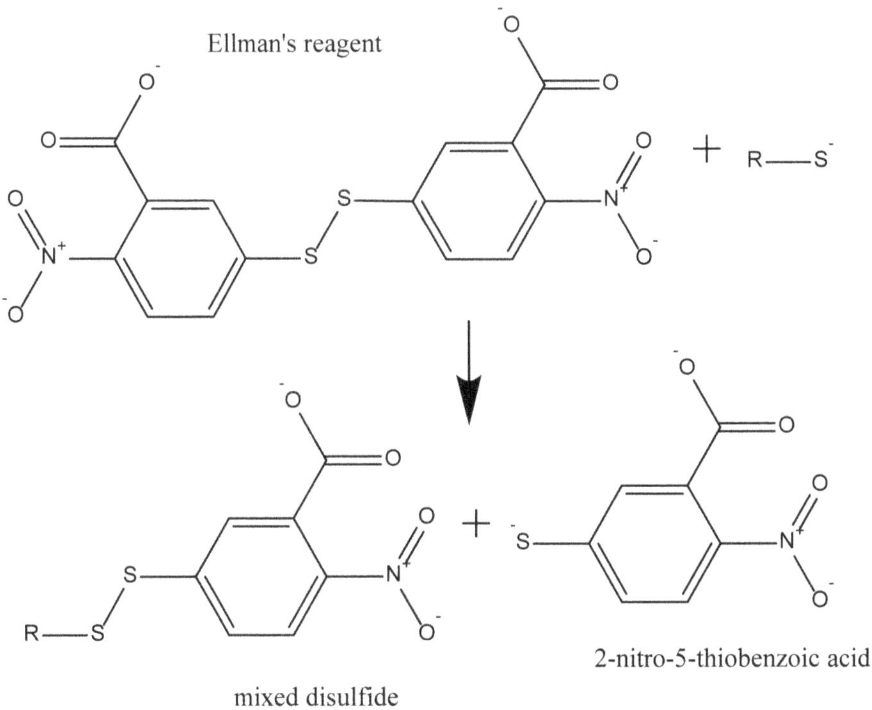

Figure 3.6 Reaction of Ellman's reagent with free thiol group.

heated, alkali treated, or lyophilized, the lysine residues may be conjugated, which affects the sensitivity of the test.[41] There would also be protein-to-protein variations because of the structure, method of protein isolation, and nativity. OPA tests cannot be applied to peptides containing tyrosine and tryptophan residues as they exhibit strong native fluorescence.

3.3.3.3 *Spectrophotometry*

Some protein–flavor reactions result in fluorescence. For example, when allyl isothiocyanate and sulfhydryl groups from a protein react, dithiocarbamate esters are formed, which give fluorescence that can be measured using a spectrophotometer.[33,42] The allyl isothiocyanate (AITC)/NH-group reaction producing thiourea derivatives has been detected by measuring the absorption at 236 nm.[43] The limitation of using this method is that not all reaction products form fluorescent end products.

3.3.3.4 *Fluorescence Quenching*

Three methods (fluorescence quenching, equilibrium dialysis, and headspace-water equilibrium) have been compared by Keppler *et al.*[33]

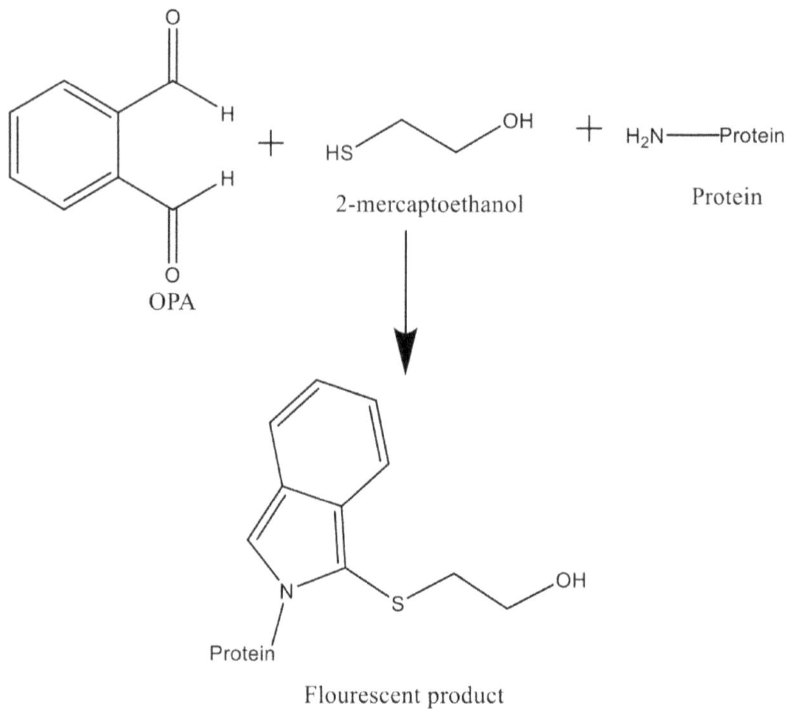

Figure 3.7 Reaction of OPA with free amines.

for determining the overall binding kinetics of AITC with BLG. This study concluded that the three methodologies were comparable and reproducible in the presence of high and low ligand concentrations for fluorescence quenching and equilibrium dialysis, respectively.[33] Fluorescence quenching spectrophotometrically measures fluorescence from the three-fluorescently emitting amino acids (*i.e.*, tryptophan, tyrosine, and phenylalanine) with absorbance at 280 nm and emission at 350 nm. Essentially, these three amino acids lose their fluorescence once they undergo a chemical reaction. Note that the method cannot determine which of these three amino acids undergoes a reaction or the type of reaction. The method simply indicates that some type of reaction has occurred that inhibits fluorescence.

3.4 Direct Methods

3.4.1 Mass Spectrometry

Advancements in MS have enabled researchers to detect and characterize peptides and protein reactions with small molecules.[44] This methodology provides a direct measure of the reaction products formed. It has also made possible the definition of the stoichiometry of the protein/adduct formation and the characterization of the sites of modification. MS, when coupled with computational and functional studies, helps explain the effects of such reactions on the functionality of a protein and its related consequences.[45]

The choice of the specific type of MS used depends on the type of modification and the resolution desired in the analysis. Most of the MS approaches fall into two categories: intact protein analysis (top-down approach) and analysis of peptides formed by enzymatically digested proteins (bottom-up approach).[46] Both of these approaches have their own advantages and disadvantages, which have been summarized in a review.[47] The top-down approach analyzes the intact protein without further sample preparation. The advantages of the top-down approach include complete sequence coverage, assessment of protein heterogeneity, and identification of posttranslational modifications (PTM). Since proteins are large molecules, appropriate ionization techniques and detectors with appropriate resolving power and mass accuracy are critical considerations. Intact protein analysis became possible only after the invention and development of matrix-assisted laser desorption ionization mass spectrometry (MALDI-MS) and electrospray ionization mass spectrometry (ESI-MS). These techniques made the ionization of proteins possible without enzymatic digestion. In MALDI, the ions are singly charged, and therefore, the resulting spectra are less complicated than in ESI-MS, where the proteins would be multiply charged. A deconvolution step is then needed to convert the ESI-MS to a single-charged entity for easy interpretation. Fourier-transform ion cyclotron resonance (FTICR) mass analyzer, orthogonal acceleration time-of-flight (oaTOF), quadrupole ion trap, triple quadrupole, and a hybrid quadrupole/oaTOF (qTOF) are detectors used for this approach, with FTICR having the highest resolution, mass accuracy, and peak capacity.[48] The resolving powers of MALDI-TOF and ESI-q MS instruments are 600 and 450 for a protein of 24 kDa, respectively.[49] This would not be sufficient for detecting the PTM of small molecules (flavor compounds), and these analyzers have limited opportunities for tandem MS.

The development of hybrid ESI qTOF MS[50,51] and MALDI qTOF MS[52,53] paved the way for protein structural characterization. The hybrid has better resolution, sensitivity, and an extended mass range in comparison to the linear mass analyzer. The mass resolving power of a hybrid qTOF mass analyzer is approximately 2300 for a protein of 24 kDa (trypsinogen) and 10 500 for a protein of size 5.7 kDa (insulin), with a mass accuracy of 20 ppm in comparison to 100 ppm for the linear mass analyzer.[49]

An example of the data obtained is shown in Figure 3.8.[54] The deconvoluted spectra of BLG with no flavor added (Figure 3.8, left) show a peak of molecular mass of 18 277 Da as the major peak, which is the mass of the unreacted BLG B variant. There were major peaks with masses of 18

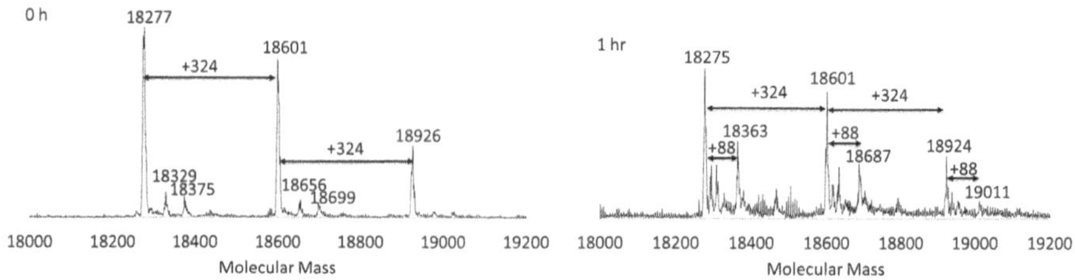

Figure 3.8 Deconvoluted spectra of pure BLG (left) and BLG + benzaldehyde (right). Masses + 88, *i.e.*, at 18 363, 18 687, and 19 011, are from the covalent addition of benzaldehyde to the base BLG 1 h after benzaldehyde addition at ambient temperature. Reproduced from ref. 54 with permission from American Chemical Society, Copyright 2020.

601 and 18 926 Da. These peaks (18 601 = BLG + lactose, 18 926 = BLG + 2 lactoses) are assumed to be Maillard reaction adducts of lactose (molecular weight 324 Da) with the main protein peak (18 276 ± 1 amu Da). While the sample was not heated during the isolation process in order to avoid these reactions, the Maillard reaction appeared to occur at ambient temperatures. Previous studies have reported lactose adducts, but their samples had been heated.[55–57]

Upon the addition of flavor compounds to the BLG solution, new masses were observed that were not present in the BLG with no flavor added. The sample with added benzaldehyde (6 h reaction time at ambient temperature) had an adduct peak with a delta shift of 88 Da [106 (molecular weight of benzaldehyde) − 18 (molecular weight of water lost during Schiff's base) = 88 Da]. This new mass corresponds to the molecular weight of Schiff's base adduct of BLG (or BLG + lactose) and benzaldehyde.

The identification of PTMs, glycosylation sites, and disulfide linkages within and outside the protein source is most often analyzed using a bottom-up approach. In this approach, the protein sample undergoes denaturation, reduction, and alkylation of the Cys residues before the enzymatic digestion step, and then, it is analyzed using MALDI MS, Orbitrap,[58] or ESI-MS.[44,47] The hydrolysate can also be sent through an LC for the separation of peptides before ESI-MS. The Orbitrap in combination with ESI provides good sensitivity, resolution, and mass accuracy of about 1–2 ppm.[59,60] It is desirable to introduce the sample *via* an LC. The limitation of the bottom-top approach is the need for a series of complicated sample preparations, protein sequence coverage depending on the MS, and loss of labile PTMs and artifactual peptides from enzymatic digestion. However, it is currently the best approach for the identification of PTMs in proteomics.

An example of the data obtained is shown in Figure 3.9 and Table 3.2. The PTM for the BLG protein with allyl isothiocyanate as flavoring showed a mass shift of 99 Da (molecular weight of allyl isothiocyanate) at the lysine groups. The authors have reported that the single allyl isothiocyanate

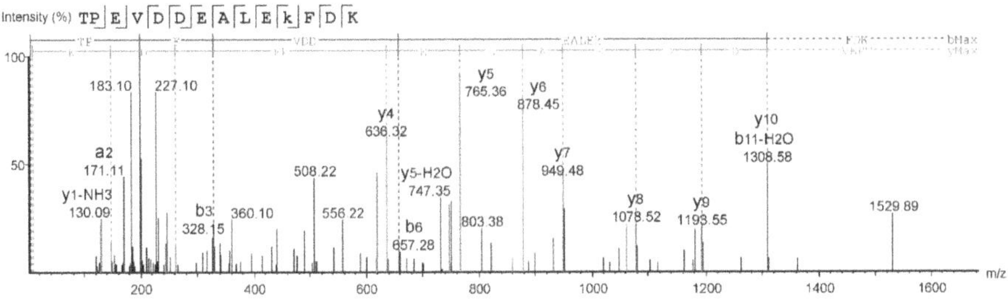

Figure 3.9 Tandem mass spectra of modified peptide TPEVDDEALEkFDK (positions 141–154, modified *m/z* = 578.9367) through the allyl isothiocyanate-Michael addition reaction with the lysine group in BLG protein. Reproduced from ref. 61 with permission from University of Minnesota, Copyright 2020.

Table 3.2 Modified peptides detected by LC-ESI-MS/MS from a tryptic digest of protein to which allyl isothiocyanate was added as flavoring at 6 h.

Peptides	Position	Modified *m/z*	Mass without PTM	Modified mass	Assignment
LIVTQTMK(+99.01)	17–24	516.7936	932.54	1031.5508	Lys Michael addition
TK(+99.01)IPAVFK	92–99	501.7945	902.56	1001.5732	Lys Michael addition
VLVLDTDYK(+99.01)K	108–117	431.5696	1192.67	1291.6846	Lys Michael addition
VLVLDTDYKK(+99.01)	108–117	646.8512	1192.67	1291.6846	Lys Michael addition
TPEVDDEALEK (+99.01)FDK	141–154	578.9367	1635.74	1733.7817	Lys Michael addition
TPEVDDEALEKFDK (+99.01)	141–154	578.9365	1634.77	1733.7817	Lys Michael addition
LK(+99.01)ALPMHIR	156–164	393.235	1077.65	1176.6624	Lys Michael addition

mass shift of 99.01 Da (Michael addition adduct) is present in the protein sample on the lysine groups preferentially. The reference protein sequence that was used for the identification is MKCLLLALALTCGAQALIVTQTMKGL DIQKVAGTWYSLAMAASDISLLDAQSAPLRVYVEELKPTPEGDLEILLQKWENG ECAQKKIIAEKTKIPAVFKIDALNENKVLVLDTDYKKYLLFCMENSAEPEQSLAC QCLVRTPEVDDEALEKFDKALKALPMHIRLSFNPTQLEEQCHI.

The PTM that was reported on the lysine groups is present in positions 24, 93, 116, 117, 151, 154, and 157. The modifications could not be confirmed on the other lysine groups present in the protein. The example MS/MS spectra for the modified lysine site at 151 are shown in Figure 3.9.

Using this methodology, PTM for other flavor compounds and proteins could be found.

3.5 Conclusions

Advances in analytical instrumentation have enabled researchers to observe, study, and, thus, better understand the reactions that occur between proteins and flavor components. All of the early methods used to measure these interactions measured total interaction – the sum of reversible and irreversible (covalent) binding. The evolution of the method from simple observation of sensory changes to the most recent ITEX headspace concentration technique has continuously improved the data obtained. However, the recognition that determining the binding mechanism (reversible *vs.* irreversible) is important in addressing the problem the industry faces in solving flavoring issues with protein has demonstrated a need for more sophisticated analytical techniques – methods that differentiate binding mechanisms.

The development of sophisticated MS-based techniques to measure the extent of binding and differentiate the type of binding has provided valuable data. Proteomic approaches are adding additional data by providing the exact location of binding. A weakness of the new MS methodology is that it does not provide good quantitative data and becomes complicated as the proteins become cross-linked, resulting in masses beyond the MS range. The approach using radiolabeled flavor molecules may provide a starting point to resolve the limitations we are observing with the MS method. The two methods provide complementary data. Another advancement in studying the protein–flavor interaction is using the present studies and constructing mathematical models to predict the reactions.[62]

In the future, with the advancement in instrumentation that would have increased sensitivity and resolution, we would realize that there is a lot to be understood about protein flavor reactions and that we have just scratched the surface.

Abbreviations

AITC	Ally isothiocyanate
BLG	β-Lactoglobulin
ESI-MS	Electrospray ionization mass spectrometry
FTICR	Fourier-transform ion cyclotron resonance
GC	Gas chromatography
HPLC	High-performance liquid chromatography
IGC	Inverse gas chromatography
ITEX	In-tube extraction

MALDI-MS	Matrix-assisted laser desorption ionization mass spectrometry
MS	Mass spectrometry
oaTOF	Orthogonal acceleration time-of-flight
OPA	*o*-Phthaldialdehyde
PDMS	Polydimethylsiloxane
PTM	Posttranslational modifications
qTOF	Quadrupole time-of-flight
SPME	Solid-phase microextraction

References

1. H. G. Maier, *Angew. Chem. Int. Ed. Engl.*, 1970, **9**, 917–926.
2. J. Solms, F. Osman-Ismail and M. Beyeler, *Can. Inst. Food Sci. Technol. J.*, 1973, **6**, A10–A16.
3. K. Wang and S. D. Arntfield, *Flavour Fragr. J.*, 2017, **32**, 92–101.
4. H. A. Gremli, *J. Am. Oil Chem. Soc.*, 1974, **51**, 95–97.
5. S. Damodaran and J. E. Kinsella, *J. Agric. Food Chem.*, 1981, **29**, 1249–1253.
6. S. Damodaran and J. E. Kinsella, *J. Agric. Food Chem.*, 1981, **29**, 1253–1257.
7. Ti. E. O'Neill, in *Flavor-Food Interactions*, ed. R. McGorrin and J. Leland, American Chemical Society (ACS), Washington, DC, 50th edn, 1996, vol. **119**, pp. 59–70.
8. T. O'neill and J. E. E. Kinsella, *J. Food Sci.*, 1988, **53**, 906–909.
9. C. Druaux, S. Lubbers, C. Charpentier and A. Voilley, *Food Chem.*, 1995, **53**, 203–207.
10. J. Schaefer, in *Handbuch der Aroma Frischung*, ed. H. Maarse, Berlin, 1981, p. 44.
11. K. L. Franzen and J. E. Kinsella, *J. Agric. Food Chem.*, 1974, **22**, 675–678.
12. C. L. Arthur and J. Pawliszyn, *Anal. Chem.*, 1990, **62**, 2145–2148.
13. IOFI Working Group on Methods of Analysis, *Flavour Fragr. J.*, 2010, **25**, 404–406.
14. D. D. Roberts, P. Pollien and C. Milo, *J. Agric. Food Chem.*, 2000, **48**, 2430–2437.
15. M. Fabre, V. Aubry and E. Guichard, *J. Agric. Food Chem.*, 2002, **50**, 1497–1501.
16. M. P. Gianelli, M. Flores and F. Toldrá, *J. Agric. Food Chem.*, 2003, **51**, 6828–6834.
17. M. Pérez-Juan, M. Flores and F. Toldrá, *J. Agric. Food Chem.*, 2006, **54**, 4802–4808.
18. M. Pérez-Juan, M. Flores and F. Toldrá, *Food Res. Int.*, 2007, **40**, 687–693.

19. M. Pérez-Juan, M. Flores and F. Toldrá, *Food Chem.*, 2008, **108**, 1226–1233.
20. J. Kühn, X.-Q. Zhu, T. Considine and H. Singh, *J. Agric. Food Chem.*, 2007, **55**, 3599–3604.
21. J. Laaks, M. A. Jochmann, B. Schilling and T. C. Schmidt, *Anal. Chem.*, 2010, **82**, 7641–7648.
22. K. Wang and S. D. Arntfield, *Food Chem.*, 2014, **157**, 364–372.
23. K. Wang and S. D. Arntfield, *Food Res. Int.*, 2015, **77**, 1–9.
24. K. Wang and S. D. Arntfield, *Food Hydrocolloids*, 2015, **43**, 410–417.
25. K. Wang and S. D. Arntfield, *Food Hydrocolloids*, 2016, **61**, 567–577.
26. K. Wang and S. D. Arntfield, *Food Chem.*, 2016, **211**, 235–242.
27. Q. X. Zhou and K. R. Cadwallader, *J. Agric. Food Chem.*, 2004, **52**, 6271–6277.
28. Q. Zhou, S.-Y. Lee and K. R. Cadwallader, *J. Agric. Food Chem.*, 2006, **54**, 5516–5520.
29. Q. Zhou and K. R. Cadwallader, *J. Agric. Food Chem.*, 2006, **54**, 1838–1843.
30. B. M. King and J. Solms, *J. Agric. Food Chem.*, 1979, **27**, 1331–1334.
31. I. Andriot, I. Marin, G. Feron, P. Relkin and É. Guichard, *Le Lait, INRA Ed.*, 1999, **79**, 577–586.
32. V. Anantharamkrishnan, T. Hoye and G. Reineccius, *J. Agric. Food Chem.*, 2020, **68**, 6395–6402.
33. J. K. Keppler, T. Koudelka, K. Palani, M. C. Stuhldreier, F. Temps, A. Tholey and K. Schwarz, *J. Biomol. Struct. Dyn.*, 2014, **32**, 1103–1117.
34. S. Arai, M. Noguchi, M. Kaji, H. Kato and M. Fujimaki, *Agric. Biol. Chem.*, 1970, **34**, 1420–1423.
35. J. J. Kehoe, A. Brodkorb, D. Mollé, E. Yokoyama, M.-H. Famelart, S. Bouhallab, E. R. Morris and T. Croguennec, *J. Agric. Food Chem.*, 2007, **55**, 7107–7113.
36. J. Russell and D. L. Rabenstein, *Anal. Biochem.*, 1996, **242**, 136–144.
37. C. K. Riener, G. Kada and H. J. Gruber, *Anal. Bioanal. Chem.*, 2002, **373**, 266–276.
38. K. Rade-Kukic, C. Schmitt and H. M. Rawel, *Food Hydrocolloids*, 2011, **25**, 694–706.
39. D. Zhu, A. Saul, S. Huang, L. B. Martin, L. H. Miller and K. M. Rausch, *Vaccine*, 2009, **27**, 6054–6059.
40. M. Roth, *Anal. Chem.*, 1943, **277**, 23.
41. J. R. Benson and P. E. Hare, *Proc. Natl. Acad. Sci. U. S. A.*, 1975, **72**, 619–622.

42. A. T. Dinkova-Kostova, W. D. Holtzclaw, R. N. Cole, K. Itoh, N. Wakabayashi, Y. Katoh, M. Yamamoto and P. Talalay, *Proc. Natl. Acad. Sci. U. S. A.*, 2002, **99**, 11908–11913.
43. R. J. Blagrove and L. C. Gruen, *Microchim. Acta.*, 1971, **59**, 639–643.
44. V. Anantharamkrishnan and G. A. Reineccius, *J. Agric. Food Chem.*, 2020, **68**, 13833–13843.
45. M. Carini and M. Orioli, *Biomarkers Antioxid. Def. Oxidative Damage Princ. Pract. Appl.*, 2010, 173–197.
46. G. E. Reid and S. A. McLuckey, *J. Mass Spectrom.*, 2002, **37**, 663–675.
47. C. A. Srebalus Barnes and A. Lim, *Mass Spectrom. Rev.*, 2007, **26**, 370–388.
48. B. Bogdanov and R. D. Smith, *Mass Spectrom. Rev.*, 2005, **24**, 168–200.
49. J. C. Rouse, J. E. McClellan, H. K. Patel, M. A. Jankowski and T. J. Porter, *Methods Mol. Biol.*, 2005, **308**, 435–460.
50. H. R. Morris, T. Paxton, A. Dell, J. Langhorne, M. Berg, R. S. Bordoli, J. Hoyes and R. H. Bateman, *Rapid Commun. Mass Spectrom.*, 1996, **10**, 889–896.
51. A. Shevchenko, I. Chernushevich, W. Ens, K. G. Standing, B. Thomson, M. Wilm and M. Mann, *Rapid Commun. Mass Spectrom.*, 1997, **11**, 1015–1024.
52. A. V. Loboda, A. N. Krutchinsky, M. Bromirski, W. Ens and K. G. Standing, *Rapid Commun. Mass Spectrom.*, 2000, **14**, 1047–1057.
53. A. V. Loboda, S. Ackloo and I. V. Chernushevich, *Rapid Commun. Mass Spectrom.*, 2003, **17**, 2508–2516.
54. V. Anantharamkrishnan and G. A. Reineccius, *J. Agric. Food Chem.*, 2020, **68**, 13121–13130.
55. J. Meltretter, C. M. Becker and M. Pischetsrieder, *J. Agric. Food Chem.*, 2008, **56**, 5165–5171.
56. F. Fenaille, V. Parisod, P. Visani, S. Populaire, J. C. Tabet and P. A. Guy, *Int. Dairy J.*, 2006, **16**, 728–739.
57. R. A. Siciliano, M. F. Mazzeo, S. Arena, G. Renzone and A. Scaloni, *Food Res. Int.*, 2013, **54**, 988–1000.
58. A. Makarov, *Anal. Chem.*, 2000, **72**, 1156–1162.
59. M. Hardman and A. A. Makarov, *Anal. Chem.*, 2003, **75**, 1699–1705.
60. J. R. Yates, D. Cociorva, L. Liao and V. Zabrouskov, *Anal. Chem.*, 2006, **78**, 493–500.
61. V. Anantharamkrishnan, PhD thesis, University of Minnesota, 2020.
62. J. Zhang, D. Kang, W. Zhang and J. M. Lorenzo, *Trends Food Sci. Technol.*, 2021, **111**, 405–425.

CHAPTER 4

Flavor–Protein Binding on Flavor Delivery

GARY A. REINECCIUS*[a] AND
VAIDHY ANANTHARAMKRISHNAN*[b]

[a]Department of Food Science and Nutrition, University of Minnesota, 1334, Eckles Avenue, St. Paul, Minnesota 55108, USA; [b]Research and Development, Lux Flavors, Chennai 600114, India
*E-mail: greinecc@umn.edu, anant040@umn.edu

4.1 Introduction

It was observed and noted in publications as early as 1920[1] that the sensory properties of a flavoring are dependent upon the food it is applied to; there are interactions between a food and an added flavoring that may change the flavor character. Tools were lacking to approach this problem very well until the advent of gas chromatography in the late 1960s. As gas chromatography became common in research labs, flavor interactions with major food components (*i.e.*, lipids, carbohydrates, and proteins) became a popular area of study.[2]

The interest in flavor interactions, specifically with proteins, was initially driven not only by scientific curiosity but also by a market need for a functional plant protein, and soy met most requirements except for two flavoring challenges. First, researchers found that flavor compounds inherent in soybeans (undesirable in a finished food product) were extremely difficult to remove from protein isolates. Various processes that would be expected to remove volatiles were applied to soy products

Food Chemistry, Function and Analysis No. 41
Flavour and Consumer Perception of Food Proteins
Edited by Jing Zhao and Changqi Liu

Published by the Royal Society of Chemistry, www.rsc.org

(*e.g.*, solvent extraction or high vacuum treatment) with little success.[3] This observation strongly suggested that chemical bonding was occurring between the flavor off-notes and the soy protein.[4] Second, the addition of desirable flavorings to foods containing soy protein isolates changed their character and intensity on their addition.[5] The first thought was that there was some unextracted lipid in the protein isolates, and this lipid was postulated to "absorb" parts or all of the added flavoring, thereby changing the sensory character. Earlier research pointed out how lipids reduce flavor intensity and may change its character. However, lipid-free soy isolates prepared in the laboratory were found to exhibit the same deleterious effects on added flavoring as commercial soy isolates. Thus, the difficulty in removing undesirable indigenous odorants from soy isolates and the change in the character of added flavorings were attributed to chemical interactions between the soy protein and volatile aroma compounds.

The observation that flavorings added to a soy protein solution changed in sensory character but did not totally disappear suggested that the bonding must vary across chemical compounds. Questions also arose as to the presence/strength of the bonds that may be formed: were they dependent upon the flavor compound, the type/source of protein, and/or the flavor–protein source, protein state (denatured or hydrolyzed), and system environment (*e.g.*, pH, salt content, moisture, and water activity)? These questions spawned research on flavor–protein interactions that continues even today. We have come to realize that reactions between flavor components and proteins are multifaceted and very complex. This chapter will provide a current review of how flavoring materials will chemically react with food proteins and what is known about factors that influence the rate and types of these reactions. It should be noted that some excellent reviews are available.[6–8]

4.2 Types of Flavor–Protein Reactions

Today, it is generally acknowledged that the interactions taking place between proteins and flavor compounds may be relatively weak, ionic, hydrophilic, or hydrophobic bonds, or stronger and potentially more permanent bonds, *i.e.*, covalent bonds. Researchers often refer to these "weak" bonds as "reversible" bonds (as they will be referred to in this chapter), where the flavor compounds can be recovered from a protein isolate by a gentle technique such as high vacuum distillation or may be free to come to some equilibrium in testing (headspace or solution). The covalent reactions that occur between proteins and flavor compounds do not reach equilibrium but may progress rapidly during heat treatment and continue slowly throughout the product shelf life (such bonds are typically referred to, and will be in this chapter, as irreversible bonds).

The reversible bonds are problematic when trying to remove inherent off-notes to make a bland protein isolate and when flavoring a finished product, as these bonds will selectively bind part or all of the added flavoring to leave the flavor imbalanced and often unacceptable. This binding can often be managed by a flavor company by allowing a food product to "age" overnight to allow for the binding and then adjusting the flavor formulation to correct for the binding.

The irreversible bonds are often the most challenging to deal with as they do not come to equilibrium, so the flavoring can be reformulated to adjust for binding: there is no solution that can potentially be done for the reversible types of interactions. This chapter is formatted to discuss reversible and irreversible interactions as separate topics, as the mechanisms of bond formation and maintenance are very different, as are the analytical methodologies used to study them. The research methodologies applied in research to study both types of interactions are discussed in the previous chapter.

4.3 Reversible Bonds

4.3.1 Effect of Protein Type on the Formation of Reversible Bonds

The largest body of research in this topic area has been devoted to three protein isolate sources: soy protein, dairy protein, and, most recently, pea protein. As was noted earlier, studies on flavor–protein interaction began with soy due to a market need for a bland, versatile plant protein. Fairly quickly, research also turned to dairy proteins, most commonly whey protein or the major whey protein, β-lactoglobulin. Over time, other proteins (canola, pea, wheat gluten, α-lactalbumin, bovine serum albumin, whey protein, caseins, fava bean, egg, fish, gelatin, myoglobin, pork muscles, and collagen) have also become the subject of similar research.[7–10] What was learned in each of these studies on different proteins depends upon the state of the protein (*e.g.*, denatured or hydrolyzed), the flavor compounds studied, the reaction environment (*e.g.*, a_W, pH, temperature, salts, or alcohol), and the methodology that was used to determine interactions (*e.g.*, measure total, reversible, or irreversible binding). Most studies in this area have depended upon equilibrium methods that measure the total reaction (*i.e.*, the sum of reversible and irreversible reactions) (see Chapter 2). When dealing with sensory issues, the sum of binding by both mechanisms is relevant: both result in sensory flavor loss. However, it is desirable to be able to distinguish between these two types of reactions because we can potentially correct reversible reactions through flavor formulation, while irreversible reactions cannot be corrected.

There are no detailed studies comparing flavor interactions broadly across protein sources. Thus, broad comparisons are complicated as different analytical methods are often used to measure interaction and different model flavor compounds are focused upon.[8] However, there has been enough work done on soy and dairy to make some comparisons between these protein sources using 2-nonanone as the model flavorant. One should note that due to the choice of the model flavor compound, *i.e.*, that it will not undergo irreversible reactions, which can be very important, one might find very different results for other aroma compounds that will undergo irreversible reactions. However, at the time of publication, 2-nonanone provided the most data for comparison. Guichard[11] has provided a summary and discussion of these data (see Table 4.1). The binding constants obtained for 2-nonanone by different authors varied substantially. Yet, the binding constants for bovine serum albumin were much higher than those of β-lactoglobulin or α-lactalbumin. Also, β-conglycinin (soy protein) showed a five-fold higher

Table 4.1 Binding constants between 2-nonanone and select dairy and soy proteins. Reproduced from ref. 11 with permission from Taylor & Francis, Copyright 2002.

Ref.	Protein	Method	No. of binding sites per monomer	K_a affinity const. [M^{-1}]	nK_a global affinity const. [M^{-1}]
12	Whole soy	ED[a]	5.5	570	3140
	β-Conglycinin (MW: 100 000)	ED	1.8	3050	5490
	β-Conglycinin (MW: 160 000)	ED	2.8		8780
	Glycinin (MW: 100 000)	ED	3.1	540	1670
	Glycinin (MW: 320 000)	ED	9	9	5340
13	β-Lactoglobulin	ED	1	2440	2440
14	Bovine serum albumin	LLP[b]	5–6	1800	9000–10 800
15	Soy protein native	ED	4	930	3720
	Soy protein partly denatured	ED	4	1240	4960
	Soy protein succinylated	ED	2	850	1700
16	β-Lactoglobulin		1	969	969
17	β-Lactoglobulin		1	833	833
18	β-Lactoglobulin	AC[c]			3629
19	Bovine serum albumin		15	14 100	211 500
	β-Lactoglobulin		13	121	1573
	α-Lactalbumin		33	10	330
20	Glycinin (20 °C)	LLP	71	891	63 261
	β-Conglycinin (28 °C)		46	254	11 684

[a]ED, equilibrium membrane dialysis

[b]LLP, liquid–liquid partition

[c]AC, affinity chromatography

affinity for 2-nonanone than the other soy proteins. This soy protein fraction had stronger binding than β-lactoglobulin but lower than that of bovine serum albumin. Despite the variation in data, the differences between some proteins or protein fractions were much greater than the differences found between authors for the same protein.

Recent work by Wang and Arntfield[21] provided an excellent comparison of the binding of aldehydes and ketones to pea, canola, and wheat gluten proteins. They found that the binding strength of aldehydes was in the order of canola > wheat gluten > salt-extracted pea. The order for binding of ketones was pea > wheat > anola. The difference in order is likely due to the aldehydes being bound through both covalent and hydrophobic bonds, while the ketones were primarily bound through hydrophobic mechanisms. This is the type of data that is needed across our commonly used and new/novel protein isolates being developed today.

While much detail is missing on what determines the reactivity of a protein with a flavor molecule, generalizations can be made that the amino acid composition of a protein and how it is assembled (protein structure) are the primary determinants of flavor binding. The amino acids provide reactive side chain groups such as free amines (lysine and arginine) or free thiols or sulfides (cysteine or cystine) to form irreversible bonding. Hydrophobic, electrically charged, or hydrophilic side chain amino acids determine the resultant protein folding (protein structure) to produce and may create unique "pockets" for the reversible flavor interactions or shield sites from reaction (see Figure 4.1).

Reaction conditions such as temperature, ionic strength, pH, and presence/concentration of salts or other chemicals (*e.g.*, ethanol) will influence the conformation of a protein, thereby influencing the potential for flavor binding. The effects of these variables on reversible flavor binding are discussed below.

4.3.2 Effect of the Chemical Properties of a Flavor Molecule on Flavor Binding

As noted earlier, the amount of flavor binding that occurs depends upon the characteristics of the binding site, the binding mechanism (hydrophobic, ionic, electrostatic, covalent, *etc.*), and the chemical characteristics of the flavor molecule. As hydrophobic interaction is the most common form of flavor–protein interaction, numerous studies have evaluated the effect of flavor compound hydrophobicity on binding. Researchers have generally found that binding increases with the hydrophobicity of the reactant flavor molecule as long as it is in the same chemical class as the flavorant (see Figure 4.2).[11] This generalization does not necessarily apply broadly across chemical classes as the functional group can add unique binding properties, *e.g.*, an acid group will offer ionic bonding opportunities.

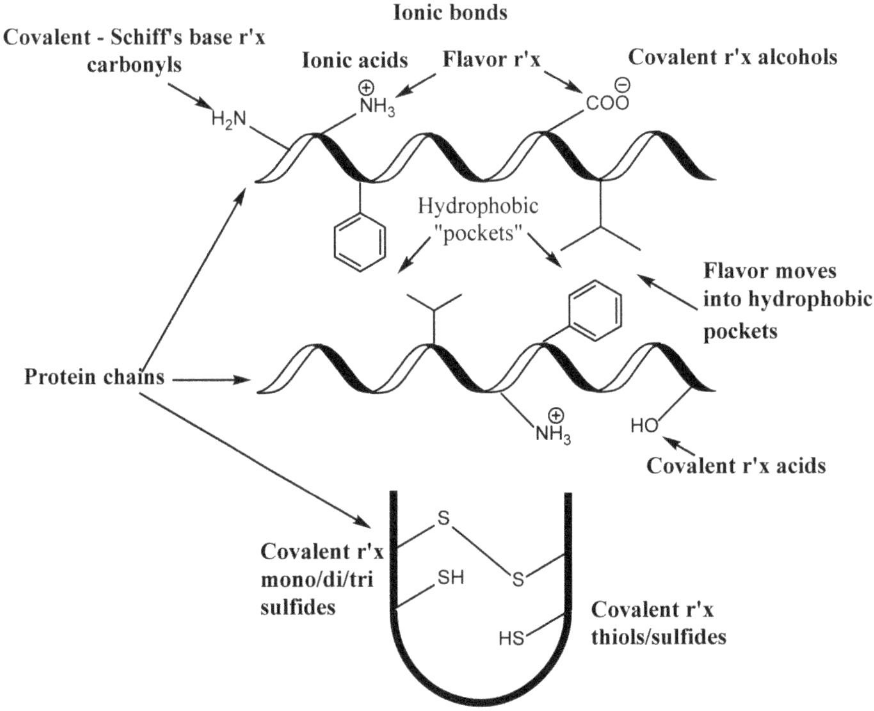

Figure 4.1 Opportunities for flavor reactions (r'x) with a protein.

Of the functional groups, flavor compounds containing a carbonyl group (specifically an aldehyde) have generally been found to be protein bound to the greatest extent. This high level of reactivity is generally due to aldehyde binding by both reversible (hydrophobic) and irreversible (covalent bond) interactions. The vast majority of flavor molecules will bind only through reversible reactions. Beyond the functional group, researchers have found that shifting the position of the functional group within a flavor molecule, *e.g.*, a ketone, can also influence binding. For example, Franzen and Kinsella[23] and Damodaran and Kinsella[14] found that shifting the position of a ketone group to the interior of a molecule resulted in less binding. They suggest that this was due to steric hindrance for the hydrophobic entity as a whole (see Table 4.2).

The effect of flavor molecule stereochemistry may be very important in determining binding. For example, Dufour and Haertle[24] found that β-lactoglobulin binds with β-ionone but not with α-ionone. These two isomers differ only in the position of the cyclohexane double bond. Lübke *et al.*[25] demonstrated a similar effect with retinol binding. The binding of retinol with β-lactoglobulin induced no change in the protein conformation, whereas the binding of α-ionone did have an effect.

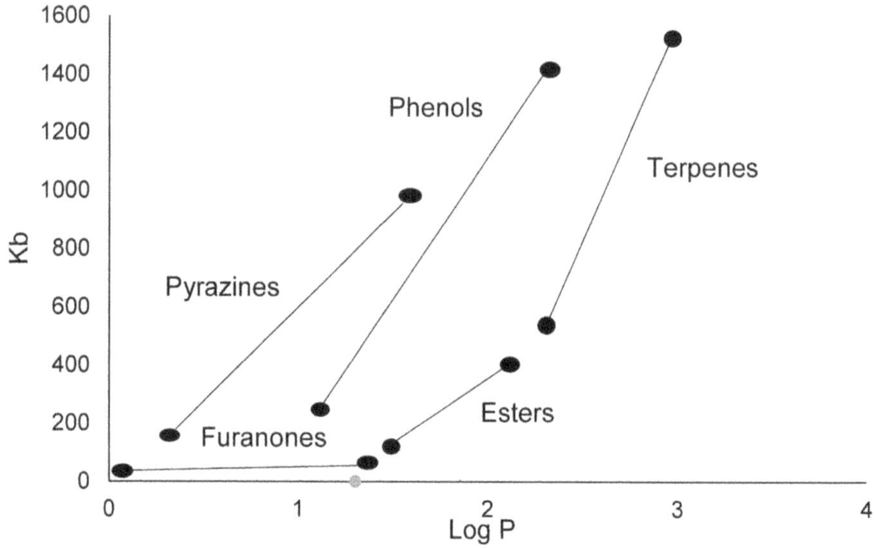

Figure 4.2 Binding constants K_b in relation to the hydrophobicity of the analyzed substances. Reproduced from ref. 22 with permission from EDP Sciences, Copyright 2000.

4.3.3 Effect of Protein Hydrolysis on Flavor Binding

There is limited information in the literature on the effect of protein hydrolysis on flavor binding. Arai *et al.*[26] chose to study how the binding of two volatiles identified in soy as off-flavor components (hexanal and hexanol) is affected by soy hydrolysis. They considered the reactions of these two compounds with native soy protein isolate, partially denatured protein isolate (90 °C for 1 h), hydrolyzed soy protein, and finally, the combination of free amino acids representing the soy isolate (see Figure 4.3). They measured binding by their ability to recover these two compounds by solvent extraction from a solution of soy isolate. They found that there was less binding of either compound when the protein was subjected to enzymatic action (proteolysis). They proposed that hydrolysis of the protein decreases the binding properties proportionally to the molecular weight. *It is of interest that there was no binding/interaction with a representative blend of soy amino acids.*

4.3.4 Effect of Protein Heat Denaturation on Flavor Binding

This area of investigation is very relevant, as most foods are thermally processed at some stage of manufacture to ensure safety. Thus, flavor binding to partially denatured proteins is generally of greater relevance to the flavor of food products than binding by native proteins.

Table 4.2 Binding constants of the experimental flavor compounds with β-lactoglobulin. Reproduced from ref. 18 with permission from Elsevier, Copyright 1998.

Flavor compound	Binding constant K_b [M^{-1}]
β-Ionone	19 143
α-Ionone	13 456
Γ-Undecalactone	9924
β-Damascenone	6073
2-Nonenal	4433
2-Nonanone	3629
Γ-Decalactone	3230
1-Nonene-3-ol	2055
2-Octenal	1579
2-Octanone	1287
Me Benzoate	1005
1-Octen-3-ol	699
2-Heptenal	576
2-Heptanone	465
γ-Octalactone	450
Benzaldehyde	341
1-Hepten-3-ol	250
Γ-Octalactone	231
2-Hexenal	218
2-Hexenone	163
2-Pentenal	61
Pentanone	42
1-Penten-3-ol	23

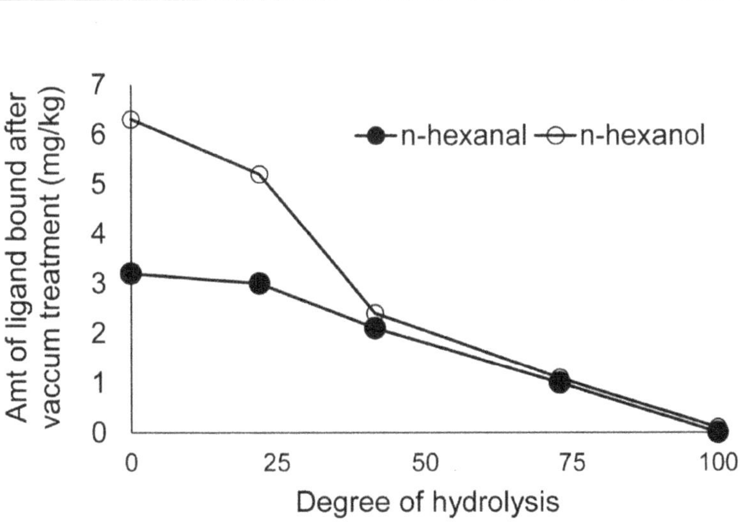

Figure 4.3 Interaction of *n*-hexanal (o) and *n*-hexanol (•) with soy protein of different degrees of hydrolysis (0–100%). Reproduced from ref. 26 with permission from Oxford University Press, Copyright 1970.

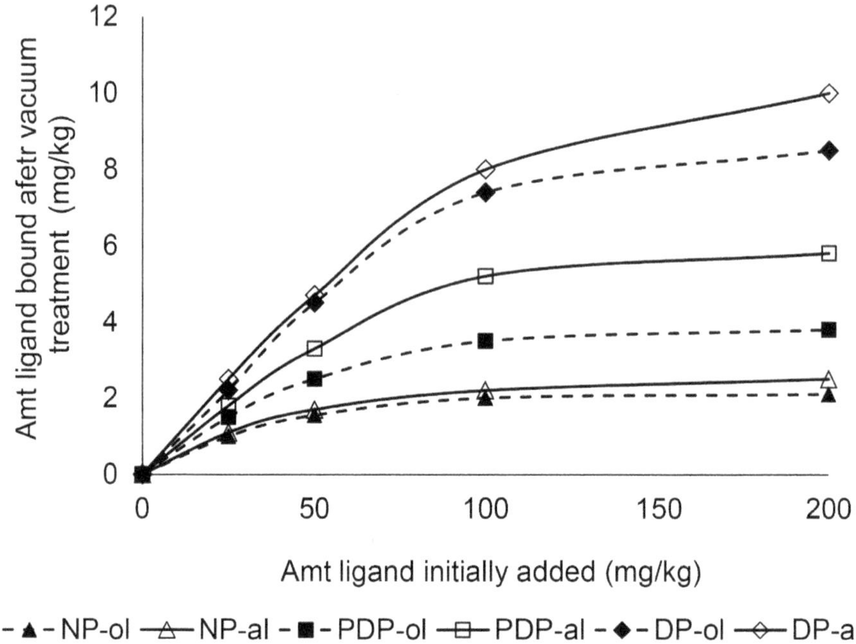

Figure 4.4 The binding of hexanol (-ol) and hexanal (-al) to soy protein. DP, denatured protein (protein suspended in ethanol and heated for 24 h at boiling); NP, native protein; PDP, partially denatured protein (protein heated at 90 °C for 1 h). Adapted from ref. 4 with permission from Oxford University Press, Copyright 1970.

Arai *et al.*[4] studied how protein denaturation (soy) affects the binding of hexanol and hexanal. As shown in Figure 4.4, Arai *et al.*[26] found that binding increased as denaturation increased. Increased binding on denaturation was attributed to an opening of the protein structure, allowing more exposed hydrophobic binding sites for flavor binding.

The effect of thermal denaturation on flavor binding has yielded both consistent and inconsistent results. For example, Ng *et al.*[27,28] reported that the denaturation of fava bean decreased the sensory intensity of vanillin in solution. Similarly, Hansen and Booker[29] found that the amount of benzaldehyde bound by β-lactoglobulin increased from 38% to 63% after the product was pasteurized (70 °C, 30 min).

To the contrary, O'Neill and Kinsella[30] found 2-nonanone binding to decrease when β-lactoglobulin was heated to 75 °C for 10 and 20 min. This discrepancy may be (partially) explained by recently published work on the covalent bonding of flavorings. Anantharamkrishnan and Reineccius[31] found benzaldehyde to react covalently very rapidly with β-lactoglobulin even at ambient temperature, whereas ketones were not reactive at ambient temperature. The extra losses observed with benzaldehyde may be due to a sum of reversible and irreversible bonding, while the binding of 2-nonanone would occur only by reversible bonding (no covalent bond formation).

Two other research groups looked at the effect of protein denaturation on the binding of vanillin to whey protein isolate[32] and bovine serum albumin.[33] In each of these cases, the heat-denatured proteins had higher levels of free vanillin than their denatured counterparts. It should be noted that vanillin, unlike benzaldehyde, does not rapidly form covalent bonds with whey proteins.[31] One may propose that the denaturation of these proteins may have favored covalent bond formation but not hydrophobic bonding. Fox and Mulvihill[34] found that vanillin binding to sodium caseinate does not appear to be influenced by heat treatment. This may be expected as casein has few secondary structures and high heat stability.

There is also some literature on the effect of heat treatment on the binding of butanol and hexanol to soy proteins.[35] The binding of these alcohols to heated soy proteins was reduced by denaturation. Similar results were observed by Semenova *et al.*[36] for the binding of hexyl acetate to the 11S globulin in soy. Thus, there does not seem to be a consistent trend in the effect of protein denaturation on flavor binding.

4.3.5 Environment Effects on Flavor–Protein Binding

4.3.5.1 Reaction System Relative Humidity

The vast majority of studies on protein-flavor reactions have been carried out in liquid solutions. Proteins are commonly used in the formulation of low/intermediate moisture foods, and thus, Zhou and Cadwallader[37] chose to provide data on the interaction of flavor molecules with soy protein as a function of relative humidity (RH). They used inverse gas chromatography (IGC) for this analysis, using protein as the stationary phase and flavor molecules in the mobile phase. On the basis of their thermodynamic data, 1-hexanol and hexanal had higher binding affinities than hexane at 0% RH. At 30% RH, the binding affinity decreased compared to 0% RH. The authors attributed this decrease in binding to competition by water for potential binding sites.

The number of flavor compounds studied was expanded upon in a later study by Zhou and Cadwallader.[38] As would be expected, the chemical structure of the flavor compound strongly influenced the binding potential. The binding of nonpolar flavor compounds was attributed primarily to nonspecific van der Waals dispersion forces and was not affected by the RH of the IGC environment. Polar flavor compounds (esters, ketones, aldehydes, and alcohols) exhibited both specific (hydrogen bonding and dipole forces) and nonspecific interactions. The binding of polar volatiles was reduced by the adsorbed water at very low RH (approaching 0%), while at higher RH (30–50%), there was no effect of humidity on the binding of model volatiles except for alcohols.

4.3.5.2 Reaction System pH

One would expect pH to have an influence on flavor–protein binding. Changes in pH would alter both the ionization and the net charge on a protein molecule and lead to structural changes in the protein that may or may not favor binding. pH also has a major effect on the rate of covalent bond formation between proteins and aldehydes (Schiff base formation). Schiff base formation is strongly enhanced by high pHs.

Jouenne and Crouzet[39] reported binding to increase from pH 3 to 9 and then decrease rapidly until pH 11 for the binding of aroma compounds (three methyl ketones, three ethyl esters, and two terpenes) by whey proteins. Ruth and Villeneuve[40] also found increased binding of 20 aroma compounds (except α-pinene and alcohols) to β-lactoglobulin when increasing the pH from 3 to 9.

On the contrary, Wang and Arntfield[41] found opposite results in the effect of pH on protein-flavor binding. Working with ketones, the overall binding to pea protein decreased in the following order: pH 5 > pH 7 > pH 9 > pH 11 > pH 3. They explained that the "high binding at pH 5 was due to the physical sorption of flavor on the aggregated protein surface upon protein isoelectric precipitation." They explained that the heavily denatured/unfolded state of protein at pH extremes (*i.e.*, pH 3 and 11) may have caused a substantial loss of hydrophobic area/binding sites, thereby accounting for the reduction in flavor binding.

4.3.5.3 Addition of Salts and Ethanol to Reaction System

Androit *et al.*[42] used two methods for studying how the presence of salt or ethanol influenced the binding of benzaldehyde to β-lactoglobulin. A static headspace technique was used to measure weak interactions, and liquid chromatography (LC) using C14-labeled benzaldehyde was used to measure covalent interactions. The two methods gave very similar data for the binding of benzaldehyde by β-lactoglobulin (see Table 4.3).

The addition of salt (50 mmol L^{-1}) decreased binding. The decrease in binding was shown to be dependent on pH (increased binding at pH 6 *vs.* 3) but independent of the protein concentration. The addition of ethanol at the concentration used (13% with salt) had no effect on binding. The effect of NaCl on binding was explained by an aggregation of β-lactoglobulin.

4.3.5.4 Effect of Reaction System Temperature (Storage Range)

There have been several studies on the effect of temperature (in the storage range) on flavor–protein binding. Studies by Damodaran and Kinsella[15] found no effect of temperature on the interactions between 2-nonanone and soy proteins within a small temperature interval (25 and 45 °C) (see Figure 4.5). Similarly, Druaux *et al.*[43] found no change in the affinity of bovine serum albumin for γ-decalactone between 20 and 30 °C. However, Damodaran and Kinsella[15] found the affinity constant of 2-nonanone to soy

Table 4.3 The effect of added salt or ethanol on the binding between benzaldehyde and β-lactoglobulin. Reproduced from ref. 42 with permission from EDPSciences, Copyright 1999.

Medium	Conc. protein (%)	pH	Bound benzaldehyde (%)
NaCl[a]	3	3	12
NaCl[a]	3	6	16
Water			25
NaCl[a]	6.9	6	18
NaCl[a] + EtOH[b]	6.9	6	20
Water[c]	6.9	6	26
NaCl[a,c]	6.9	6	11
(NaCl[a] + EtOH[b])[c]	6.9	6	10

[a]50 mmol L^{-1}.

[b]13%.

[c]Protein was heated at 76 or 55 °C.

protein increased substantially when reducing the temperature to 5 °C, as did the number of binding sites. Druaux *et al.*[43] likewise found affinity for γ-decalactone to increase when they used 10 °C *vs.* 20–30 °C. The authors suggested that the temperature-induced changes in binding affinity were likely due to alterations in the tertiary and quaternary structures of soy protein at low temperatures.

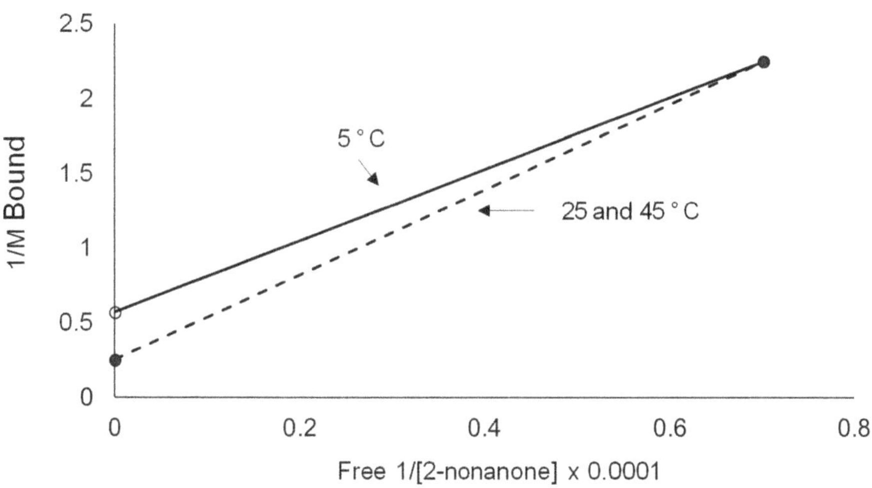

Figure 4.5 Effect of temperature on the binding of 2-nonanone to soy protein. Double-reciprocal plot of 2-nonanone binding to soy protein at 5 °C (solid line) and 25 and 45 °C (dashed line). Adapted from ref. 15 with permission from American Chemical Society, Copyright 1981.

4.4 Irreversible Bonds (Covalent Bonding)

Some of the earliest work on irreversible bonding between flavors and proteins was summarized by Maier.[44] Maier[44] noted that aldehydes were found to be the most reactive of the common flavor chemicals, and he postulated that this was due to their reactivity with the amine group on lysine or SH group on cysteine, forming a Schiff base. While there was not a lot of work published in this area at the time, Maier[44] commented that the formation of esters, hemiacetals, acetals, or ketals in food systems is highly unlikely.

Gremli[5] was also interested in distinguishing between reversible and irreversible bonding. He chose to look into the issue of flavor–protein binding using two methods of chemical analysis: the headspace equilibrium method and the high vacuum transfer system method. He proposed that a headspace equilibrium method would reflect all bonding reactions between soy protein (reversible or irreversible) and flavor volatiles, while a high vacuum transfer method would recover all volatiles other than those "irreversibly" bound. He studied the reaction of model alcohols, aldehydes, and ketones with a soy protein isolate (Proline D, Central Soya). He found no interaction (binding) among the alcohols tested, but both aldehydes and ketones were found to react reversibly and irreversibly, with the generalization that longer carbon chains interacted to a greater extent than short-chain compounds. Gremli[5] used his data to predict that at the flavor dosage levels common in flavor applications, *ca.* 70% of the added heptanal and 60% of the 2-nonanone would be bound due to flavor–protein reactions.

Numerous other studies have been conducted in this topic area using various methods to access flavor–protein binding. Studies focused on determining why lemon is a difficult flavor to use in protein-containing products showed that all of the added citral was completely bound in a 5% casein or soy protein solution.[45,46] This would result in the complete loss of lemon flavor. Hansen and Heinis[47] reported vanillin interaction with whey proteins and casein using sensory evaluation. They clearly demonstrated the impact of flavor–protein binding on the sensory perception of vanillin in a dairy product. They did not use any analytical methods for determining the binding mechanism but commented on the possible role of Schiff base formation, *i.e.*, covalent reactions between flavor and protein. While the methods available to researchers at the time could not document the formation of covalent bonds, evidence was mounting that the formation of covalent bonds between flavor components and proteins was of substantial importance.

Over time, methods have been developed to specifically measure irreversible binding (see Chapter 2), and proof of reaction has become possible. The remainder of this chapter will be devoted to what has been learned by using more specific methodologies to measure only irreversible binding.

4.4.1 Protein–Flavor Covalent Reactions

If one considers the types of functional groups residing on amino acid side chains and the broad array of functional groups comprising flavor molecules, one might expect various covalent reactions to occur. The amino acid content, protein structure and concentration, flavor composition, and reaction environment (*e.g.*, pH, temperature, and water activity) directly influence the reactions and rates of reactions that will occur.

Covalent bond reactions that are of the greatest interest in flavor–protein reactions are carbonyl reactions (*e.g.*, Schiff base and Michael addition) and disulfide reactions.[48] Most of the reactions of aldehydes, ketones, carboxamides, and carboxylic acid derivatives, which are popular functional groups of flavor molecules, directly involve carbonyl chemistry.[49] Of the carbonyls, aldehydes are more reactive than ketones because of steric effects in the latter. The additional substituent increases the steric restrictions for the nucleophilic approach and causes greater steric interaction in the tetrahedral adduct as the hybridization changes from trigonal to tetrahedral.

Aldehydes and ketones may also react with alcohols to form reversible addition products (hemiacetals). Hemiacetals may then undergo dehydration, and the addition of another alcohol molecule will result in the formation of an acetal. These reactions commonly occur in concentrated flavorings using ethanol or propylene glycol as a solvents, but a food environment typically does not favor their stability.

While several of the amino acid side chain functional groups could be reactive, there are specific groups (*e.g.*, $-NH_2$, –SS–, and –SH) where covalent linkages are most likely to take place with flavor components.[50,51] The products of these covalent reactions are salts, amides, esters, and aldols.[52] As noted earlier, the most common reaction for forming covalent bonding is between a flavor molecule containing an aldehyde functional group and a lysine group from the protein.[53]

Due to the historical lack of definitive methods to distinguish between covalent and non-covalent bond formation, there is little information in the literature documenting the occurrence of this reaction. Often times, covalent reactions were assumed to occur, but the method could not absolutely demonstrate that they did occur or determine where on the protein molecule the reaction occurred. This section will discuss what is known about the loss of desirable flavor components due to irreversible flavor–protein reactions.

4.4.2 Effect of Protein Type on the Formation of Flavor Bonding

While there is minimal information on reversible binding across protein sources, there is even less information on the irreversible binding of flavors across protein sources. As will be noted later in this chapter, flavor–protein reactions outside the food area offer some useful data.

4.4.3 Effect of the Chemical Properties of a Flavor Molecule on Flavor Binding

Meynier *et al.*[54] reported on covalent reactions between hexanal or *trans*-2-hexenal and sodium caseinate. They found that the proportion of unreacted Lys residues decreased upon the addition of either flavor compound, but the His residues decreased only with the addition of *trans*-2-hexenal. Earlier, Aliaz and Giron[55] found similar results for the reaction of 2-octenal with bovine serum albumin. They monitored the covalent reaction of 2-octenal using amino acid analysis. They found that after 4 h at 37 °C in an aqueous solution, 50% of the His in the bovine serum albumin covalently reacted with the added 2-octenal. Both research groups noted that the turbidity of the reacting solution increased over time, and this was proposed to be due to protein crosslinking by the aroma compounds. These results are consistent with other studies, *e.g.*, Yamaki *et al.*,[56] who found that the addition of 2-octenal to α-crystallin resulted in 45% of the Lys and 60% of the His residues reacting.

While aldehydes are considered the most reactive of the flavor chemicals, sulfur compounds have also been shown to be reactive. Sulfur-containing flavor chemicals are relevant, particularly for savory flavors. Sulfur-containing aroma compounds have been shown to react with sulfhydryl and disulfide groups on a protein.[57,58] Flavor molecules with a higher number of sulfhydryl groups, *e.g.*, diallyl disulfide and dimethyl trisulfide, can react with cysteine and result in disulfide interchange reactions. These reactions are thermodynamically spontaneous and, therefore, favorable, as free energy ($\Delta G = -RT \ln K$) is negative.[59]

Mottram *et al.*[58] demonstrated this reactivity by taking a commercially available savory flavoring containing thiol and disulfide aroma compounds and heating it to 100 °C with egg albumin protein. This heating resulted in a significant reduction in the concentration of the sulfur-containing flavor compounds in the flavoring (Chapter 2, Likens-Nickerson method). The results indicated that there was an interchange of thiol and disulfide groups with sulfhydryl and disulfide groups of proteins, *e.g.*, cysteine, as there was over a 100-fold reduction in bis(2-furanylmethyl)disulfide concentration when heated with albumin protein. There was a smaller reduction in the concentration of disulfides when casein, a protein with a low concentration of sulfhydryl groups, was used. During the reaction, the disulfides were reduced to thiol compounds by a redox reaction.

Some data on the influence of functional groups, specifically on covalent binding, has recently been reported.[31] Covalent reactions among 47 different flavor compounds from 13 different functional groups with β-lactoglobulin have been studied using liquid chromatography-mass spectrometry (LC-MS).[60] Anantharamkrishnan, Hoye, and Reineccius[31] showed that aroma compounds containing aldehyde, sulfur (especially thiols), and furan functional groups were the most reactive flavor compounds to form covalent bonds with β-lactoglobulin. Sulfur-containing molecules (*e.g.*, propanethiol) and dimethyl trisulfide reacted with the cysteine groups in β-lactoglobulin to form covalent linkages. The thiol-containing groups on the flavor compound cleaved one or both of the electrophilic disulfide linkages in β-lactoglobulin. Flavor compounds containing a ketone functional group did not form covalent bonds, but diacetyl, a diketone, formed covalent bonds with β-lactoglobulin. It reacted with the guanidine group in arginine residues to form dinitrogen alicyclic compounds. Hydrocarbons, alcohols, acids, esters, lactones, and pyrans did not form covalent adducts with β-lactoglobulin.

One should note that this study was conducted for short reaction times (48 h maximum, generally at ambient temperature), and, therefore, the reactivity of some compounds might not have progressed to detectable levels within the study reaction time. An example of this is likely vanillin. Anantharamkrishnan, Hoye, and Reineccius[31] found no evidence of the reaction of vanillin with β-lactoglobulin while other researchers have found a loss in vanillin as evidenced by sensory data (largely reversible binding) and brown color formation (irreversible binding). Research considering longer storage/reaction periods needs to be conducted to make the results more appropriate for shelf-life issues.

4.4.4 Environment Effects

The effects of reaction environment (pH, temperature, and water activity) on the occurrence of covalent bond formation between select flavor compounds and a model food protein (β-lactoglobulin) were reported by Ananthramkrishnan and Reineccius,[31] and are summarized in the following sections. Data on the covalent reaction of allyl isothiocyanate (AITC) under different environmental conditions have also been reported.[61] The researcher's interest in AITC reactions was not related to its flavor but the loss of a desired bioactive compound.

4.4.4.1 pH

One would expect pH to have a considerable effect on these interactions/reactions. pH would affect the reaction rate by inducing structural changes to the protein (*e.g.*, denaturation), making reaction sites more or less available depending upon how the pH changed protein structure and/or altered the charge on the protein-reactive sites. For example, the ionization of the amine group on lysine would reduce the amine–carbonyl reaction rate.

Liquid chromatographic/mass spectrometric data showed that covalent adducts were formed more slowly at low pHs (3) than at basic pHs for some compounds (*e.g.*, citral, AITC, and dimethyl trisulfide).[62] No reactivity was observed between citral and β-lactoglobulin at pH 3, but substantial reactivity was found at pHs 7 and 8 (see Figure 4.6). The lack of observable

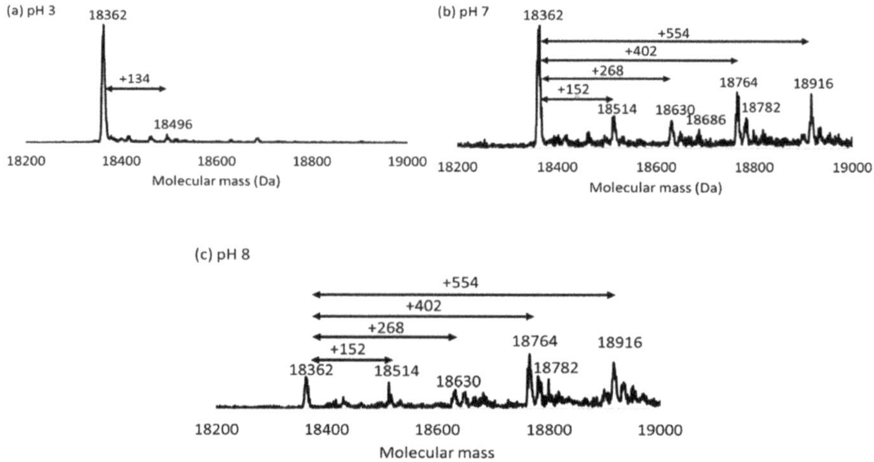

Figure 4.6 Deconvoluted electrospray ionization (ESI) spectra of β-lactoglobulin + citral at pHs 3, 7, and 8. Masses + 134, 152, 268, 403, and 554 are from the covalent addition of citral molecules to β-lactoglobulin 1 h after addition at ambient temperature. Reproduced from ref. 62 with permission from American Chemical Society, Copyright 2020.

reactions at pH 3 is likely due to the short reaction times monitored in the study.

4.4.4.2 *Temperature*

Foods will undergo a substantial range of heat treatments depending on the food and the reason for the heat treatment (prolongs shelf life, kills pathogens, or provides sterility). Additionally, food products may be stored at very different temperatures (*e.g.*, ambient, refrigerator, or freezer). Due to the wide variation in potential heat treatment and storage temperatures across foods, the effect of temperature on flavor reactions is of substantial interest. Unfortunately, there is very little in the literature related to covalent bond formation between flavor compounds and proteins. Ananthramkrishnan and Reineccius[62] conducted a brief study considering only a limited range of storage temperatures (5, 20, and 45 °C) for periods up to 24 h. As one would expect, higher temperatures (45 °C) led to the formation of products that were not observed at lower temperatures (4 and 20 °C) (see Figure 4.7).

Chobpattana *et al.*[33] studied the reaction of vanillin with amino acids (L-cysteine, L-lysine, and phenylalanine) and three peptides (aspartame, glutathione, and pentalysine) at temperatures ranging from 55 to 75 °C for 5–25 h, and caseinate and bovine serum albumin[63] under high-temperature–short-time pasteurization temperatures (68 °C for 30 min or 75 °C for 15 min). They found vanillin to react with the amino acids and peptides under these very severe conditions *via* covalent reactions. There was a

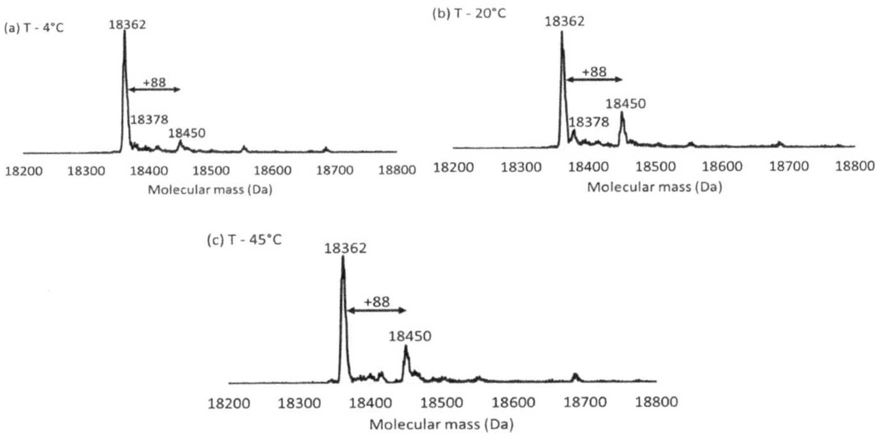

Figure 4.7 Deconvoluted electrospray ionization (ESI) mass spectra of β-lactoglobulin incubated with benzaldehyde at different temperatures (a) 4 °C, (b) 20 °C, and (c) 45 °C. Reproduced from ref. 62 with permission from American Chemical Society, Copyright 2020.

limited reaction with the proteins studied due to the much-reduced heating times in that study. They attributed losses of vanillin primarily to hydrophobic/hydrophilic bonding with caseinate and bovine serum albumin rather than covalent bonding.[63]

4.4.4.3 *Water Activity*

Ananthramkrishnan and Reineccius[62] conducted a brief study on the effect of water activity (0.11–0.75) on reaction rates. They found that increasing a_W's led to an increase in the formation of adducts for AITC. There were no observable differences in adduct formation as a function of a_W for benzaldehyde, citral, and dimethyl disulfide. However, this lack of observed effect may be due to the rate of the reaction being too slow to be detected in the timeframe of this study. These studies on environmental effects need to be repeated using longer reaction times to get relevant data encompassing real-world storage times.

4.5 Related Research

4.5.1 Metabolic Disruption by Protein: Small Molecule Reactions

There have been numerous publications on proteins and their covalent reactions with small molecules (*e.g.*, flavor compounds) in the medical and toxicology fields. Some examples of this literature follow. They have provided directions on how to monitor covalent flavor reactions in foods.[64]

Matrix-assisted laser desorption ionization time-of-flight mass spectrometry (MALDI-TOF MS) and liquid chromatography–electrospray ionization mass spectrometry (LC-ESI MS) were used to prove that 2(*E*),4(*E*)-decadienal (DDE) covalently bonded to Lys residues of cytochrome *c*, ribonuclease A, and β-lactoglobulin proteins. Two of these proteins (cytochrome *c* and ribonuclease A) serve important functions in the body, so their loss/reduction due to small-molecule reactions can have very negative effects on health. All three are small proteins and, thus, lend themselves to study by the noted MS techniques with high accuracy. It was observed that 2(*E*),4(*E*)-decadienal formed measurable Lys Schiff base adducts only 10 min after mixing with cytochrome *c* and ribonuclease A, but the adducts are relatively unstable and can reverse in 24 h. However, β-lactoglobulin reactions with 2(*E*),4(*E*)-decadienal were NOT reversible: the Lys Schiff adducts increased and proceeded to form the more stable Lys pyridinium adducts and Cys Michael adducts.[65] While flavor chemists think of 2(*E*),4(*E*)-decadienal as a desirable flavor compound generated by the thermal oxidation of lipids, it has been shown to potentially have negative health effects.

A study by Zhu *et al.*[66] found that 4-oxo-2-nonenal (ONE), a lipid oxidation product, cross links proteins by conjugate addition of its C=C with the side chain nucleophiles such as sulfhydryl or imidazole groups of β-lactoglobulin to form a ketoaldehyde. This ketoaldehyde further reacts to undergo Paal–Knorr condensation with the primary amine from Lys residues. This phenomenon was monitored using sodium dodecyl sulfate-polyacrylamide gel electrophoresis (SDS-PAGE), MALDI-TOF MS, and LC-ESI MS after tryptic and chymotryptic digestion.[67] This study also showed that the presence of Glutathione made ONE more reactive with cytochrome *c* and ribonuclease A, but it inhibited the modification by 4-hydroxy-2-nonenal (HNE).

A review article by Sayre *et al.*[68] noted that although simple Michael and Schiff base adducts are formed initially, only some of these adducts survive proteolysis in subsequent analysis. The adducts that survive are ONE and HNE Michael Cys and His adducts. Schiff base adducts are reversible but can be stabilized by reducing the sample with $NaBH_4$ to survive proteolysis and then be detected by LC-ESI MS.[68] Thus, samples are commonly reduced to provide better data on these reactions.

In a study where HNE was added to β-lactoglobulin and human hemoglobin protein, >99% of the adducts were formed *via* Michael addition, and only trace amounts of adducts were formed by Schiff base reactions (using LC-ESI MS, gas chromatography/mass spectrometry, and spectroscopic protein carbonyl assays). Michael addition reactions can proceed to cross-linking or result in a product with more than one aldehyde reacting at the same amino acid residue.[69] Another study bolstered the finding that Michael adducts are the predominant modification mechanism between

HNE and low-density lipoproteins. The authors suggested that the Michael adducts could be a reliable marker for atherosclerosis.[70]

4.5.2 Loss of Bioactivity *via* Reaction with Proteins

The reactivity of AITC (a character impact flavor compound in the Mustard family) with proteins has been studied due to its positive health benefits; its consumption has been linked to anti-carcinogenic, antimicrobial, anticancer, and cryoprotective activities.[71] AITC exerts its anticancer activity by binding to certain proteins, such as tubulins.[72,73] The reaction between AITC and lysine and phenolic groups of tyrosine residues of the mustard 12S protein was found to increase with pH, temperature, and reaction time. The reaction was complete when the AITC to protein molar ratio was 100 : 1.[74] AITC reacts slowly with insulin, bovine serum albumin, and lysosome protein to cleave the disulfide bonds from the cysteine residues that lead to polymerization.[75] It also reacted with the arginine and lysine residues to form thiourea-like adducts. After reaction with AITC, the digestibility of bovine serum albumin protein with trypsin and chymotrypsin decreased but did not change with pepsin. This was postulated to be due to the nature of the enzyme, *i.e.*, trypsin and chymotrypsin attack peptides with basic and aromatic amino acids, but pepsin has a much broader range of attack sites.

Benzyl isothiocyanate (BITC) (also found in the mustard family), a widely used compound in chemotherapy and cancer chemopreventive action, disturbs lipid metabolism in rats following *in vivo* long-term treatment and *in vitro* adipocytes.[76] This was postulated to be due to adduct formation. At very high concentrations (50 mg g^{-1} protein), BITC has also been shown to decrease the nutritional quality of egg proteins when tested by a protein quality rat bioassay. BITC reduced the lysine content and its availability and significantly affected the bio-utilization of nitrogen and the deposition of energy.[77]

4.6 Summary

The recognition that the composition of a food influences how an added flavoring is perceived by a person is over 100 years old. Our ability to study interactions between a food and a flavoring (*i.e.*, having adequate research techniques) is now nearly 60 years old. During this 60-year period, we have learned a great deal about how flavor compounds may form reversible bonds with proteins. Researchers have shown that amino acid composition and subsequent protein structure affect interactions. While many types of reversible bonding can occur, it is generally recognized that hydrophobic bonding between flavor compounds and hydrophobic pockets in a protein plays a major role in influencing flavor perception. There are effects of pH, water activity, denaturation, temperature, and added salts or solvents (*e.g.*, ethanol) that may modify the extent of these reversible reactions. Despite the effort to understand these interactions, little has been done, or likely can be done, to mitigate these effects. We can only work toward understanding and strategizing how to manage their interactions, especially at a time when numerous new plant-based proteins are potentially coming to the market.

We lack studies across proteins utilizing the same equipment/methodologies, which would enable comparisons to be made as to whether a protein has more or less affinity for certain types of flavors. The primary solution to this problem rests with flavor companies, which must be able to formulate flavors for a given protein. As noted in this review, reversible binding comes to an equilibrium and can be managed *via* the flavor formulation. The industry needs a good compilation of flavor-binding properties for a broad array of flavor compounds and protein types to help in designing flavor systems.

The problems faced through irreversible binding are a greater challenge. These interactions do not come to equilibrium but are continuous, occurring at a high velocity during thermal processing and slowly throughout the shelf life. The reaction comes to an end when there are either no reactive flavoring components remaining or all of the reactive sites on a protein have chemically reacted. We know very little about these reactions at this time; only recently have the analytical techniques become available to conduct such research. We need to understand what flavor chemicals will react and how reaction rates vary with flavor compound, protein amino acid composition and structure, the food environment (pH, water activity, composition, and temperature profile), and thermal processing (*e.g.*, pasteurization and sterilization). We are not likely able to change nature to make aldehydes (for example) nonreactive with amines (*e.g.*, Lys), but we should know where we can use a certain flavor–protein combination and how long we can expect the product to have an acceptable flavor. We may have to live with either the short shelf life of certain products (*e.g.*, lemon-flavored high protein products) or find/design flavors that have less reactive chemicals.

There is still much to be done.

Abbreviations

AITC	Allyl isothiocyanate
BITC	Benzyl isothiocyanate
DDE	2(*E*),4(*E*)-Decadienal
HNE	4-Hydroxy-2-nonenal
LC-ESI MS	Liquid chromatography-electrospray ionization mass spectrometry
MALDI-TOF MS	Matrix-assisted laser desorption ionization time-of-flight mass spectrometry

ONE	4-Oxo-2-nonenal
SDS-PAGE	Sodium dodecyl sulfate-polyacrylamide gel electrophoresis

References

1. H. Meyer and K. H. Gottlieb-Billroth, *Z. Physiol. Chem.*, 1920, **112**, 55.
2. J. Solms, F. Osman-Ismail and M. Beyeler, *Can. Inst. Food Sci. Technol. J*, 1973, **6**, A10–A16.
3. J. C. Cowan, J. J. Rackis and W. J. Wolf, *J. Am. Oil Chem. Soc.*, 1973, **50**, a426–a444.
4. S. Arai, M. Noguchi, M. Yamashita, H. Kato and M. Fujimaki, *Agric. Biol. Chem.*, 1970, **34**, 1569–1573.
5. H. A. Gremli, *J. Am. Oil Chem. Soc.*, 1974, **51**, 95–97.
6. E. Guichard, in *Interaction of aroma compounds with food matrices*, Elsevier Ltd, 2015.
7. J. Kühn, T. Considine and H. Singh, *J. Food Sci.*, 2006, **71**, R72–R82.
8. K. Wang and S. D. Arntfield, *Flavour Fragr. J.*, 2017, **32**, 92–101.
9. K. Wang, PhD thesis, University of Manitoba, 2015.
10. L. Y. Heng, PhD thesis, Wageningen Universiteit, 2005.
11. E. Guichard, *Food Rev. Int.*, 2002, **18**, 49–70.
12. T. E. O'Neill and J. E. Kinsella, *J. Food Sci.*, 1987, **52**, 98–101.
13. T. E. O'Neill and J. E. Kinsella, *J. Agric. Food Chem.*, 1987, **35**, 770–774.
14. S. Damodaran and J. E. Kinsella, *J. Agric. Food Chem.*, 1980, **28**, 571–578.
15. S. Damodaran and J. E. Kinsella, *J. Agric. Food Chem.*, 1981, **29**, 1249–1253.
16. M. Espinoza-Diaz, PhD thesis, *Etude de la rétention et de la ibération des composés d'arôme dans... ou non de β-lactoglobuline*, Dijon, 1991.
17. E. Charles, M. Lambert, S. Brondeur, P. Courthaudon and E. Guichard, in *Flavor Release*, ed. D. Roberts and A. J. Taylor, American Chemical Society, Washington D.C., 2000, pp. 342–354.
18. K. Sostmann and E. Guichard, *Food Chem.*, 1998, **62**, 509–513.
19. A. Jasinski and E. Kilara, *Milchwissenschaft*, 1985, **40**, 596–599.
20. S. F. O'keefe, A. P. Resurreccion, L. A. Wilson and P. A. Murphy, *J. Food Sci.*, 1991, **56**, 802–806.
21. K. Wang and S. D. Arntfield, *Food Chem.*, 2014, **157**, 364–372.
22. J. Reiners, S. Nicklaus and E. Guichard, *Lait*, 2000, **80**, 347–360.
23. K. L. Franzen* and J. E. Kinsella, *J. Agric. Chem.*, 1974, **22**, 675–678.
24. E. Dufour and T. Haertlé, *J. Agric. Chem.*, 1990, **38**, 1691–1695.
25. M. Lübke, E. Guichard and J. L. Le Quéré, in *Flavor Release*, ed. D. Roberts and A. J. Taylor, Washington D.C., 2000, pp. 282–292.

26. S. Arai, M. Noguchi, M. Kaji, H. Kato and M. Fujimaki, *Agric. Biol. Chem.*, 1970, **34**, 1420–1423.
27. P. K. W. Ng, E. Hoehn and W. Bushuk, *J. Food Sci.*, 1989, **54**, 105–107.
28. P. K. W. Ng, E. Hoehn and W. Bushuk, *J. Food Sci.*, 1989, **54**, 324–325.
29. A. P. Hansen and D. C. Booker, *ACS Symp. Ser.*, 1996, **633**, 75–89.
30. T. O'neill and J. E. E. Kinsella, *J. Food Sci.*, 1988, **53**, 906–909.
31. V. Anantharamkrishnan, T. Hoye and G. Reineccius, *J. Agric. Food Chem.*, 2020, **68**, 6395–6402.
32. V. L. Mcneill and K. A. Schmidt, *J. Food Sci.*, 1993, **58**, 1142–1144.
33. W. Chobpattana, I. J. Jeon and J. S. Smith, *J. Agric. Food Chem.*, 2000, **48**, 3885–3889.
34. P. F. Fox and D. M. Mulvihill, *J. Dairy Res.*, 1982, **49**, 679–693.
35. S. Chung and R. Villota, *J. Food Sci.*, 1989, **54**, 1604–1606.
36. M. G. Semenova, A. S. Antipova, T. A. Misharina and R. V. Golovnya, *Food Hydrocolloids*, 2002, **16**, 557–564.
37. Q. Zhou and K. R. Cadwallader, *J. Agric. Food Chem.*, 2004, **52**, 6271–6277.
38. Q. Zhou and K. R. Cadwallader, *J. Agric. Food Chem.*, 2006, **54**, 1838–1843.
39. E. Jouenne and J. Crouzet, *J. Agric. Food Chem.*, 2000, **48**, 1273–1277.
40. S. M. Van Ruth and E. Villeneuve, *Food Chem.*, 2002, **79**, 157–164.
41. K. Wang and S. D. Arntfield, *Food Res. Int.*, 2015, **77**, 1–9.
42. I. Andriot, I. Marin, G. Feron, P. Relkin and E. Guichard, *Lait*, 1999, **79**, 577–586.
43. C. Druaux, S. Lubbers, C. Charpentier and A. Voilley, *Food Chem.*, 1995, **53**, 203–207.
44. H. G. Maier, *Angew. Chemie Int. Ed. English*, 1970, **9**, 917–926.
45. H. Friedrich and B. A. Gubler, *Lebensm. Wiss. Technol.*, 1978, **11**, 215–218.
46. B. A. Gubler, in *Flavour 91*, ed. P. Schrier, Berlin, 1981, p. 717.
47. A. P. Hansen and J. J. Heinis, *J. Dairy Sci.*, 1991, **74**, 2936–2940.
48. N. V. Bhagavan and C.-E. Ha, in *Essentials of Medical Biochemistry*, Elsevier, 2011, pp. 29–38.
49. F. A. Carey and R. J. Sundberg, in *Advanced Organic Chemistry*, Springer, US, 2007, pp. 119–251.
50. G. Reineccius and H. B. Heath, in *Flavor chemistry and technology*, Taylor & Francis, 2006.
51. S. M. Van Ruth and J. P. Roozen, in *Food Flavour Technology: Second Edition*, Wiley-Blackwell, Oxford, UK, 2nd edn, 2010, pp. 190–206.
52. O. E. Mills and J. Solms, *Leb. Wiss.Technol.*, 1984, **17**, 331–335.

53. M. Preininger, in *Ingredient Interactions Effects on Food Quality*, ed. A. G. Gaonkar and A. McPherson, CRC Press, Boca Raton, FL, 2nd edn, 2006, pp. 477–542.
54. A. Meynier, V. Rampon, M. Dalgalarrondo and C. Genot, *Int. Dairy J.*, 2004, **14**, 681–690.
55. M. Alaiz and J. Girón, *J. Agric. Food Chem.*, 1994, **42**, 2094–2098.
56. S. Yamaki, T. Kato and K. Kikugawa, *Chem. Pharm. Bull.*, 1992, **40**, 2138–2142.
57. D. S. Mottram and I. C. C. Nobrega, in *Flavor Release*, ed. D. Roberts and A. J. Taylor, American Chemical Society, Washingon D.C., 2000, pp. 274–281.
58. D. S. Mottram, A. Dodson and C. Szauman-Szumski, *J. Agric. Food Chem.*, 1996, **44**, 2349–2351.
59. S. Damodaran and J. E. Kinsella, *J. Agric. Food Chem.*, 1981, **29**, 1253–1257.
60. V. Anantharamkrishnan, PhD thesis, University of Minnesota, 2020.
61. K. Rade-Kukic, C. Schmitt and H. M. Rawel, *Food Hydrocolloids*, 2011, **25**, 694–706.
62. V. Anantharamkrishnan and G. A. Reineccius, *J. Agric. Food Chem.*, 2020, **68**, 13833–13843.
63. W. Chobpattana, I. J. Jeon, J. S. Smith and T. M. Loughin, *Food Chem. Toxicol.*, 2002, **67**, 973–977.
64. V. Anantharamkrishnan and G. A. Reineccius, *J. Agric. Food Chem.*, 2020, **68**, 13121–13130.
65. X. Zhu, X. Tang, J. Zhang, G. P. Tochtrop, V. E. Anderson and L. M. Sayre, *Chem. Res. Toxicol.*, 2010, **23**, 467–473.
66. L. M. Zhu, X. Tang, X. Zhang, J. Tochtrop, G. P. Anderson and V. E. Sayre, *Chem. Res. Toxicol.*, 2010, **23**, 467–473.
67. X. Zhu, M. M. Gallogly, J. J. Mieyal, V. E. Anderson and L. M. Sayre, *Chem. Res. Toxicol.*, 2009, **22**, 1050–1059.
68. L. M. Sayre, D. Lin, Q. Yuan, X. Zhu and X. Tang, *Drug Metab. Rev.*, 2006, **38**, 651–675.
69. B. A. Bruenner, A. D. Jones and J. B. German, *Chem. Res. Toxicol.*, 1995, **8**, 552–559.
70. K. Uchida, S. Toyokuni, K. Nishikawa, S. Kawakishi, H. Oda, H. Hiai and E. R. Stadtman1, *Biochemistry*, 1994, **33**, 12487–12494.
71. Y. Zhang, *Mol. Nutr. Food Res.*, 2010, **54**, 127–135.
72. L. Mi, A. J. Di Pasqua and F.-L. Chung, *Carcinogenesis*, 2011, **32**, 1405–1413.
73. L. Mi, Z. Xiao, B. L. Hood, S. Dakshanamurthy, X. Wang, S. Govind, T. P. Conrads, T. D. Veenstra and F.-L. Chung, *J. Biol. Chem.*, 2008, **283**, 22136–22146.

74. N. V. K. K. Murthy and M. S. N. Rao, *J. Agric. Food Chem.*, 1986, **34**, 448–452.
75. S. Kawakishi and T. Kaneko, *J. Agric. Food Chem.*, 1987, **35**, 85–88.
76. M. Okulicz and I. Hertig, *J. Physiol. Biochem.*, 2013, **69**, 75–84.
77. M. Hernández-Triana, J. Kroll, J. Proll, J. Noack and K. J. Petzke, *J. Nutr. Biochem.*, 1996, **7**, 322–326.

CHAPTER 5

The Flavour of Dairy Proteins

ROBERT J. MCGORRIN*

Department of Food Science and Technology, Oregon State University, Corvallis, Oregon 97331, USA
*E-mail: Robert.mcgorrin@oregonstate.edu

5.1 Introduction

Dairy proteins are widely used as ingredients in foods and beverages, and their consumption is growing. Dairy protein ingredients are added to foods to impart emulsification and stabilization properties and to provide nutritional value. Commercially available dried and fluid dairy proteins include milk protein concentrate (MPC) and milk protein isolate (MPI), whey protein concentrate (WPC) and whey protein isolate (WPI), rennet casein, acid casein, sodium caseinate, and concentrated serum protein.[1] Non-fat dry milk (NDM), skimmed milk powder (SMP), and dry whey products are the major dried milk ingredients produced in the United States and a significant source of dairy protein.[2] Other major global manufacturers of dairy proteins are New Zealand and Ireland. The market demand for dairy protein ingredients is expanding due to product innovations in the protein bar and dairy beverage categories.[3] MPCs are increasingly being utilized as components of high-protein beverages for muscle building, nutritional supplementation, and clinical/medical nutrition applications.[4]

Flavour quality is a major factor that drives consumer acceptance and preference for food products.[5] The flavour chemistry of dairy proteins in relation to the sensory contributions of volatile and non-volatile compounds

Food Chemistry, Function and Analysis No. 41
Flavour and Consumer Perception of Food Proteins
Edited by Jing Zhao and Changqi Liu

Published by the Royal Society of Chemistry, www.rsc.org

derived from them is an area of active research. Over 100 volatile aroma compounds have been identified that provide sensory contributions to the flavour of milk, cheese, and dairy products.[6–8] Dairy-related flavour and off-flavour compounds are formed *via* a concerted series of chemical and biochemical pathways, including heat-induced reactions, lipid oxidation, glycolysis, lipolysis, and proteolysis of milk-based products. The flavour chemistry of the key odour-active components in bovine milk, cheese, and other fermented dairy products has been comprehensively reviewed; many of these flavour-active compounds originate from dairy proteins.[6–10]

Although dairy proteins and dairy ingredients typically have neutral taste profiles, challenges arise when developing products with very high protein levels. These sensory challenges are primarily caused by intrinsic undesirable flavour contributions from the protein raw materials or off-flavours that develop during their processing and storage.[11] The resulting sensory characteristics tend to be highly variable. Research studies indicate that consumers can discern sensory differences in rehydrated dairy protein ingredients and that these flavours carry through into product applications.[12] Consequently, the incorporation of a dairy protein into a food product may amplify undesirable off-flavours inherent in the protein ingredient. The aroma, sensory, and off-flavour challenges with dairy protein ingredients have been recently reviewed.[10,13–15]

Alternately, interactions between added proteins and flavours are known to modify the flavour profile of a food product due to associations, altered release, or binding with other desired flavour components.[16] The interactions of dairy proteins with flavour components are another area of active research.[17–20] In these situations, aroma imbalance and flavour fading are common issues. Flavour fade is the sensed loss of flavour affected by the non-specific binding of flavour compounds with the protein. As fade does not occur equally among the added flavourants, the flavour profile may shift, be perceived as imbalanced, or decrease in overall intensity.[13,21]

This chapter focuses on the flavour chemistry of bovine dairy protein concentrates and isolates that are used as ingredients in foods and beverages. The characteristics of major odour and taste compounds, the source and influencing factors of flavour compound formation, and the potential methods of eliminating undesirable off-flavours in dairy protein products are described. Equally important is the impact of processing and storage conditions on the flavour and flavour stability of protein ingredients. Insights from sensory-directed analysis and evaluation of dairy protein products are also discussed.

5.2 Sources of Flavours in Milk and Dairy Products

The characteristic flavours of dairy products are generated by chemical and fermentation pathways that occur during the processing and storage of milk. Typical flavour volatiles in milk originate from the cow's metabolism with influence from its diet, such as feed, green forages, or silage.[22] The resulting balance of volatile (aromatic) and non-volatile (tastant) flavour compounds is derived from the fat, carbohydrate, or protein components in milk. Dairy flavour compounds can be produced *via* four separate pathways.[6,10,23]

- **Lipid:** Flavour constituents derived from unsaturated fatty acids in milkfat *via* lipolysis or oxidation to produce aldehydes, lactones, hydroxyl acids, and methyl ketones. The lipolysis of milkfat triglycerides produces significant quantities of C_4–C_{10} free fatty acids that are potent flavour contributors in cultured dairy products.
- **Carbohydrate:** Flavour compounds derived from lactose fermentation by lactic acid bacteria (*Lactococcus* ssp.) or citrate metabolism generate lactic acid and the important flavour compounds diacetyl (2,3-butanedione) and acetoin (3-hydroxy-2-butanone). These are significant in cultured dairy products, such as buttermilk, sour cream, and yoghurt.
- **Protein:** The proteolytic breakdown of casein and other milk proteins provides tastants (free amino acids and peptides), aroma compounds, including Strecker aldehydes (phenylacetaldehyde and iso-valeraldehyde), and sulphur compounds produced from cysteine and methionine (methanethiol, dimethyl disulphide, and methional).
- **Thermal generation:** Non-enzymatic browning/Maillard reactions between sugars (lactose) and amino acids (from proteolysis of casein) generate a wide variety of aroma compounds, including pyrazines, pyridines, and thiazoles. The thermal degradation pathways of lactose/glucose produce caramelized sugar and maple-like aromas from the formation of 4-hydroxy-2,5-dimethyl-3(2*H*)-furanone (HDMF, furaneol), 3-hydroxy-4,5-dimethyl-2(5*H*)-furanone (sotolon), and other furanone compounds.

5.3 Dairy Protein Flavours

From a sensory perspective, intact dairy proteins provide texture to foods without discernible flavours or aromas.[14] This is primarily because proteins are macromolecular amino acid polymers that do not interact with aroma or taste receptors. However, proteins can contribute to flavour in several ways.

Proteins develop volatile aroma compounds as minor breakdown products from enzymatic or chemical pathways or are partially hydrolyzed to provide taste sensations from small peptides and individual amino acids, as outlined above.[6,10,23] In some cases, proteolysis can produce bitter and metallic tastes.[24]

While dairy protein ingredients (*e.g.*, WPC and MPC) should ideally impart a bland, mild, or dairy-like flavour, they often exhibit off-flavour characteristics that are contributed by minor impurities from non-protein constituents. For example, MPC contains traces of residual lipids that are carried over from the milkfat separation process. Small amounts of lipids can be oxidized or fragmented to form unsaturated aldehydes, ketones, lactones, and other potent aromatics.[10] While these off-flavour components are not formed from proteins, they are often a dominant contributor to the characteristic aroma or flavour of the dairy protein ingredient. The aroma compounds that contribute to these undesirable non-dairy flavour attributes have a high impact due to their relatively low sensory thresholds and are significant contributors to the overall flavour.[11] Aroma compound analysis can be utilized to identify potential sources and develop methods to prevent or remove undesirable off-flavour components.

5.4 Flavours of Major Dairy Protein Ingredients

The key flavour compounds that have been identified in dairy protein ingredients share similar characteristics with those identified in milk and fermented dairy products.[25] Desirable dairy flavour qualities such as sweet aromatic or cooked/milky attributes are associated with fresh milk or whey.[25] The delicate flavour of milk and cheese is the result of the appropriate prevalence and concentrations of a wide range of odorants to obtain the correct flavour balance.[6,10,26] If the flavour-active compounds are not appropriately balanced, then the desired flavour outcome is not attained and may be perceived as flawed. Minor residual carry-over components (fats and lactose) in dairy protein ingredients contribute characteristic but objectionable off-flavours/taints depending on their relative amounts. Undesirable non-dairy flavour characteristics include cardboard and corn chip/tortilla, which are not generally associated with fresh fluid milk or dairy products.[5,11,27] During the preparation of protein concentrates and isolates, the processes of pasteurization, drying, concentration, filtration, *etc.* can produce chemical changes resulting in desirable and undesirable aromas and flavours in the finished ingredient. The protein flavour quality is influenced by the time/temperature consistency of the process, the freshness of the milk, storage time, and conditions. The flavour characteristics of the principal dairy protein ingredients are discussed individually in the following sections.

5.4.1 Milk Concentrates: NDM and SMP

NDM and SMP are prepared by removing water from pasteurized skimmed milk. Both are classified for use as ingredients according to the heat treatment used in their manufacture. There are three main classifications: low-heat, medium-heat, and high-heat. NDM and SMP contain a maximum of 5% moisture (by weight) and 1.5% or less milkfat. The difference between them, as defined by two different sets of regulations and authorities, is that SMP must contain a 34% minimum milk protein content, whereas NDM has no standardized protein level.[28] NDM is defined by the United States Food and Drug Administration, while SMP is defined by *Codex Alimentarius*.

The aroma of NDM and SMP, produced with different heat treatments, reflects the sensory impact of volatile compounds formed from oxidation, lipolysis, protein degradation, and sugar caramelization of fluid milk components.[25,29,30] Higher temperatures during the production of NDM will intensify the oxidized, lipolyzed, and caramelized aroma volatiles generated *via* these flavour pathways. The overall levels of flavour compounds identified in NDM powder are very low, and their compositions are complex. In particular, volatile breakdown products from small quantities of residual fat have a large influence on the flavour of non-fat milk.

Free fatty acids and lactones, which were present at relatively high levels, are considered to be fundamental flavour contributors to NDM and SMP.[25,30] Oxidation of unsaturated fatty acids produces a complex mixture of aldehydes, ketones, and alcohols that impart off-flavours and limit shelf life and storage stability.[31] Lipid oxidation products impart specific off-flavours to whole milk and SMP, which are described as painty, metallic, fishy, and grassy.[31] Pasteurization and drying of milk promote the autoxidation of unsaturated fatty acids, resulting in elevated levels of volatile aldehydes (*E,E*)-2,4-decadienal (fried fatty), ketones (1-octen-3-one, mushroom, metallic), and lactones δ-decalactone (sweet, cream). Excessive amounts of free fatty acids, including butyric (rancid cheese), hexanoic (goaty), and octanoic (waxy, soapy) acids, are strong contributors to spoiled and heated milk aroma.

Heated milk flavours are also contributed by volatile sulphur compounds, including methanethiol (cooked corn), dimethyl trisulphide (cabbage), and methional (boiled potato).[30] *o*-Aminoacetophenone (foxy, corn tortilla) is a prominent aroma compound in stale dry milk[32] and stale flavour fractions of sterilized concentrated milk.[33] It is likely produced by the degradation of kynurenine, a metabolite of the amino acid tryptophan. Heterocyclic compounds derived from milk protein, such as indoles or thiazoles, indirectly influence the flavour of non-fat milk. Vanillin (vanilla, sweet) has a high odour intensity in milk and originates from cow rumen breakdown of feed lignins to coniferyl alcohol. The latter compound is conveyed through the milk and may subsequently oxidize to vanillin during the heat treatment of the milk.[34]

Caramelized flavours are generated in fluid milk subjected to high-heat treatments; these sugar degradation products include HDMF (furaneol; caramel and sweet), 3-hydroxy-2-methyl-4*H*-pyran-4-one (maltol; burnt sugar), and 3-hydroxy-4,5-dimethyl-2(5*H*)-furanone (sotolon; sugar odour). Maillard reactions between amino acids and sugars generate cooked, malty, and stale flavours and lead to the formation of the popcorn-odour compounds—2-acetyl-1-pyrroline, 2-acetyl-2-thiazoline, and 2-acetylthiazole—in heat-treated milk.

Although many of the above flavour compounds are present in fresh milk, their concentration levels increase with temperature treatment and storage time when they shift to off-flavours.[35,36] During the milk drying process, these same thermally and oxidatively generated aroma components are formed in different proportions and contribute varying intensities to the overall flavour depending on their individual sensory thresholds. For example, high heat-treated NDM exhibits strong aroma intensity, principally contributed by HDMF, butanoic acid, methional, *o*-aminoacetophenone, *trans*-4,5-epoxy-(*E*)-2-decenal, sotolon, and vanillin.[30,37] A flavour study of high heat-treated SMP identified methyl 2-methyl-3-furyl disulphide and bis(2-methyl-3-furyl) disulphide as the key character impact compounds that provide brothy and vitamin-like flavour attributes.[38] These aroma compounds are generated by the thermal degradation of vitamin B1 (thiamin) during processing. Different flavour intensities can develop in NDM with extended storage. A stored sample of NDM had particularly high odour intensities for methional, *o*-aminoacetophenone, C_2–C_{12} fatty acids, and skatole (faecal).[39] A comparison of sensory-significant volatile aroma-active compounds identified in NDM, SMP, and other dry dairy protein ingredients is presented in Table 5.1.

A lexicon of 17 flavour descriptors for SMP and dried dairy ingredients was developed by a trained sensory panel in parallel with instrumental volatile analysis.[40] A sensory lexicon is a defined, established sensory language that enables the sharing of standardized descriptions of food products and ingredients among different audiences. This lexicon was applied to characterize the overall flavour and flavour variability. Significant sensory attributes for NDM and SMP have been classified as sweet, cooked/heated, animal, cardboard, and astringent characteristics.[41] Many flavours are common to all dried dairy ingredients, while others are specific to particular products. Differences in sensory attributes have been reported for SMP produced from milk obtained from dairy cows on pasture and non-pasture diets.[22,42] The ability to show correlations between sensory lexicons, flavour chemical structures, and mechanisms of their formation improves the overall knowledge of NDM and SMP aroma and flavour and can be used to make ongoing improvements in product quality.

The aroma profile of milk powder is significantly influenced by seasonality, resulting in varying elevated quantities of dimethyl sulphide, pentanal, hexanal, and butanoic acid.[43] Whole milk powder manufactured during summer had significantly higher levels of dimethyl sulphide and the "green" smelling aldehydes hexanal and pentanal in comparison to milk dried in autumn and winter. However, butanoic acid levels were significantly different between production in autumn and spring.

Table 5.1 Significant aroma-active compounds identified in NDM, SMP, MPC, MPI, acid casein, rennet casein, micellular casein (MCC), sweet whey powder (SWP), WPC, WPI, and whey protein hydrolysate (WPH).

Aroma compound	Odour descriptor	NDM[a]	SMP[b]	MPC[c]	MPI[d]	Acid casein	Rennet casein	MCC	SWP	WPC[e]	WPI[f]	WPH
Acetic acid	Vinegar		25[g]	**44,45**[h]	**45**	**45**						
Propionic acid	Garbage				**45**	**45**						
Butyric acid	Vomit, cheesy	**30**	**25**	**45**	**45**	**45**	**45**		**59**	**27,64**	**27**	
Hexanoic acid	Goaty	**30**	**25**						**59**			
Octanoic acid	Goaty	**30**	**25**						**59**			
3-Methylbutanal	Malty			44	44			**44**				**81**
2-Methylbutanal	Malty, chocolate											**81**
2-Methyl-1-propanol	Burning plastic			**45**	**45**	45	45					
Hexanal	Green		25	**44,45**			**45**		**59**			
Phenylacetaldehyde	Honey, floral		25				**45**		**59**			
1-Octen-3-one	Mushroom	30	25	**44,45**	45	45	45	44,45	**59**	27,**64**	27	**81**
(*E*,*Z*)-2,4-Decadienal	Fatty, corn chips		25	45	45				**59**			
(*E*,*E*)-2,4-Decadienal	Fatty, corn chips	30	25	45	45		**45**		**59**			
(*E*,*Z*)-2,6-Nonadienal	Cucumber	30								**27,64**	**27**	**81**
Nonanal	Green, fatty	**30**		**44,45**	**45**	**45**	**45**	44,**45**	**59**			
(*E*)-2-Nonenal	Cardboard	**30**	25			**45**	45	44		**27,64**	**27**	**81**
o-Aminoacetophenone	Tortilla, corn chips	**30**	25	45	**45**	**45**		**45**		**27**	**27**	81
Decanal	Fatty									**27,64**	**27**	
4-Methyl octanoic acid	Waxy, soapy									27,**64**	27	
Decanoic acid	Fatty, stale									**27,64**	**27**	
Dimethyl sulphide	Sulphur											**81**
Dimethyl disulphide	Garlic, rubbery			45		**45**	**45**		59	27,64	27	
Dimethyl trisulphide	Cabbage			44,45	**45**	45	**45**	44,**45**	**59**	**27,64**	**27**	**81**

(*continued*)

Table 5.1 (*continued*)

Aroma compound	Odour descriptor	NDM[a]	SMP[b]	MPC[c]	MPI[d]	Acid casein	Rennet casein	MCC	SWP	WPC[e]	WPI[f]	WPH
Methyl-2-methy-3-furyl disulphide	Cooked, canned corn		38									81
Bis(2-methyl-3-furyl) disulphide	Vitamin B_1		38									
2-Methyl-3-furanthiol	Brothy, burnt, nutty									**27**,64	27	81
Methional	Potato	**30**	**25**	**44,45**	**45**	45	45		**59**	27,**64**	**27**	**81**
Benzothiazole	Carpet, clay			**45**	**45**		45	45				
2-Acetyl-1-pyrroline	Popcorn	30	**25**	**45**	**45**		45	**45**	**59**	**27,64**	**27**	81
2-Methoxy phenol	Smoky, fatty			45	**45**	**45**		45		**27,64**		81
γ-Nonalactone	Peach, floral		25							27,**64**		81
δ-Decalactone	Coconut, waxy	**30**	25						**59**	27,**64**	27	81
δ–Dodecalactone	Coconut, sweet				**45**			45	59			
Maltol	Cotton candy	**30**		**44,45**								
4-Hydroxy-2,5-dimethyl-3(2*H*)-furanone (Furaneol)	Caramelized, burnt sugar	**30**	**25**									
3-Hydroxy-4,5-dimethyl-2(5*H*)-furanone (sotolon)	Maple, spicy	30	**25**	45	**45**		**45**	**45**	**59**	**27**,64	**27**	81

[a]Non-fat dry milk (NDM), medium heat treatment.

[b]Skimmed milk powder (SMP), stored for 6 months at 21 °C.

[c]Milk protein concentrate (MPC), ref. 44 = 85% milk protein; ref. 45 = 70, 80, 85% milk protein.

[d]Milk protein isolate (MPI), > 90% milk protein.

(*continued*)

Table 5.1 *(continued)*

Aroma compound	Odour descriptor	NDM[a]	SMP[b]	MPC[c]	MPI[d]	Acid casein	Rennet casein	MCC	SWP	WPC[e]	WPI[f]	WPH

[e]Whey protein concentrate (WPC), 80% whey protein.

[f]Whey protein isolate (WPI), > 90% whey protein.

[g]Numbers indicate literature references to identification of specific aroma compounds in the dairy protein ingredient.

[h]Bold font indicates that this aroma compound is a significant contributor to the flavour characteristic.

5.4.2 Milk Protein Concentrates and Isolates: MPC and MPI

MPC and MPI are comparatively new categories of dried dairy ingredients.[44] MPC is a complete dairy protein ingredient consisting of 40–89% concentrations of caseins and whey proteins, while MPI is further concentrated to contain 90% or greater protein content. MPCs/MPIs are manufactured from fluid skimmed milk using membrane filtration to concentrate the caseins and whey proteins, with subsequent spray drying. One current application for MPCs is Greek yoghurt. Commercially available MPCs contain 45, 70, 80, and 85% protein.[1]

The flavours of concentrated milk proteins have diverse profiles. MPCs with lower protein contents (≤70%) have similar flavour characteristics to fluid milk: cooked/milky, cereal, sweet aromatic, and a sweet taste. As the MPC/MPI protein contents increase (>80%), the off-flavour profile intensifies to display more cardboard, potato/brothy, brothy/animal, tortilla/corn chip, soapy, and fatty overtones and a higher astringency.[14] Significant aroma-active compounds identified in MPCs include *o*-aminoacetophenone, nonanal, 1-octen-3-one, dimethyl trisulphide, 2-acetyl-1-pyrroline, heptanal, methional, 1-hexen-3-one, hexanal, dimethyl disulphide, butanoic acid, and acetic acid.[44,45] Tortilla flavour is a defining flavour of milk protein ingredients, and many of the odour-active volatile compounds found in corn tortillas, corn chips, and taco shells are also odour-active compounds in milk proteins.[45] While *o*-aminoacetophenone is a significant contributor to the tortilla flavour of dried milk proteins, research studies have suggested that a mixture of other odour-active compounds plays a supporting role. These include *E,E* and *E,Z*-2,4-decadienal, acetic acid, 3-methylbutanal, 2-acetyl-1-pyrroline, *E*-2-nonenal, 2-methylbutanal, *E*-2-decenal, methional, β-ionone, hexanal, 1-octen-3-ol, and 2-ethyl-3,5-dimethylpyrazine, which were also previously identified in corn tortillas, chips, and taco shells.[45]

Spray-drying parameters for MPC 70 and NDM were shown to directly impact the flavour quality and shelf life of these dried dairy ingredients. During the spray drying process, increasing the inlet temperature from 160 °C to 260 °C and the solids concentration from 30 to 40% enhanced the sweet aromatic attributes and reduced the undesirable cardboard flavour and associated volatile lipid oxidation products. Although the dairy particulates received greater heat treatment during drying, the overall flavour quality was observed to improve at elevated inlet temperatures.[46]

With ageing, MPC and MPI develop higher flavour intensities of cardboard, animal, burnt sugar, and tortilla/corn chip sensory characteristics. Storage temperature is directly coupled with the flavour quality of MPC. High temperature (40 °C) storage reduced the flavour shelf life of MPC 45 by 50% and MPC 80 by 75% compared to storage at 25 °C. Based upon sensory and protein solubility/functional changes, the shelf life of MPC at temperatures greater than 40 °C is less than three months. The MPC shelf life improves significantly to 6 months at 25 °C and can surpass 12 months at 3 °C. Especially during the shipping and storage of MPC and MPI, the ambient temperature should be controlled to avoid a drastic reduction in the protein flavour shelf life.[45]

5.4.3 Casein Proteins

Caseins possess unique solubilizing and heat stability/thermoplastic properties and are used in a multitude of ingredient applications. Rennet casein, a prominent dairy protein ingredient, has been widely used in analogue cheese, meal-replacement beverages, and protein bar products. Rennet casein is manufactured by the addition of chymosin to skimmed milk to coagulate the curd, followed by washing (to remove whey proteins, lactose, and minerals) and drying steps.[47] Acid caseins are traditionally produced by acid precipitation from skimmed milk at pH 4.6, followed by spray or roller drying. The soluble form, sodium caseinate, is produced when casein curd (acid precipitated) is treated with sodium alkali to reach a final pH of 6.7 and fully dissolved prior to spray drying.[47] Potassium and calcium caseinates are prepared similarly, with the corresponding alkali solubilization of acid casein.[47]

5.4.3.1 Rennet Casein

Rennet casein as a dried ingredient possesses an intense and unpleasant off-aroma/flavour, which is described as glue-like, animal/wet dog, mucilage, and gelatin.[48] Initial volatile flavour analyses by gas chromatography/mass spectrometry (GC/MS) found that no single compound contributed the typical gluey flavour, which likely resulted from a synergistic mixture of aldehydes, furans, dimethyl disulphide, and benzothiazole.[48] While present at sub-threshold levels, *o*-aminoacetophenone contributed characteristics of the gluey odour of rennet casein as part of the volatile mixture. *o*-Aminoacetophenone was also identified as an off-flavour compound in NDM and a prevalent component of stored milk powders.[32,39] Subsequent model system aroma studies using descriptive sensory analysis confirmed that a combination of volatile components, including hexanoic acid, indole, guiacol, and *p*-cresol, may be responsible for the prevalent animal/wet dog off-flavour of rennet casein.[48] A more recent flavour study of rennet casein concluded that it was characterized by high aroma intensity, cardboard, and animal flavours, with accompanying high levels of volatiles identified as phenylacetaldehyde (honey), sotolon (sweet), *E*-2-heptenal, 2,4-decadienal (fatty/corn chips), 1-hexen-3-one (burnt), hexanal (green), dimethyl disulphide (sulphur/cabbage), dimethyl trisulphide (cabbage/garlic), and 2-pentyl furan.[45] In this recent study, hexanoic acid was not identified as a significant contributor to the off-flavour. Compound concentrations and relative abundances were highly variable, presumably because of differences in ingredient manufacturing and processing.

5.4.3.2 *Acid Casein*

The flavour of acid casein, similar to rennet casein, has been described as animal/wet dog and gluey. Cardboard, tortilla, musty, and burnt feather off-flavour descriptors are also prevalent in caseinates,[14] reflecting flavour volatiles generated *via* oxidative, Maillard, and Strecker degradation pathways. Acid casein was characterized by high aroma intensity, sour aromatics, cardboard, animal flavour, and tortilla/corn chip flavours. Aroma volatiles identified in acid casein at high concentrations included *o*-aminoacetophenone (tortilla/corn chip), *E*-2-heptenal, and *E*-2-nonenal (cardboard).[45] To receive optimum consumer acceptance as food ingredients, rennet, acid casein, and caseinates should ideally contribute minimal off-flavours.

5.4.3.3 *Micellular Casein*

Micellar casein (MCC) can be manufactured by membrane fractionation and spray drying of fluid milk. The advantage of this alternative process is that casein has a more neutral flavour. The outcome is a dairy protein with reduced off-flavours, accompanied by some of the functional and nutritional attributes observed in rennet casein.[14] A flavour evaluation of freshly prepared MCC determined that the liquid protein concentrate had lower aroma intensity and no-cooked flavours compared with spray-dried MCC. The dried MCC had a slight cardboard flavour, an undesirable corn chip flavour, and a higher abundance of volatile aldehydes.[44] The improved flavour of MCC retentates from membrane fractionation offers the potential to provide lower protein-related off-flavours in protein-fortified food and beverage products. An added benefit is that it avoids the usual difficulties with the rehydration of high-protein powder concentrates.

5.4.4 Whey Proteins

Dried whey proteins are utilized as food ingredients for their functional and nutritional properties.[49] Their precursor is liquid whey, a by-product of cheesemaking, which is processed into dried whey powder, WPC (35–80% protein), and WPI (>90% protein). Although a large portion of dried whey protein produced in the United States is the byproduct of cheddar and mozzarella cheese manufacture, many other whey sources exist.[1] Whey is classified either as sweet whey (rennet coagulation) or acid whey (acid coagulation).

While many of the volatile compounds identified in heat-treated milk (Section 5.4.1) occur in whey protein ingredients, their flavour characteristics and intensities will be different due to different proportions of flavour components generated from the lower protein/higher lactose contents and residual traces of milkfat. Predictably, the type of cheese influences the flavour of the whey by-product.[50,51] Cheddar whey flavour is characterized by

2,3-butanedione/buttery notes and acid wheys (acid casein, cottage cheese, and Greek yoghurt) with a sour aromatic flavour.[52] Acid casein whey exhibits a soapy flavour, while acid and Greek yoghurt wheys have a characteristic potato flavour.[52] Dried whey has a different flavour profile than liquid whey because of the additional concentration and spray drying steps. This further processing can reduce the concentrations of certain flavour compounds and generate new ones, thereby altering the flavour profile. Therefore, the wide variability in flavour profiles among commercial dried whey products can be attributed to different milk supplies, whey sources, and processing parameters. Off-flavours (brothy, diacetyl, sourness, and bitterness) of dried whey are often limiting factors in its usage as an ingredient in bland food products.[27] The flavour attributes of whey protein ingredients have been recently reviewed.[53,54]

In contrast to sweet whey, acid whey contains higher concentrations of organic acids, which contribute to an increased sour taste accompanied by lipid oxidation, sour aromatics, and cardboard flavours.[50,52] To promote future utilization of sour whey, a currently active area of research is the application of nanofiltration and other membrane-based filtration processes to extract proteins and lactose from acid whey.[55]

The freshness of the liquid whey prior to spray drying has a major impact on the flavour quality of the dried whey protein. Cardboard and cabbage flavours were shown to increase in spray-dried WPC and WPI products when the storage time of the liquid whey was extended prior to drying.[56,57] Combined sensory and flavour analysis studies determined that lipid oxidation products (hexanal, heptanal, and octanal) and sulphur degradation products (dimethyl disulphide and dimethyl trisulphide) increased in spray-dried products with increased liquid whey storage time, whereas 2,3-butanedione decreased. The flavour shelf stability decreased in spray-dried products with longer liquid whey storage times. Results from this study determined that for optimum quality and shelf life, liquid whey should be held for less than 12 h before spray drying.[56]

5.4.4.1 *Sweet Whey Powder*

Liquid sweet whey is dehydrated or spray-dried into a dry powder ingredient. Liquid sweet whey possesses sweet, aromatic, and cooked milk flavours with cardboard and metallic off-flavours.[50,58] Fluid whey can be dehydrated by reverse osmosis or spray drying/evaporation to produce an SWP containing 11–14% protein and 63–75% lactose.[1]

The flavour profile of SWP is different from that of liquid whey due to the additional concentration and spray drying steps involved in its processing. These additional steps reduce or generate unique flavour compounds, thereby changing the flavour profile of SWP.[59] Solvent-assisted flavour evaporation in combination with GC/olfactometry and GC/MS was used for the identification of aroma compounds in SWP from cheddar whey.[59] Of the most aroma-intense free fatty acids identified, butanoic, hexanoic, and octanoic acids contributed cheesy and rancid odours, while butanoic acid showed very high aroma intensity. 3-Methylbutanoic acid provided a sweaty odour with a high OSME GC/olfactometry aroma intensity score, while pentanoic and dodecanoic acids added a pungent odour, and octanoic acid provided an animal-like odour. The most aroma-prominent neutral volatile compounds identified were hexanal, heptanal, nonanal, phenylacetaldehyde, 1-octen-3-one, methional, 2,6-dimethylpyrazine, 2,5-dimethylpyrazine, 2,3-dimethylpyrazine, 2,3,5-trimethylpyrazine, furfuryl alcohol, *p*-cresol, 2-acetylpyrrole, maltol, furaneol, and C_{10}–C_{12} lactones.[59] The SWP flavour of cheddar whey is characterized by cooked/milky and oxidized aromatics and a sweet taste.[60] High lactose concentrations in SWP promote sugar caramelization and Maillard reaction products, resulting in Strecker degradation compounds, furans, and pyrroles, with predominant caramelized and cooked flavours accompanied by oxidized notes.[60] The source of SWP flavour results from compounds generated by the starter culture during cheese making, the type of SWP manufacturing process, and a foundation of flavour components originating from the milk.

5.4.4.2 Whey Protein Concentrate

WPC is obtained by removing non-protein constituents from pasteurized sweet whey to achieve a 34–80% protein content in the dried product.[1] The whey protein fraction in SWP is concentrated by removing lactose using physical separation techniques (membrane filtration), thus creating 34% total solids in WPC (WPC 34).[61] Further concentration by ultrafiltration boosts the whey protein content to 80% (WPC 80).[54,61,62]

The levels of lactose and residual milkfat significantly influence the flavour profile of dried whey proteins. The sensory attributes of WPC are generally characterized as sweet aromatic, milky, cardboard, and fatty.[63] However, the milky and sweet flavours are dependent on the relative amounts of protein and lactose. For example, the relatively lower protein/higher lactose of WPC 34 has more cooked/milky, sweet aromatic, and sweet taste than the higher protein/lower lactose of WPC 80.[60] Generally, as the level of whey protein increases, the cooked milky and sweet aromatic flavours decrease. Lactose is removed as the protein and fat levels are concentrated by diafiltration to produce WPC 80. Because its fat content is 6.5% on a dry weight basis, WPC 80 develops off-flavours during storage due to lipid oxidation initiated during peroxide bleaching and drying steps.[27,64–66]

WPCs exhibit sweet aromatic, cardboard/wet paper, animal/wet dog, soapy, brothy, cucumber, and cooked milky flavours, accompanied by a bitter taste and an astringent sensation.[27] The key volatile flavour compounds in WPC 80 are butanoic acid (cheesy), 2-acetyl-1-pyrroline (popcorn), 2-methyl-3-furanthiol (brothy/burnt), 4-hydroxy-2,5-dimethyl-3-(2*H*)-furanone (maple/spicy), 2-nonenal (cardboard/old books), (*E,Z*)-2,6-nonadienal (cucumber), and (*E,Z*)-2,4-decadienal (fatty/oxidized).[27] A representative listing of sensory-significant volatile compounds in WPC 80 is shown in Table 5.2.

Liquid wheys from cheddar and mozzarella have dissimilar aroma compounds, and cheddar whey is more prone to the formation of lipid oxidation volatiles than mozzarella whey.[68] Therefore, it is not unexpected that the flavour profiles of WPCs prepared from the two cheese types are different. Cheddar WPC had a higher concentration of decanal, heptanal, and acetic acid than mozzarella WPC. These differences were attributed to the dissimilar starter cultures that were used.[69]

During extended storage, WPC develops a recognizable off-flavour that limits its application in food products with subtle flavours. From a comparison of sensory studies for various wheys, cardboard flavours increased while sweet aromatic and buttery flavours decreased with storage. Proportional quantities of measured volatile compounds in whey changed with storage, corresponding to the sensory results, and aldehydes produced from lipid oxidation increased in WPC with storage time.[70] Cardboard flavour is a common off-flavour caused by lipid oxidation in liquid whey, dried whey, and MPCs.[44] A direct flavour comparison between liquid and spray-dried whey proteins, as well as other liquid milk proteins, showed a distinct increase in cardboard off-flavour in the spray-dried forms.[44] Because spray drying is a thermal process, it is likely to change the nature of the fat on the surface of the particles and enhance the rate of lipid oxidation, leading to cardboard and other lipid-derived off-flavours.

Table 5.2 A comparison of sensory-significant volatile flavour compounds in WPC 80 and WPI.

Compound	Odour descriptor	WPC 80[a]	WPI[b]
Butyric acid	Cheesy, rancid	X	X
2-Acetyl-1-pyrroline	Popcorn	XX	XX
Dimethyl disulphide	Garlic, rubbery	Trace	XX
Dimethyl trisulphide[c]	Cabbage	X	XX
2-Methyl-3-furanthiol	Brothy, burnt	XX	X
Methional	Potato	X	XX
Hexanal	Green grass	XX	X
(*E*)-2-Octenal	Citrus, fatty	X	—
(*E*)-2-Nonenal	Cardboard	XX	Trace
(*E,Z*)-2,6-Nonadienal	Cucumber	XX	X
δ-Decalactone	Coconut	XX	—
o-Aminoacetophenone	Grape	Trace	Trace
3-Hydroxy-4,5-dimethyl-2(5*H*)-furanone	Maple, spicy	XX	X
4-Hydroxy-2,5-dimethyl-3(2*H*)-furanone	Burnt sugar	XX	X

X = compound is present.

XX = significant flavour impact.

[a]Aroma-active compounds identified in WPC 80 (80% protein) in ref. 27.

[b]Aroma-active compounds identified in WPI (90% protein) in ref. 27.

[c]First reported as the source of a cabbage off-flavour in WPI in ref. 67.

5.4.4.3 Whey Protein Isolate

Dried whey with a protein concentration above 90% of the total solids is denoted as WPI. As a functional ingredient and high-quality protein, WPI has useful properties for emulsification, water binding, viscosity building, gelling, whipping/foaming, and as a protein supplement.[1] In food applications, WPI is often used in baked goods, beverages, dairy, processed meat, snack, and confection products.

A typical WPI process involves whey fluid clarification, fat separation, pasteurization, bleaching, microfiltration or ion exchange, ultrafiltration, diafiltration, and spray drying.[5,71] In the manufacture of WPI, residual fat is removed from SWP by microfiltration or anion exchange chromatography.[11,72] Compared to WPC, WPI has a much lower concentration of fat (ranging from 0.15% to 0.60% on a dry-weight basis). The removal of residual fat is a necessary step to achieve a concentration of protein above 90% of the total solids.[27,73,74] The reduced fat content in WPI corresponds to a reduction in lipid oxidation and off-flavours,[27] and WPI powders contain fewer lipid oxidation-derived aroma compounds compared to WPC 80 powders. Although lipid content influences off-flavour formation, the increased binding interactions of oxidation-derived volatile compounds with WPI protein may also diminish the off-flavour of WPI compared to WPC 80.[75]

The source of whey (cheese type, protein coagulant, and starter culture) used to prepare WPI directly affects its sensory and flavour properties. WPIs from cheddar and cottage cheese wheys were determined to have the highest cardboard flavour intensities with the highest levels of aldehyde concentrations derived from lipid oxidation.[73] WPI from cottage cheese wheys (prepared as a by-product of acid precipitation of casein) had a characteristic brothy flavour, sour taste, and astringency. The sour and astringent flavour characteristics of cottage cheese WPI limit its use in many foods, except for high-acid applications, such as kefir, yoghurt, and clear acidic beverages. Because of differences in cheese-making processes and starter cultures, WPIs prepared from mozzarella, cheddar, and rennet sweet wheys exhibit distinct flavour characteristics and different levels of lipid oxidation-derived aroma volatiles.[73] These differences provide an opportunity to tailor which WPI manufacturing sources are optimal to use as dairy protein ingredients for specific product applications so that the WPI flavour characteristics are congruent with the desired sensory attributes of the product.

Aroma-impact compounds in WPI as compared to WPC 80 (80% protein) were evaluated using descriptive sensory techniques in conjunction with GC/MS and GC/olfactometry analysis of aroma-active volatile compounds.[27,73] A comparison of sensory-significant flavour compounds in WPI and WPC 80 is presented in Table 5.2. WPI exhibited sweet aromatic, cardboard/wet paper, animal/wet dog, soapy, brothy, cucumber, and cooked/milky flavours, along with a bitter taste and an astringent sensation.

For rennet and mozzarella WPI, the aroma-active volatiles produced from lipid oxidation (pentanal, hexanal, heptanal, octanal, nonanal, decanal, 2- and 3-methyl butanal, *E*-2-nonenal, and methional) were not appreciably different; however, mozzarella WPI was higher in diacetyl, acetic acid, 2-pentyl furan, *Z*-4-heptenal, and 1-octen-3-one than rennet whey WPI. These volatiles (lipid oxidation and Strecker degradation compounds) are attributed to the presence of the starter culture. Because rennet whey manufacture does not include starter culture addition, the overall quantity of flavour and oxidative volatiles in rennet WPI was lower than those in cheddar, mozzarella, and cottage cheese WPI.[75] Starter cultures have been shown to promote lipid oxidation in liquid whey,[65,68,69] explaining the higher relative quantities of volatile oxidation products in other WPIs compared to rennet WPI.

5.4.4.4 Whey Protein Hydrolysates

Dairy protein hydrolysates are an emerging new category of nutritional and functional ingredients derived from milk proteins.[14,53] Whey protein is hydrolyzed into smaller peptides that are more rapidly digested and absorbed, have improved heat stability and solubility, are less allergenic, and impart unique functionality.[76] Dairy ingredient processors have become increasingly specialized in the production of whey protein hydrolysates (WPHs).[77] The process begins with sweet whey, which is concentrated by cross-flow filtration and then enzymatically hydrolyzed to produce a mixture of peptides and free amino acids for enhanced nutritional and functional benefits.

Commercially available 80% protein WPHs are utilized for nutritional/medical supplements, sports nutrition applications, hypoallergenic infant formulae, and elder nutrition.[77] In the latter application, whey is an ideal quality protein to help restore muscle loss in older adults following inactivity during illness or extended hospitalization.[78] Hydrolysis of proteins expands their functional properties, including heat stability and solubility, and improves their digestibility, a specific benefit for the use of hydrolysates in infant formulae.[79] To improve functionality, WPHs are commonly formulated into meal replacement bars to inhibit bar hardening over shelf life.[80] Enzymatic hydrolysis of milk, casein, or whey proteins can generate bioactive peptides with enhanced solubility and heat stability benefits.[53,81]

The presence of strong off-flavours limits the use of protein hydrolysates in food product formulations, particularly beverages. The flavour of WPHs differs from WPC and WPI depending on the type of enzyme used, the degree of hydrolysis, and the heating conditions.[82,83] Product developers have significant challenges masking the inherent brothiness, free fatty acid aromatics, and distinctive bitter tastes in protein hydrolysates. The degree and intensity of bitterness in WPH increase with smaller oligopeptides—three to six amino acids in length—and L-amino acids with hydrophobic side chains.[76,82] Therefore, the concentration of specific types of peptides is more significant to WPH bitterness than the degree of protein hydrolysis.[76,84] Bitter peptides in WPH, identified by sensory-guided fractionation of peptides with two-dimensional reverse-phase liquid chromatography and MS, were shown to arise by hydrolysis of α-lactalbumin, β-lactoglobulin, serum albumin, and β-casein.[76] Bitterness inhibitors have been effective in helping to reduce the bitter taste of WPH in product applications. A recent study identified bitter taste inhibitors (sweeteners and nucleotides) of different peptide compositions for product applications with WPH.[85]

Besides its bitter taste, WPH has off-flavour challenges. While WPC typically has a milky, sweet, and mild milkfat odour, extensively hydrolyzed WPH possesses cheese-like, acrid, and sour olfactory characteristics. The corresponding volatile compounds with high flavour dilution factors in WPH were identified as methional, phenylacetaldehyde, and 3-methylbutanoic acid.[86] The concentration of these Strecker aldehyde aroma compounds increases with the corresponding hydrolysis of whey proteins to produce free methionine, phenylalanine, and leucine amino acids. In another sensory study of WPH, high levels of cooked/sulphur, potato/brothy, tortilla, and animal flavours were described as resulting from extended heating and proteolysis.[82] Metallic, cabbage, cereal, and burnt sensory characteristics are included in a lexicon developed for WPH and casein hydrolysates.[82,87] These findings confirmed that the volatile components in WPH are principally derived from protein degradation. These compounds include dimethyl sulphide (sulphur), dimethyl trisulphide (cabbage), and methional (potato), which contribute characteristic flavours to WPH and other whey protein ingredients.[67,82]

Products formulated with hydrolysates exhibit intense aromas and flavours, requiring that they be sensory evaluated at lower levels compared to other dairy proteins.[53] Dairy ingredient producers are continuously making improvements to enzymes and process conditions to reduce the bitter flavour profile for sports and nutritional bar applications.

Oligopeptides generated during the preparation of enzymatic hydrolysates of WPI 84 were recognized as important flavour precursors of the Maillard reaction, which generates pyrazine flavour compounds that impart roasted and baked aromas to heated foods. A recent study[81] confirmed the significant role of oligopeptides in pyrazine formation compared to free amino acids, even in model systems with high levels of amino acids. Hydrolysates obtained with different peptidase enzymes (pepsin, chymotrypsin, and thermolysin) and heated with glucose yielded a larger variety of pyrazines and Strecker aldehydes. A range of 15 C_1–C_7-substituted pyrazines were identified in the various hydrolyzed WPI–glucose model systems after dry heating (180 °C, 90 min); however, the most abundant pyrazine formed was 2,5(6)-dimethylpyrazine (cocoa, nutty, and roasted).[88]

5.5 Off-Flavours from Dairy Protein Ingredients

Off-flavours are undesirable odours and tastes imparted to food, typically by oxidation, microbial deterioration, overprocessing, and extensive or improper storage conditions.[89] In particular, food products and beverages fortified with dairy proteins have substantial hurdles to overcome in order to deliver balanced and appealing flavours. Dairy protein ingredients may have inherent undesirable off-aromas or off-tastes that develop during their processing or storage. Moreover, proteins can participate in volatile interactions or chemical binding with specific flavour components, such as aldehydes, that modify flavour release, thereby altering the overall flavour profile of food products.[16,18] The net result is often an aroma imbalance and fading. The occurrence and formation of off-flavours in dairy protein ingredients were recently reviewed.[15] This section will highlight the principal types of off-flavours contributed by dairy ingredients.

5.5.1 Off-aromas (Taints)

The characteristic aroma compounds that contribute a positive flavour impact at low levels can become off-flavours when present in foods at higher concentrations. Among these are volatile sulphur compounds—important contributors to cheese flavour—which manifest their impact as off-flavours in dairy protein concentrates when they occur at elevated concentrations due to their low sensory detection thresholds.[90] For example, while dimethyl sulphide provides a subtle "corn-like" flavour characteristic at low levels, it contributes a highly undesirable "cooked vegetable" or "cabbage-like" off-odour when present at concentrations significantly above its sensory threshold (30–45 ppb).[91] The heating of dairy proteins during drying or concentration can accelerate the release of sulphur compounds from cysteine and methionine constituents, thereby generating volatile sulfhydryl compounds (methanethiol, dimethyl sulphide, and dimethyl trisulphide), resulting in intense eggy and sulphide off-flavours.[92] Methanethiol was recently identified as the initial volatile sulphur compound produced in skimmed milk exposed to high-intensity fluorescent light and caused by the photooxidation of methionine in the presence of riboflavin.[93] An earlier study attributed the source of a cardboard off-flavour in ultra-high-temperature (UHT) milk and milk proteins exposed to sunlight or fluorescent light to a related volatile compound, dimethyl disulphide.[94] Because methanethiol is known to be readily oxidized to dimethyl disulphide, this compound may be formed as a secondary reaction product in milk.[90] The characteristic sulphuric/rubber off-flavour in spray-dried SMP is contributed by benzothiazole at low ppb concentration.[95]

- **Cooked/milky aromatics associated with cooked/heated milk.** Caramelized flavours develop from Maillard reactions between lactose degradation products and amino acids in fluid milk subjected to heat treatments greater than traditional high-temperature–short-time pasteurization temperatures.[53] This pathway results in the formation of pyrazines, furans, and other Maillard products that impart cooked, malty, and brown off-flavours.[96]
- **Oxidized flavour associated with rancid or oxidized oils.** Heat treatments promote the autoxidation of unsaturated fatty acids in milk. Concentrations of many volatile aldehydes, methyl ketones, and alcohols increase with extended pasteurization time and temperature. Volatile lipid oxidation products identified in NDM were nonanal, (*E*)-2-nonenal, (*E*,*E*)-2,4-nonadienal, and (*E*,*Z*)-2,6-nonadienal, which provide fatty, hay, stale, and cucumber off-odour notes, respectively.[30,39] Because they were present below their sensory thresholds, these volatile oxidation products were not detected by sensory analysis of dry milk powders. However, several of these compounds occur as indicators of oxidized off-flavour in WPC and WPI.[27,54]
- **Grassy/hay/cardboard.** (*E*)-2-Nonenal (hay) and β-ionone aromatics are correlated with the off-odour of dried grasses.[30] At higher concentrations, (*E*)-2-nonenal is associated with the odour of cardboard.[97]
- **Cowy/barny.** *p*-Cresol, skatole, and indole aromatics are linked to the odour of cow faeces or urine.

Examples of high-impact off-flavour compounds that have been identified in dry dairy protein ingredients are shown in Figure 5.1.

5.5.2 Off-tastes

5.5.2.1 Astringency

An astringent mouth sensation is a common occurrence with whey protein beverages, which may limit their consumer acceptability.[98] Astringency and chalky off-tastes have been observed in studies of pasteurized skimmed milk treated with proteases; however, they have also been noted in UHT-pasteurized milk, NDM, and cheddar cheese.[99] Astringency is a mouth-drying and puckering sensation caused by the precipitation or aggregation of proteins through interaction with salivary proteins. Whey proteins exhibit high levels of astringency at acidic conditions typical of high-acid (pH 3.4) whey protein beverages;[98] however, astringency is also perceptible in pH-neutral whey beverages[100] and dairy products containing

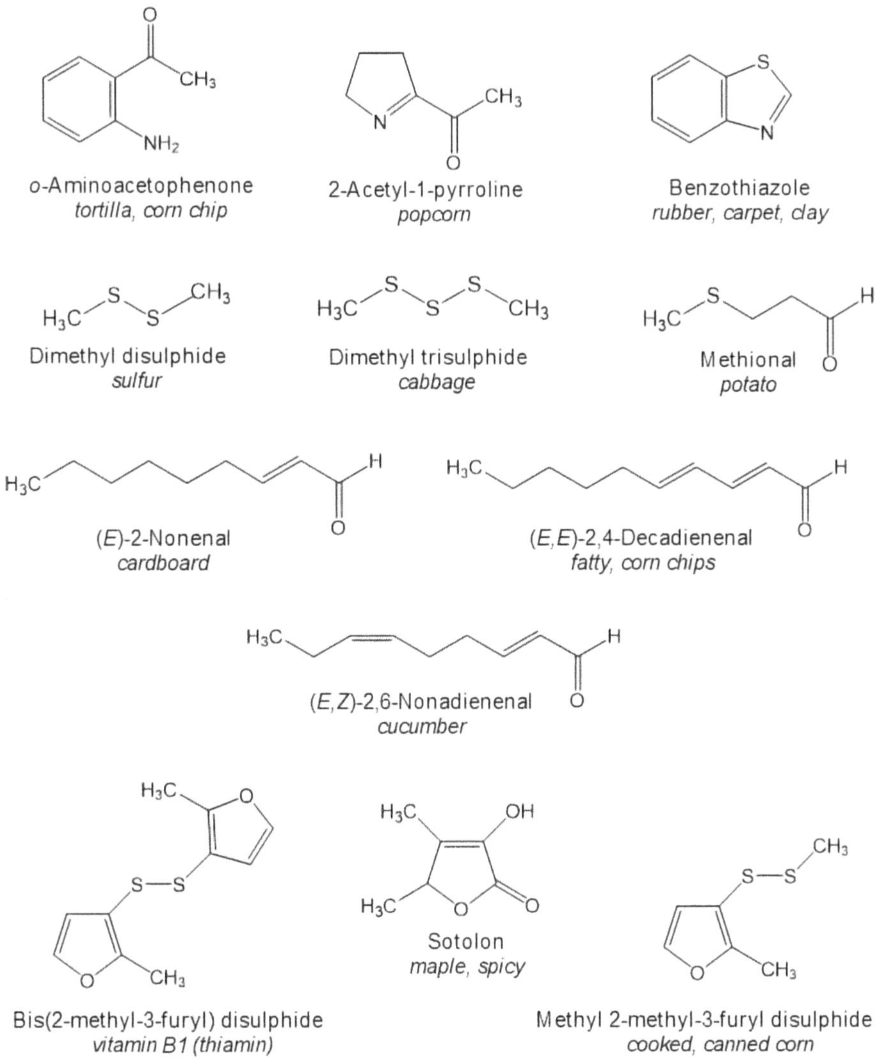

Figure 5.1 Representative high-impact off-flavour compounds identified in dairy protein ingredients.

γ-caseins.[99] The precise molecular mechanism for whey protein astringency remains unclear, although factors such as low pH, protein level, electrostatic interactions, and denaturation are known to change the intensity of astringency.[98] If the complexation between whey proteins and salivary proteins can be reduced or prevented, the astringency should be diminished. Conditions that cause astringency seem to be inherent in the whey protein and the acidity of the solution.[98] Additional research studies are required to understand how to reduce or eliminate whey protein ingredient astringency so that food manufacturers can improve consumer liking of whey-containing products in order to deliver their nutritional benefits.

5.5.2.2 Bitterness

The hydrolysis of dairy proteins most often produces bitter-tasting sensory attributes that restrict the use of these ingredients because of their unfavourable consumer acceptance. The investigation of the bitterness intensity of whey protein hydrolysates has been directly related to the degree of hydrolysis.[76] The breakdown of protein is well known to generate bitterness in dairy protein-based products such as cheddar, Gouda, Swiss, and camembert cheeses, casein hydrolysates, and other food protein ingredients such as soy protein hydrolysates. Structure-activity relationships of peptides with bitterness perception have been extensively studied, and the degree of bitterness is particularly influenced by hydrophobic amino acids.

5.6 Conclusion

The flavour profiles of concentrated dairy proteins are very complex. Ideally, these proteins should be bland and flavourless to provide maximum applicability as food ingredients. However, off-flavours are readily produced from dairy proteins by proteolytic breakdown, lipolysis, lipid oxidation, lactose fermentation, and thermal degradation (Maillard reactions). In many instances, combinations of multiple flavour-generation pathways can give rise to tortilla/corn chip, cardboard, cooked milk, and/or cabbage-like sensory perceptions that are correlated with specific volatile flavour compounds. The freshness, storage conditions, and treatment of milk and liquid whey proteins during concentration and drying steps can directly affect the intensity and character of dairy-based off-flavours that carry over into finished products.

The goals of this chapter were to impart current research knowledge about dairy protein flavour chemistry, to build awareness of the factors and types of flavour formation pathways that milk protein ingredients undergo, and to provide insights into unique off-flavour compounds that are generated during processing or developed during storage. A better understanding of how these dairy off-flavours arise can improve the quality, stability, and expanded usage of dairy proteins in food and beverage applications.

Abbreviations

GC/MS	Gas chromatography/mass spectrometry
HDMF	4-Hydroxy-2,5-dimethyl-3(2*H*)-furanone

MPC	Milk protein concentrate
MPI	Milk protein isolate
NDM	Non-fat dry milk
SMP	Skimmed milk powder
SWP	Sweet whey powder
WPH	Whey protein hydrolysate
MCC	Micellar casein
WPI	Whey protein isolate

References

1. U.S. Dairy Export Council, Commercial milk powders, milk proteins, and whey proteins manufactured in U.S.A, https://www.thinkusadairy.org/products/milk-proteins. Accessed 17 January 2023.
2. USDA 2022, U.S. Total milk powder, skim milk powder, and dry whey production. United States Department of Agriculture, Online, https://www.nass.usda.gov/Publications/Todays_Reports/reports/dary0822.pdf. Accessed 17 January 2023.
3. M. Eckert and P. Riker, *Food Technol.*, 2007, **61**(3), 20–26.
4. G. S. Meena, A. K. Singh, N. R. Panjagari and S. Arora, *J. Food Sci. Technol.*, 2017, **54**(10), 3010–3024.
5. M. A. Drake, R. E. Miracle and J. M. Wright, in *Milk Proteins: From Expression to Food*, ed. A. Thompson, M. Boland and H. Singh, Academic Press, San Diego, CA, 2009, pp. 429–448.
6. T. H. Parliment and R. J. McGorrin, in *Flavour Chemistry: Industrial and Academic Research*, ed. S. J. Risch and C. T. Ho, ACS Symp. Ser. 756, American Chemical Society, Washington, DC, 2000, pp. 44–71.
7. R. J. McGorrin, in *Food Flavours and Chemistry: Advances of the New Millennium*, ed. A. M. Spanier, F. Shahidi, T. H. Parliment, C. J. Mussinan, C. T. Ho and E. T. Contis, Royal Society of Chemistry, Cambridge, 2001, pp. 67–84.
8. T. K. Singh, M. A. Drake and K. R. Cadwallader, *Compr. Rev. Food Sci. Food Saf.*, 2003, **2**, 139–162.
9. M. A. Drake, K. R. Cadwallader and M. E. Carunchia-Whetstine, in *Flavour of Dairy Products*, ed. K. R. Cadwallader, M. A. Drake and R. J. McGorrin, ACS Symp. Ser. 971, American Chemical Society, Washington DC, 2007, pp. 51–77.
10. K. R. Cadwallader and T. K. Singh, in *Advanced Dairy Chemistry*, ed. P. L. H. McSweeney and P. F. Fox, Springer, Boston, MA, 3rd edn, 2009, ch. 14, vol. **3**, pp. 631–690.
11. T. J. Smith, R. E. Campbell and M. A. Drake, in *Advanced Dairy Chemistry, Proteins: Applied Aspects*, ed. P. L. McSweeney and J. A. O'Mahony, Springer, London, 4th edn, 2016, vol. **1B**, pp. 197–223.

12. M. A. Drake, in *The Wonders of Whey . . .Catch the Power*, Proceedings or the 4th International Whey Conference, American Dairy Products Institute, Elmhurst, IL, 2006, pp. 292–300.
13. K. R. Cadwallader, in *Handbook of Functional Beverages and Human Health*, ed. F. Shahidi and C. Alasalvar, CRC Press, Boca Raton, FL, 2016, vol. **11**, pp. 27–34.
14. M. A. Drake, R. E. Miracle and J. M. Wright, in *Milk Proteins: From Expression to Food*, ed. M. Boland, H. Singh and A. Thompson, Elsevier Science and Technology, New York, NY, 2nd edn, 2014, ch. 16, pp. 473–492.
15. X. Su, M. Tortorice, S. Ryo, X. Li, K. Waterman, A. Hagen and Y. Yin, *Molecules*, 2020, **25**, 569.
16. R. J. McGorrin, in *Protein + flavour = a formulation challenge*, Proceedings of the 2018 Protein Trends and Technologies Seminar, Global Food Forums, Itasca, IL, 2018, May, 22–23.
17. J. Kühn, T. Considine and H. Singh, *J. Food Sci.*, 2006, **71**, R72–R82.
18. T. E. O'Neill, in *Flavour-Food Interactions*, ed. R. J. McGorrin and J. V. Leland, ACS Symp. Ser. 633, American Chemical Society, Washington, DC, 1996, pp. 59–74.
19. A. P. Hansen and D. C. Booker, in *Flavour-Food Interactions*, ed. R. J. McGorrin and J. V. Leland, ACS Symp. Ser. 633, American Chemical Society, Washington, DC, 1996, ch. 7, pp. 75–89.
20. A. Amman and K. Schroen, *J. Agric. Food Chem.*, 2018, **66**, 9869–9881.
21. I. Suppavorasatit and K. R. Cadwallader, in *Chemistry, Texture and Flavour of Soy*, ed. K. R. Cadwallader and S. K. Chang, ACS Symp. Ser. 1059, American Chemical Society, Washington, DC, 2010, pp. 339–359.
22. H. Faulkner, T. F. O'Callaghan, S. McAuliffe, D. Hennessy, C. Stanton, M. G. O'Sullivan, J. P. Kerry and K. N. Kilcawley, *J. Dairy Sci.*, 2018, **101**, 1034–1047.
23. J. Adda, in *Developments in Food Flavours*, ed. G. G. Birch and M. G. Lindley, Elsevier Applied Science, London, 1986, pp. 151–172.
24. L. Lemieux and R. E. Simard, *Lait*, 1992, **72**, 335–382.
25. M. E. Carunchia Whetstine and M. A. Drake, in *Flavour of Dairy Products*, ed. K. R. Cadwallader, M. A. Drake and R. J. McGorrin, ACS Symp. Ser. 971, American Chemical Society, Washington, DC, 2007, ch. 13, pp. 217–251.
26. R. J. McGorrin, in *Flavour of Dairy Products*, ed. K. R. Cadwallader, M. A. Drake and R. J. McGorrin, ACS Symp. Ser. 971, American Chemical Society, Washington, DC, 2007, ch. 2, pp. 23–49.
27. M. E. Carunchia Whetstine, A. E. Croissant and M. A. Drake, *J. Dairy Sci.*, 2005, **88**, 3826–3839.

28. U.S. Code of Federal Regulations, CFR 21, Sect. 131.125, Nonfat dry milk, https://www.accessdata.fda.gov/scripts/cdrh/cfdocs/cfcfr/CFRSearch.cfm?fr=131.125. Accessed 17 January 2023.
29. H. Shiratsuchi, M. Shimoda, K. Imayoshi, K. Noda and Y. Osajima, *J. Agric. Food Chem.*, 1994, **42**, 994–998.
30. Y. Karagul-Yuceer, M. A. Drake and K. R. Cadwallader, *J. Agric. Food Chem.*, 2001, **49**, 2948–2953.
31. H. J. Clarke, M. G. O'Sullivan, J. P. Kerry and K. N. Kilcawley, *Antioxidants*, 2020, **9**, 338.
32. O. W. Parks, D. P. Schwartz and M. Keeney, *Nature*, 1964, **202**, 185–187.
33. R. G. Arnold, L. M. Libbey and E. A. Day, *J. Food Sci.*, 1966, **31**, 566–573.
34. W. Y. Cobb, S. Patton and H. Grill, *J. Dairy Sci.*, 1963, **46**, 566–567.
35. R. Bassette, D. Y. C. Fung, V. R. Mantha and E. H. Marth, *CRC Crit. Rev. Food Sci. Nutr.*, 1986, **24**, 1–52.
36. A. P. Lee, D. M. Barbano and M. A. Drake, *J. Dairy Sci.*, 2016, **99**, 9659–9667.
37. Y. Karagul-Yuceer, M. A. Drake and K. R. Cadwallader, in *Freshness and Shelf Life of Foods*, ed. K. R. Cadwallader and H. Weenen, ACS Symp. Ser. 836, American Chemical Society, Washington DC, 2003, pp. 108–123.
38. N. Kobayashi, Y. Mizota, K. Kumazawa and O. Nishimura, *J. Agric. Food Chem.*, 2008, **56**, 2469–2476.
39. Y. Karagül-Yüceer, K. R. Cadwallader and M. A. Drake, *J. Agric. Food Chem.*, 2002, **50**, 305–312.
40. M. A. Drake, Y. Karagul-Yuceer, K. R. Cadwallader, G. V. Civille and P. S. Tong, *J Sens. Stud.*, 2003, **18**, 199–216.
41. A. K. Abdalla, K. Smith and J. Lucey, *Int. J. Dairy Sci.*, 2017, **12**, 190–196.
42. Z. Cheng, M. G. O'Sullivan, J. P. Kerry, M. A. Drake, S. Miao, D. Kaibo and K. N. Kilcawley, *Food Res. Int.*, 2020, **138 (A)**, 109749.
43. A. Biolatto, G. Grigioni, M. Irurueta, A. M. Sancho, M. Taverna and N. Pensel, *Food Chem.*, 2007, **103**, 960–967.
44. B. G. Carter, H. Patel, D. M. Barbano and M. A. Drake, *J. Dairy Sci.*, 2018, **101**, 3900–3909.
45. T. J. Smith, R. E. Campbell, Y. Jo and M. A. Drake, *J. Dairy Sci.*, 2016, **99**, 4325–4346.
46. C. W. Park, M. A. Stout and M. A. Drake, *J. Dairy Sci.*, 2016, **99**, 9598–9610.
47. G. Bylund, Casein, in *Dairy Processing Handbook*, Tetra Pak Processing Systems, Lund, Sweden, 2015, ch. 20, pp. 449–458.
48. Y. Karagül-Yüceer, K. N. Vlahovich, M. Drake and K. R. Cadwallader, *J. Agric. Food Chem.*, 2003, **51**, 6797–6801.

49. C. V. Morr and E. A. Foegeding, *Food Technol.*, 1990, **44**(4), 100–112.
50. F. J. Gallardo-Escamilla, A. L. Kelly and C. M. Delahunty, *J. Dairy Sci.*, 2005, **88**, 2689–2699.
51. I. W. Liaw, R. E. Miracle, S. M. Jervis, M. Listiyani and M. A. Drake, *J. Food Sci.*, 2011, **76**, C1188–C1194.
52. S. Smith, T. J. Smith and M. A. Drake, *J. Dairy Sci.*, 2016, **99**, 3434–3444.
53. M. Stout and M. A. Drake, in *Whey Proteins: From Milk to Medicine*, ed. H. C. Deeth and N. Bansal, Elsevier Science & Technology, New York, NY, 2018, ch. 10, pp. 377–406.
54. B. G. Carter and M. A. Drake, *J. Dairy Sci.*, 2018, **101**, 6691–6702.
55. B. E. Erickson, *Chem. Eng. News*, 2017, **95**(6), 27–30.
56. M. Whitson, R. E. Miracle, E. Bastian and M. A. Drake, *J. Dairy Sci.*, 2011, **94**, 3747–3760.
57. C. W. Park, M. Parker and M. A. Drake, *J. Dairy Sci.*, 2016, **99**, 4303–4308.
58. M. E. Carunchia Whetstine, J. D. Parker, M. A. Drake and D. K. Larick, *J. Dairy Sci.*, 2003, **86**, 439–448.
59. S. S. Mahajan, L. Goddik and M. C. Qian, *J. Dairy Sci.*, 2004, **87**, 4057–4063.
60. R. Sithole, M. R. McDaniel and L. M. Goddik, *J. Food Sci.*, 2006, **71**, C157–C163.
61. C. I. Onwulata, in *Whey Processing, Functionality, and Health Benefits*, ed. C. I. Onwulata and P. J. Huth, Wiley-Blackwell, Ames, IA, 2008, pp. 369–391.
62. M. Rosenberg, *Trends Food Sci. Technol.*, 1995, **6**, 12–15.
63. J. Evans, J. Zulewska, M. Newbold, M. A. Drake and D. M. Barbano, *J. Dairy Sci.*, 2009, **92**, 4773–4791.
64. J. Evans, J. Zulewska, M. Newbold, M. A. Drake and D. M. Barbano, *J. Dairy Sci.*, 2010, **93**, 1824–1843.
65. R. E. Campbell, R. E. Miracle and M. A. Drake, *J. Dairy Sci.*, 2011, **94**, 1185–1193.
66. M. A. D. Listiyani, R. E. Campbell, R. E. Miracle, L. O. Dean and M. A. Drake, *J. Dairy Sci.*, 2011, **94**, 4347–4359.
67. J. M. Wright, M. E. Carunchia Whetstine, R. E. Miracle and M. A. Drake, *J. Food Sci.*, 2006, **71**, C86–C90.
68. I. W. Liaw, R. E. Miracle, S. M. Jervis, M. Listiyani and M. A. Drake, *J. Food Sci.*, 2011, **76**, C1188–C1194.
69. R. E. Campbell, R. E. Miracle, P. D. Gerard and M. A. Drake, *J. Food Sci.*, 2011, **76**, S354–S361.
70. M. E. Whitson, R. E. Miracle and M. A. Drake, *J. Sens. Stud.*, 2010, **25**, 616–636.

71. A. H. Varnam and J. P. Sutherland, in *Milk and milk products*, Chapman and Hall, New York, 1994, p. 168.
72. L. M. Huffman and J. M. Harper, *J. Dairy Sci.*, 1999, **82**, 2238–2244.
73. T. J. Smith, E. A. Foegeding and M. A. Drake, *J. Food Sci.*, 2016, **81**, C849–C857.
74. M. E. Whitson, R. E. Miracle, E. Bastian and M. A. Drake, *J. Dairy Sci.*, 2011, **94**, 3747–3760.
75. R. J. Stevenson and X. D. Chen, *Food Res. Int.*, 1996, **29**, 495–504.
76. X. Liu, D. Jiang and D. G. Peterson, *J. Agric. Food Chem.*, 2014, **62**, 5719–5725.
77. Whey protein hydrolysates, Hilmar Ingredients, https://www.hilmar-ingredients.com/wp-content/uploads/2022/08/Hilmar8390_4132018.pdf. Accessed 17 January 2023.
78. S. Y. Oikawa, C. McGlory, L. K. D'Souza, A. K. Morgan, N. I. Saddler, S. K. Baker, G. Parise and S. M. Phillips, *Am. J. Clin. Nutr.*, 2018, **108**(5), 1060–1068.
79. I. A. Nnanna and C. Wu, in *Handbook of Food Products Manufacturing*, ed. Y. H. Hui, John Wiley and Sons, Hoboken, NJ, 2007, pp. 537–556.
80. J. L. Childs, M. D. Yates and M. A. Drake, *J. Food Sci.*, 2007, **72**, S425–S434.
81. A. B. Nongonierma and R. J. FitzGerald, *J. Funct. Foods*, 2015, **17**, 640–656.
82. P. P. Leksrisompong, R. E. Miracle and M. A. Drake, *J. Agric. Food Chem.*, 2010, **58**, 6318–6327.
83. S. Ziajka, W. Dzwolak and J. Zubel, *Milchwissenschaft*, 1994, **49**, 382–384.
84. J. Newman, T. Egan, N. Harbourne, D. O'Riordan, J. C. Jacquier and M. O'Sullivan, *Talanta*, 2014, **126**, 46–53.
85. P. P. Leksrisompong, P. D. Gerard, K. Lopetcharat and M. A. Drake, *J. Food Sci.*, 2012, **77**, S282–S287.
86. H. Nakada, M. Ohata, M. Hosaka, H. Ochi, F. Abe and K. Arihara, *Anim. Sci. J.*, 2018, **89**, 1348–1354.
87. J. Newman, D. O'Riordan, J. C. Jacquier and M. O'Sullivan, *J. Sens. Stud.*, 2014, **29**, 413–424.
88. G. L. L. Scalone, P. Lamichhane, T. Cucu, N. De Kimpe and B. De Meulenaer, *Food Chem.*, 2019, **278**, 533–544.
89. R. J. McGorrin, in *Flavour, Fragrance, and Odor Analysis*, ed. R. Marsili, CRC Press Taylor & Francis Group, Boca Raton, FL, 2012, pp. 207–262.
90. R. J. McGorrin, in *Volatile Sulphur Compounds in Food*, ed. M. C. Qian, X. Fan and K. Mahattanatawee, ACS Symp. Ser. 1068, American Chemical Society, Washington, DC, 2011, pp. 3–31.

91. C. J. Scarlata and S. E. Ebeler, *J. Agric. Food Chem.*, 1999, **47**, 2505–2508.
92. Z. Al-Attabi, B. R. D'Arcy and H. C. Deeth, *Crit. Rev. Food Sci. Nutr.*, 2009, **49**, 28–47.
93. M. Asaduzzaman, M. Scampicchio, F. Biasioli, P. J. Bremer and P. Silcock, *Flavour Fragr. J.*, 2020, **35**(1), 34–41.
94. M. Y. Jung, S. H. Yoon, H. O. Lee and D. B. Min, *J. Food Sci.*, 1998, **63**, 408–412.
95. N. Kobayashi, Y. Mizota, K. Kumazawa and O. Nishimura, *J. Agric. Food Chem.*, 2008, **56**, 2469–2476.
96. T. K. Singh, K. R. Cadwallader and M. A. Drake, in *Handbook of Food Products Manufacturing*, ed. Y. H. Hui, John Wiley & Sons, Hoboken, NJ, 2006, pp. 715–748.
97. M. Czerny and A. Buettner, *J. Agric. Food Chem.*, 2009, **57**, 9979–9984.
98. B. G. Carter, E. A. Foegeding and M. A. Drake, *J. Dairy Sci.*, 2020, **103**, 5793–5804.
99. V. R. Harwalkar, H. Cholette, R. C. McKellar and D. B. Emmons, *J. Dairy Sci.*, 1993, **76**, 2521–2527.
100. C. A. Withers, M. J. Lewis, M. A. Gosney and L. Methven, *J. Dairy Sci.*, 2014, **97**, 1233–1247.

CHAPTER 6

Flavour of Fish and Fish Proteins

SERKAN SELLI*[a], ONUR SEVINDIK[b], GAMZE GUCLU[a] AND JING ZHAO[c]

[a]Department of Food Engineering, Faculty of Agriculture, Cukurova University, Adana 01130, Turkey; [b]Department of Food Engineering, Faculty of Engineering, Adana Alparslan Turkes Science and Technology University, Adana 01250, Turkey; [c]School of Exercise and Nutritional Sciences, San Diego State University, San Diego, CA 92120, United States
*E-mail: sselli@cu.edu.tr

6.1 Introduction

6.1.1 Fish Production and Consumption

Fish has always been one of the most spearheaded and popular foods, from the past to the present. The global production of fish has reached above 170 million on a yearly basis, and the consumption has now peaked at about 20.5 kg per person in a year.[1] The high amount of fish consumption in the world is caused by both its rich nutritional content (free amino acids or FAAs, proteins, lipids, nucleotides, organic acids, carbohydrates, fat-soluble vitamins, *etc.*) and its unique taste and flavour. This massive amount of fish consumption also results in a high amount of fish waste. The amount, type, and proportions of the components that compose these wastes can vary according to the processing technology and the type and size of the fish.

Food Chemistry, Function and Analysis No. 41
Flavour and Consumer Perception of Food Proteins
Edited by Jing Zhao and Changqi Liu

Published by the Royal Society of Chemistry, www.rsc.org

6.1.2 Fish Production By-products

Despite its significant potential, fish waste remains unevaluated, and an important amount of it is thrown away into the environment as a pollutant. In general, fish waste contains fish and shellfish, skeletons, fins, heads, skin, and internal organs that are mostly transformed into products such as fish oil, fish meal, fertilizer, pet food, and fish silage. These waste materials contain high amounts of proteins and rich bioactive constituents[2] that can be converted into value-added products, such as fish protein hydrolysates (FPHs) that have high nutritional value and substantial functional properties[3] (Figure 6.1). Therefore, the valourization of this remarkable waste is important to both the economic values of the products and the sustainable production of foods.

6.1.3 Fish Protein and Hydrolysates

There are several remarkable fish-based products other than fishmeal, such as fish protein concentrates (FPCs),[4,5] isolates,[6] and hydrolysates.[7] Among these products, FPH possesses several health-beneficial properties, such as antiplatelet, antihypertensive, anticancer, and immunomodulatory and antioxidant activities.[8–12] FPCs have less significance regarding bioactivity levels as the proteins need *in vivo* (proteolytic release) or *in vitro* (use of enzymes) treatment to release their bioactive capabilities.[12]

Functional properties of FPH highly depend on protein sources, types of proteases, the degree of hydrolysis, reaction time, temperature, and pH, while sensory properties depend on the type of proteolytic enzyme and the effects of hydrolysis conditions.[13] Although a large number of studies have examined the optimization of enzymatic treatments to obtain higher protein recovery from fish wastes, a series of recent studies have focused on the bioactivities of protein peptides.[14] Peptides in FPHs are considered an important natural bioactive component for functional food, nutraceuticals, and pharmaceutical sectors. Antioxidant, anti-hypertension, calcium-binding, anti-diabetic, anticoagulant, anti-proliferation, and anti-obesity activities are the most prominent functional properties of peptides exhibited in FPHs.[2]

Several previous studies have emphasized the remarkable antioxidant potential of FPHs and possible techniques to increase their release in FPHs.[2,3,15] In this context, enzymatic hydrolysis is a well-known and commonly used procedure to obtain higher antioxidant activities, although the effectiveness of the hydrolysis reaction is highly dependent on several parameters, such as reaction rate, concentration of peptides and FAAs, amino acid sequences, and hydrophobicity.[16,17] Chi *et al.*[18] examined the functional properties of *Scomberomorus niphonius*, namely the Spanish mackerel, by determining the effect of the average molecular weight of collagen hydrolysates.[18] According to the study, the researchers observed an inverse correlation between the logarithm of average molecular weight

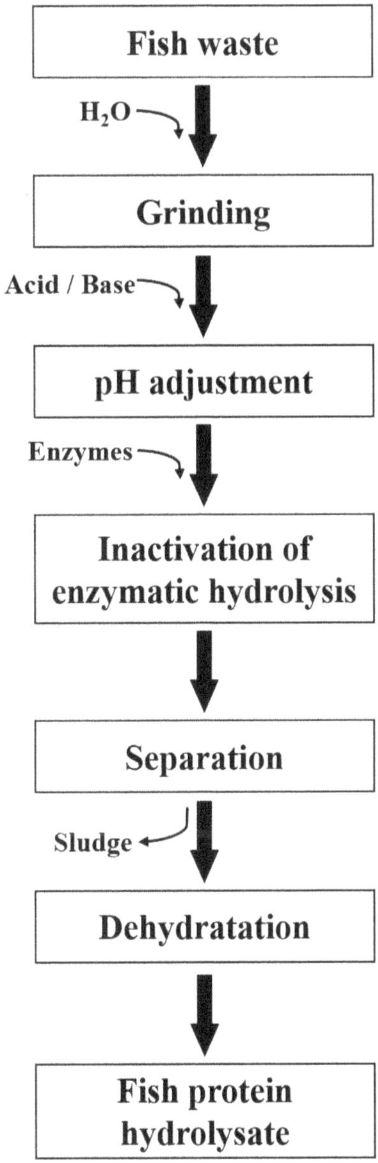

Figure 6.1 Production of FPHs from fish waste.

and the antioxidant potential of the hydrolysate fractions. Furthermore, the collagen hydrolysates demonstrated an effective antioxidant and emulsifying agent. In line with this study, García-Moreno *et al.* evaluated the use of FPHs with their high antioxidant and emulsifying potential in fish oil-in-water emulsions.[19] According to the results, researchers observed that the degree of hydrolysis is an effective parameter for both the stabilization of emulsions and their antioxidative properties. Therefore, FPHs were recommended as alternative protein emulsifiers in the stabilization of fish oil-in-water emulsions against oxidation reactions.

The majority of prior studies discussed the antioxidant potential of FPHs.[15,20,21] Indeed, FPHs possess many other bioactive properties. Among them, the anti-inflammatory activity of FPHs is an important property to be evaluated. Inflammation is a response of the body indicating the presence of an important amount of bacteria, fungi, viruses, allergens, cytokines, harmful chemicals, or foreign objects that might be the reason for several serious diseases. Therefore, cheap and accessible anti-inflammatory resources, such as marine, are gaining importance day by day. Subhan *et al.* reported that the peptides obtained from fish scale collagen exhibited remarkable anti-inflammatory activity.[22] Pro-inflammatory cytokines, such as tumour necrosis factor-alpha (TNF-α) and interleukin (IL)-6, are the main components responsible for inflammation in the body, and fish scale collagens exhibited a regulatory effect over these two key components.[23] Another anti-inflammatory fish by-product is the tripeptide (Pro-Ala-Tyr) obtained from the pectoral fin peptic hydrolysates of a salmon fish (*Salmo salar*). Similarly, Ahn *et al.* discussed the inhibitory effect of fish by-products against inflammation by inactivating proinflammatory cytokines, such as IL-6, IL-1β, and TNF-α.[20]

FPHs, the relatively new method of fish waste valourization, are facing some other challenges, such as consumer acceptance of their daily consumption. Kurniawan *et al.* determined the response of children to traditional food with the addition of Indian scad (*Decapterus russelli*) protein concentrate and flavour. According to the results, colour, texture, odour, and taste are crucial parameters to determine the most suitable and consumable FPC-added products.[24] Many essential aroma compounds are derived by enzymatic oxidative degradation, which involves lipoxygenase activity on the highly unsaturated fatty acids, providing the metallic, mushroom-like, and mild green aroma of fresh fish.[25] Although lipids play an important role as fatty acid precursors, the formation of fish flavour is additionally based upon proteins, particularly FAAs and peptides. FAAs are not only substantial precursors for the aroma but also for the formation of principal non-volatile taste-active compounds that provide a unique taste to the fish. Therefore, fish FAAs have a remarkable impact on both retronasal and orthonasal receptors when a fish is sniffed and chewed. Histidine is the most abundant FAA, particularly in fishwith darker flesh, such as mackerel and tuna.[25] It is known to be the precursor of the vital compound histamine, which can be a marker for fish quality and safety. In the presence of bacterial decarboxylase enzymes and suitable conditions, not only histamine but also putrescine and cadaverine arise *via* the decarboxylation reaction of FAAs. All these amino acid-based aroma compounds are fingerprints of the freshness and safety of a fish.

The Maillard reactions are responsible for protein-based aroma and taste substances in fish and fish products. In many aspects, the Maillard reaction positively affects the aroma quality of cooked fish and fishery products; it can be undesirable for the fish's taste, especially in some specific conditions, such as overheating. The degradation of some fish amino acids, such as lysine, arginine, cysteine, glutamic acid, aspartic acid, and methionine, which are also well-known aroma precursors, results in the production of new aroma compounds *via* the Maillard reaction and Strecker degradation, providing burnt, cooked flavour, while D-amino acids are known to be the precursors of sweet tastes in seafood.[26] In an earlier study, Maga and Lorenz evaluated the sensory characteristics of FPCs and determined that the carbonyl content of FPCs was not found to be significant regarding flavour acceptability; however, perceived off-flavour is mostly associated with the amine nature of the product.[27] As the deodorization of marine-based protein products is a popular subject due to the developed valourization techniques, some recent studies have focused on the removal of these off-flavours derived from the fish protein by adjusting the media (pH, ionic equilibrium, ion strength, *etc.*) used in the production.[28]

Therefore, this chapter emphasizes the complex structure of fish's aroma, including the generation phase (oxidative, enzymatic, microbial, and other aspects), essential non-volatiles as aroma precursors, and the characteristics and bioactive compounds of fish protein products.

6.2 Fish Flavour

6.2.1 Overview of Fish Flavour

Fish or fish products are appreciated all around the world for their nutritional value and health functions. Fish include a proper balance of lipids, FAAs, protein, nucleotides, sugars, and organic acids, contributing to a distinct and delicate flavour of seafood that makes it different from other meat flavours. The flavour characteristics of fish and their products are critical parameters for consumer acceptance and preferences.

Until now, a number of research studies and reviews have carried out sensorial tests, which could be a fast and simple way to detect quality measurements. However, human judgement may be subjective and time-varying. The quality of fish is first and rapidly assessed by sensory techniques based on the alteration in aroma, taste, appearance, and texture properties. Earlier research revealed that neither chemical methods nor sensorial analyses alone were sufficient to determine the freshness and overall quality of fish species.[28,29] Therefore, the correct quality measurement of fish requires a comprehensive monitoring of oxidative, enzymatic, and bacterial deteriorations by both chemical and sensorial evaluations collaterally. Several biochemical changes, including nucleotide deterioration, active taste inosine generation, hypoxanthine aggregation, endogenous enzyme activity, pH reduction, and the progress of volatiles leading to spoilage alterations that take place after the harvest period, may influence the overall fish quality. In this delicate post-harvest period, some new compounds, such as alcohols and aldehydes, may form by the enzymatic pathway of polyunsaturated fatty acids (PUFA), and some other compounds may degrade or disappear as a result of the biochemical reactions. Thermal processing, storage, microbial activity, and spoilage can be the main reasons for inter-reactions of proteins, peptides, amino acids, sugars, and nucleotides that have important impacts on the organoleptic characteristics of the fish species. Among these characteristics, aroma is considered as one of the most important factors of sensorial properties and directly affects consumer preferences.

6.2.2 Generation of Fish Flavour

The fish flavour is fundamentally composed of non-volatile taste-active and volatile substances mainly derived from lipids, amino acids, enzymatic, and microbial activities (Figure 6.2). Although the concentrations of aroma compounds are quite low in fresh fish, their perception is quite high when a fish is sniffed due to their low threshold values.[30,31] The increased interest of consumers in fish and fish products has created a huge demand for precise work to characterize both the pleasant properties, particularly the freshness, and the off-flavours. Therefore, several well-documented pioneering types of research and reviews focused on the generation of fish flavour and prevention methods of off-flavour formation.[32,33] Each part of the fish has its own importance, especially when the fresh fish aroma is considered. In an earlier study, Josephson and Lindsay mentioned that both fish slime and tissue are responsible for the fresh notes of fish, but slime especially plays a crucial role in the perception of the overall freshness of a harvested fish.[34]

Recent theoretical developments have revealed that microbial activity, oxidation, enzymatic, and non-enzymatic reactions are known to be the four vital features of the generation of overall fish flavour and off-flavour.[30,31,34]

6.2.2.1 Lipid Oxidation

Oxidative processes are well-known to be the principal causes of fish and other types of meat quality deterioration, such as flavour, colour, and nutritional properties. Lipids are one of the main components of fish, and their amount is quite variable because of the complex matrix of fish, including protein, water, and minerals. Lipid oxidation is defined as a chain reaction of free radicals and consists of three main stages: initiation, propagation, and termination (Figure 6.3). This reaction is a complicated and important interaction that threatens food quality, especially products containing a high amount of unsaturated fat. Important losses in quality, generation of unacceptable flavour, decreasing shelf life, loss of nutritional

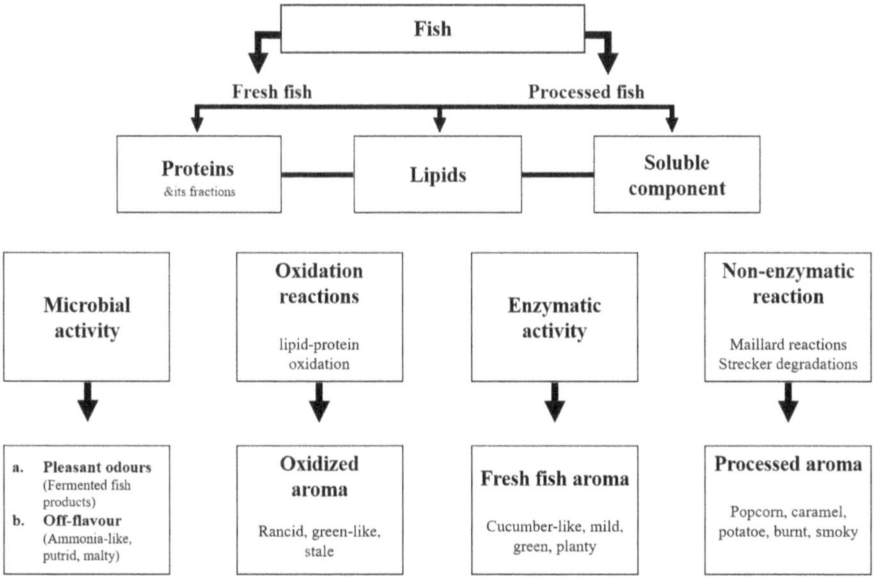

Figure 6.2 An overview of aroma production in fish.

properties, and possible production of unhealthy molecules, such as free radicals and reactive aldehydes, are some of the main results of lipid oxidation in food samples.[35,36] However, lipid oxidation will also lead to significant changes in the sensory and nutritional characteristics of fish and other food products, including odour, flavour, colour, and texture properties, which are easily detected by consumers. Extended storage under unfavourable conditions can create rancid odours described as tallowy for beef; muttony for mutton; stale, cheesy, acrylic, fishy, or oily for pork; and rancid, painty, fishy, and cod-liver oil-like for fish. This off-odour develops from the products of autoxidation of unsaturated fatty acids, such as oleic, linoleic, linolenic, and arachidonic.[37] Fish and fish products are the principal sources of PUFAs. These ingredients are extremely vulnerable to degradation processes, including oxidation. The change in the number of fish lipids also leads to changes in the amount and composition of PUFAs contained in the lipids. In this regard, the composition of the fish and its sensitivity to oxidative deterioration are influenced by factors such as geographic conditions, hunting methods, seasonal climate, species, gender, age, and nutrition.[38,39]

Fish generally possess pleasant, delicate, and mild flavours, and the main well-accepted fresh fish aromas are known to be plant-like and sweet. It is a primary source of omega-3 (n-3) PUFAs, such as docosahexaenoic acid (C22:6n3) and eicosapentaenoic acid (C20:5n3), and has many health benefits for consumers.[40] However, PUFAs are susceptible to lipid oxidation. The main factors affecting the oxidation rate in fish are the type of fatty

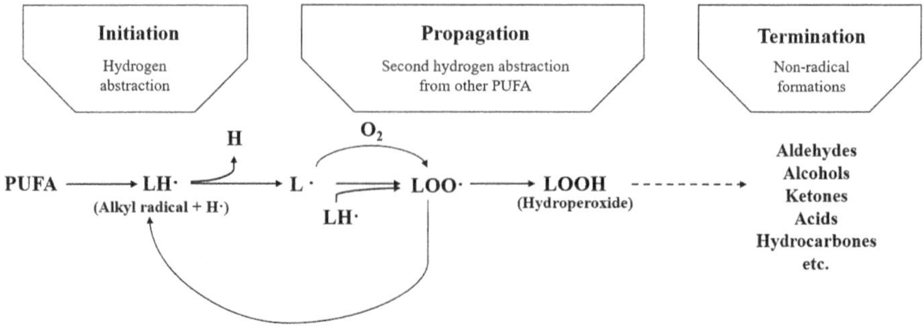

Figure 6.3 Fatty acid oxidation scheme realizing in fish and other seafood.

acids, oxygen, temperature, light, and the presence of antioxidants and prooxidants (oxidation enhancers). Owing to the rich amount of dark-coloured muscles, oily fish are usually more susceptible to oxidation than other lean fishes. The oxidation process initially increases the formation of primary lipid oxidation products, such as hydroperoxides and conjugated dienes. The further decomposition of these compounds is responsible for the rise of volatile components such as carbonyl compounds (aldehydes and ketones) and alcohols. Therefore, the development of lipid oxidation can be checked by the primary products of fatty acid oxidation (Figure 6.3), such as hydroperoxides and conjugated dienes, or secondary oxidation products, such as pentanal, hexanal, and malonaldehyde. The lipoxygenase group that catalyses the oxygenation of polyunsaturated acids containing a (*Z*,*Z*)-1,4-pentadiene moiety to produce hydroperoxides has been assumed to play a vital role in lipid oxidation and the occurrence of volatile components in different food samples. These enzymes lead to the formation of lipid-derived products such as aldehydes, ketones, carboxylic acids, and peroxides and cause the degradation of phosphatides. Among these aroma groups, several post-mortem oxidation aroma compounds form, such as short-chain alcohols, carbonyls and esters, and trimethylamine (TMA). These compounds are known to be responsible for their unpleasant smell and taste. Many studies have been conducted by different researchers on factors affecting lipid oxidation in fish and fishery products. Refsgaard *et al.* studied the lipid oxidation of farmed Atlantic salmon during storage at −10 °C or −20 °C for 34 weeks.[31] The amounts of hydroperoxides and fatty acids in the samples increased throughout the storage, and the most significant changes occurred in the sample stored at −10 °C. The hydroperoxides increased from 0 to 10 meq O_2 kg^{-1}, while free fatty acids escalated from 1% to 8.7% in 34 weeks of storage in the samples. The increase in products formed by oxidative reactions was related to a decrease in highly unsaturated fatty acid content (C20:5n3, C22:5n3, and C22:6n3) in the salmon. Additionally, volatile compounds, including aldehydes and ketones, were found in the samples. Among these, hexanal, heptanal, (*E*)-2-hexenal, (*E*,*E*)-2,4-heptadienal, and nonanal had a significant increase in the amount proportionally to the storage time. Aro *et al.* reported that lipid autoxidation in fish over storage time results in off-odour notes.[41] Indeed, oxidative procedures taking place over fish storage may cause saturated and unsaturated aldehyde accumulation, including 2,4,7-decadienal, 2,4-heptadienal, (*Z*)-4-heptenal, and hexanal, which give rancid notes. Different cooking applications, including boiling, frying, roasting, and steaming, lead to the inactivation of enzymes and promote a series of diverse physicochemical reactions, of which the oxidation of lipids is the most prominent one.[42,43] These authors studied the development of lipid oxidation in sous-vide cooked and chilled stored Atlantic mackerel by applying fluorescence microscopy along with various analyses. Results showed that sous-vide cooking followed by chilled storage resulted in primary and secondary oxidation products in the samples. Reversely, the use of natural antioxidants, such as rosemary extract that is rich in tocopherols and ascorbyl palmitate, was reported to effectively delay lipid oxidation reactions in cooked mackerel through storage at low temperatures.[43] Wang *et al.* reported that lipid oxidation is a crucial reaction that is induced when fish is cooked.[44] These authors found that the increase in the core temperature during steaming resulted in a rise in the amounts of hexanal, heptanal, octanal, and nonanal, especially at 75 and 85 °C ($P < 0.05$), due to the lipid oxidation in the large-mouth bass (*Micropterus salmoides*) fish sample, and simultaneously gave the steamed fish grassy, fatty, and fruity odour characteristics.

6.2.2.2 *Protein Oxidation*

Protein oxidation of fish and fishery products has been one of the subjects of great scientific interest because the proteins are known to be preferential targets of radical species, and protein oxidation has important effects on protein functionality, digestibility, and other quality properties. This oxidation has been much less investigated than lipid oxidation in the literature, but some studies have reported that protein molecules from fish and other muscle foods are sensitive to oxidation reactions during storage.[45] Protein oxidation has a complex mechanism because the pathways and types of products formed depend on the targets in the protein and how the reactions begin. These oxidative alterations comprise the cleavage of peptide bonds,[46] the modification of amino acid side chains, and the formation of covalent intermolecular cross-linked protein derivatives, as presented in Figure 6.4. Specific amino acid side chains and/or a chemical modification cause peptide backbone fragmentation, aggregation, loss of solubility and functionality, and decreased proteolysis sensitivity, including adverse changes in the physical properties of proteins.[21] The reactive oxygen species (ROS), including superoxide, nitric oxide, hydrogen peroxide, and hydroxyl radicals, are produced in the muscle by metals or enzymes. These ROS can catalyse oxidation reactions that probably have a negative impact on the quality parameters of fish, affecting both

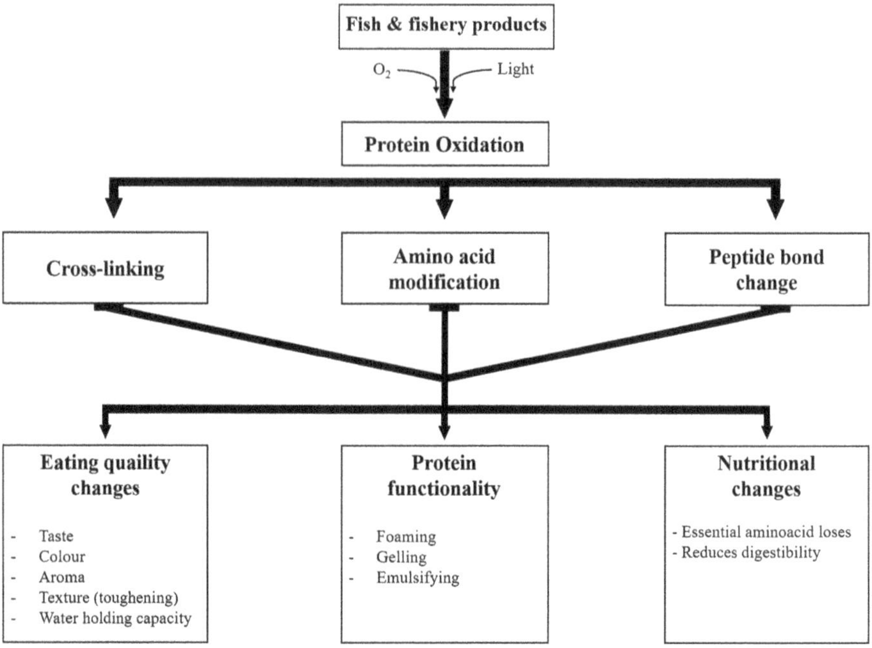

Figure 6.4 The diverse effects of protein oxidation realized in fishes.

lipids and proteins. Stadtman and Levine studied the basic mechanisms of ROS-mediated oxidation of proteins and FAAs.[47] They reported that protein oxidation leads to hydroxylation of aromatic and aliphatic amino acid side chains, nitration of aromatic amino acids, nitrosylation of sulphhydryl groups, sulphoxidation of methionine residues, chlorination of aromatic and primary amino groups, and conversion of some amino acid residues to carbonyl derivatives. The same authors also reported that the oxidation process can cause cleavage of the polypeptide chain and the creation of cross-linked protein aggregates, resulting in the loss of protein function. It is known that protein oxidation is an essential parameter resulting in possible quality loss of various muscle foods both during processing and during storage.

Besides, it is widely known that proteolytic and oxidative reactions contribute significantly to the release of aroma compounds, which directly affect the flavour and fragrance of fish and fish-like muscle foods. Thus, it is anticipated that protein oxidation will also affect the final flavour of foods containing muscles. Indeed, the oxidation of proteins is known to be a source of active carbonyl components, which are associated mostly with the Maillard reactions contributing to the formation of heterocyclic aromatic amines and some other aroma compounds. In the extant literature, the presence of these precursor compounds and their potential effects on the change of fish aroma and flavour were recently mentioned by some authors.[48] Moreover, Al-Dalali *et al.* discussed the positive relationship between lipid and protein oxidations and their certain effects on the aroma alteration of marinated raw beef meat.[49]

In post-mortem muscle tissue, when proteins are targeted by ROS, the result of this interaction is frequently carbonyl creation and decreased sulphydryl content of the protein. These alterations can considerably change the properties of fish proteins and may ultimately influence the quality of the products and lead to several health problems.[50,51] Due to protein oxidation in fish fillets, hydrophobicity, solubility, water-holding capacity, tenderness, and texture properties are the most important negatively affected quality parameters. The progress of protein oxidation in fish has been studied through the formation of diverse radical-producing model systems. Kjærsgård *et al.* investigated the realization of protein oxidation throughout the frozen storage of rainbow trout (*Oncorhynchus mykiss*) fillets.[52] The results obtained in this study with the aid of UV spectrophotometric determination of protein carbonyl groups showed that storage at −20 °C resulted in increased protein carbonylation and protein solubility changes in the rainbow trout. Protein carbonylation is undoubtedly the most investigated oxidative alteration of protein molecules, having the benefit that it can be combined with proteomic tools that allow the identification of oxidized proteins and the site of the oxidation. Pazos *et al.* elucidated the susceptibility of specific myofibrillar proteins of mackerel mince to carbonylation reactions during chilled storage and the antioxidant capacity of (+)-catechin to prevent oxidation of proteins.[53] In this study, the carbonylation of each specific protein was measured by combining the labelling of protein carbonyls by fluorescein-5-thiosemicarbazide with 1-D or 2-D gel electrophoresis techniques. As a result of this study, the addition of catechin decreased protein oxidation in a concentration-dependent manner, and the highest catechin amounts (100 and 200 ppm) displayed the highest antioxidant activity. Catechin (200 ppm) had a significant reducing effect on the carbonylation of protein spots identified as glycogen phosphorylase, pyruvate kinase muscle isozyme, and isoforms of creatine kinase. On the contrary, catechin was not active in inhibiting the oxidation of actin and an unnamed protein product similar to enolase. Timm-Heinrich *et al.* investigated the effect of both fish oil and fish meal replacement on oxidative changes in rainbow trout stored on ice for up to 12 days by measuring different quality parameters.[54] Fish samples were fed six different diets, which differed in their concentrations of marine oil and protein *vs* vegetable oil and protein. In this study, protein oxidation was determined using the measurement of protein carbonyl (spectroscopy and immuno-blot), protein sulphydryl, α-aminoadipic semialdehyde, and γ-glutamic semialdehyde. The authors reported that the differences in protein and amino acid amounts were not significant in the samples, and there were no important effects of different diets on protein oxidation reactions. However, the obtained data showed that compounds present in the marine ingredients might have affected protein oxidation. Zhang *et al.* investigated the effect of different packaging methods, including air packaging, vacuum packaging, and modified atmosphere packaging (MAP), on the protein oxidation and degradation of grouper (*Epinephelus coioides*) fillets during refrigerated storage conditions.[55] The authors reported that the protein oxidation level of the samples gradually increased as the storage time increased, and high carbon dioxide MAP played a positive role in the inhibition of myofibril degradation and oxidation for the refrigerated grouper fillet samples. Similarly, Wang *et al.* investigated the effects of different temperatures (45, 55, 65, 75, and 85 °C) on the protein oxidation of a large-mouth bass fish sample.[44] They highlighted that to ensure safety, pleasant flavour properties, and minimal protein oxidation, a core temperature between 65 and 75 °C was recommended for the steamed large-mouth bass.

Lipid and protein oxidations can cause important quality alterations in fish and fishery products. Overall, the same factors triggering lipid oxidation will also cause protein oxidation. In the extant literature, published studies reveal that protein oxidation and lipid oxidation show a strong correlation, stimulating a remarkable increase in carbonyl content. Carbonyl content is considered a marker of protein oxidation in muscle foods such as meat and fish. Berardo *et al.* provided an extensive report about the protein oxidation that occurs in dry fermented sausages by measuring the carbonyl content, while Al-Dalali *et al.* recently exhibited the impact of protein oxidation on the flavour profile by monitoring the frozen storage of raw beef meat.[49,56] In another study, Feng and Ahn demonstrated the effect of protein oxidation on the aroma profile of irradiated ready-to-eat cured turkey meat products. Researchers found a strong correlation between protein oxidation and an increase in aldehyde and ketone concentrations.[57] There exist several possible reasons for protein oxidation and the corresponding alteration of the flavour profile of muscle foods, including fragmentation of backbones, oxidation of amino acid side chains, and insertion of reactive carbonyl compounds.[58]

The main factors proposed for inhibiting oxidation reactions and aroma alteration in these products are the use of antioxidants, suitable packaging methods, and applying low temperatures in the storage period. Furthermore, in the case of seafood oil products, the microencapsulation technique can provide the necessary protection against oxidation reactions for these compounds.[59]

6.2.2.3 Enzymatic Reactions

Enzymatic techniques are commonly used to produce several food ingredients for a wider range of applications in the food sector. Enzymes are known to be biochemical catalysts that accelerate chemical reactions inside the cell structure by lowering the activation energy. They form a complex layer between highly specific active sites and the substrate. Therefore, these crucial components are widely used in the food industry to catalyse the desired reaction and convert the raw materials into high-value products.[60] Some of these high-value products produced by the traditional enzymatic processes of seafood include fermented fish, fish sauce, and fish silage. In the processing of these products, endogenous enzymes play a critical role in the formation of flavour by the partial hydrolysis of the tissue macromolecules.[61] Additionally, endogenous enzymes are responsible for the pleasant, green, mild odour of freshly harvested raw fish.[62] This pleasant odour of raw fish comes from the presence of C_6, C_8, and C_9 alcohols and carbonyls.[63–65] These compounds are produced by endogenous enzymes catalysing oxidation, *i.e.*, lipoxygenases and hydroperoxide lyases.[8]

As explained in detail in the previous section, lipid oxidation is one of the milestone reactions of aroma production in both fresh and processed fish and other seafood. Generally, long-chain PUFAs act as precursors for the formation of carbonyls, alcohols, and aldehydes. The interaction between unsaturated fatty acids in fish muscles and oxygen molecules results mostly in the formation of hydroperoxides and derivatives by complex reactions, namely lipid oxidation. Hydroperoxides are fundamentally known as odourless compounds; however, they may be subject to consecutive degradation reactions and produce several secondary products, particularly those associated with rancid odour.[66] The above-mentioned two enzymes (lipoxygenases and hydroperoxide lyases) mostly catalyse oxidation to generate aroma compounds. As the types and concentrations of lipoxygenases change according to the fish species, the difference in aroma occurs with various freshwater and saltwater fish.[34] The main enzymes responsible for the generation of aroma compounds are 12- and 15-lipoxygenases and hydroperoxide lyase. The n-9 hydroperoxides are produced by the action of 12-lipoxygenases on specific PUFAs. Following, specific hydroperoxide lyases hydrolyse the 9-hydroperoxide and form (*Z*,*Z*)-3,6-nonadienal, which can be spontaneously turned into its isomer of (*E*,*Z*)-2,6-nonadienal. These aldehydes can undergo enzyme reduction to their respective alcohol forms.[34,67] In addition to these aldehydes, C_8 alcohols and ketones also contribute to the fish aroma to a great extent. The representative compounds of these groups are 1-octen-3-ol, 1-octen-3-one, (*Z*)-1,5-octadien-3-ol, and (*Z*)-1,5-octadien-3-one contributing to mushroom, metallic, and leafy aroma notes in fish.[65,68] The generation of these compounds is again realized by the action of 12-lipoxygenases. For this reaction, the intermediate compound is 12-hydroperoxide, which will hydrolyse by hydroperoxide lyase to form C_8 alcohol. The formed alcohol will then transform into its corresponding ketone.[34]

Other C_6 aldehydes are mostly produced by 15-lipoxygenases activity with hydroperoxide lyases. The produced aldehydes differ according to the type of PUFAs degraded in the reaction, such as hexanal from n-6 PUFAs and (*Z*)-3-hexenal from n-3 PUFAs. Another enzymatic activity is the isomerization of (*Z*)-3-hexenal to (*E*)-2-hexenal.[30] Hexanal is an important aldehyde that contributes to a wide variety of food odours.[69] This compound was reported to be responsible for the grassy, green, plant-like, or rotten fish odour notes in meat and fish.[30,70,71]

In sum, aldehydes are the dominant oxidation products regarding the overall aroma of fish due to their relatively low threshold values. In particular, 3-hexanal, 2-pentanal, 4-heptanal, 2,4-heptadienal, 2,4,7-decatrienal, and 2–6-nonadienal are the aldehydes providing off-flavours *via* linoleic acid-derived autoxidation reactions.[72] Autoxidation of fish and fishery products is related to not only aldehydes derived from omega-3 PUFAs but also some ketones (1-penten-3-one, 1,5-octadien-3-one, *etc.*) and volatile alcohols (1-penten-3-ol). Specific autoxidation aroma compounds (2,4,7-decatrienal, (*E*,*Z*)-2,6-nonadienal, (*Z*)-4-heptenal, and 1-penten-3-one) provide an intense burnt-fishy note to the overall aroma of oxidized fish oil.[73]

Another important group of enzymes is the proteinases that are used to generate flavour compounds from fish and fish by-products. Proteases play a substantial role in the production of seafood flavourants, fermented fish, fish sauce, and FPH. Even though there are endogenous proteases in fish, proteases are often added as an exogenous enzyme to accelerate the protein hydrolysis in new or by-products of seafood.[74] Enzymatic hydrolysis has been used to produce seafood flavourants from seafood processing by-products. The demand for the use of seafood flavours as food additives is quite high in artificial crab, surimi, fish sausage, and cereal-based extrusion products.[75] The use of several plant proteinases, including papain, bromelain, and ficin, was reported to shorten the needed time for fish sauce fermentation. In addition, it enhanced the flavour of the product, which resulted in more preference from consumers than traditional fish sauces.[74,76] The action of proteases results in the degradation of muscle proteins and the production of their constituent amino acids and peptides. These molecules are the precursors of the distinguishable flavour of fermented fish. Amino acids also participate in the Maillard reactions and Strecker degradation when heat is applied.[77,78] A controlled Maillard reaction contributes favourably to the flavour to a certain extent. The generation of aroma by the Maillard pathway is explained in detail in further sections of this chapter.

6.2.2.4 *Microbial Activity*

Microorganisms are capable of creating several chemical classes of compounds with low boiling points and small molecular masses. Volatile compounds of microorganism origin are responsible for the characteristic aroma of fermented fish products, FPHs, and fish sauces, as well as the unpleasant odour notes of spoiled fish and fish products.

The overall aroma of fish and fish products is the result of a combination of parameters and many complex reactions. Microbial activity is considered one of the most important key factors for the generation of fish flavour and the interactions between biochemical pathways, environmental parameters, varietal differences, and post-harvest conditions. Apart from enzyme production, microbial strains are also commonly used for the production of fermented fish products, which provide them with unique characteristics, particularly aroma, taste, and texture.[79] During the fermentation process, the quality parameters of the fish species and products are altered by the effects of degradation and microorganism action. The determination of the volatile profile in the muscle of fermented fish is a significant factor in figuring out the variation in sensorial parameters of the final products. To increase fish and fishery product consumption, the development of new, suitable, and delicious tasting products is desirable for consumers. Nordvi *et al.* investigated the volatile compounds found in a novel, sliceable, fermented, and dried fish product obtained from the fatty fish Atlantic salmon (*Salmo salar*), the lean fish saithe (*Pollachius virens*) (1 : 1, w/w), *Lactobacillus sakei,* and four different milk protein-based ingredients.[80] The variations in the volatiles during the cold storage (4 °C) of vacuum-packed dried fish sausages were studied by a dynamic headspace gas chromatography–mass spectrometer. In the samples, alcohols, alkanes, esters, aldehydes, ketones, and compounds derived from amino acid metabolisms were the most dominant volatile components. Regarding the bacteria, *Lactobacillus sakei* in combination with the enzymes in the fish, promoted a diverse profile of remarkable volatiles in the fermented fish. Li *et al.* identified the volatiles in Stinky Mandarin fish (*Siniperca chuatsi*) through fermentation (with spices and salt) and storage periods using an electronic nose in combination with gas chromatography-mass spectrometry.[81] 24 aroma compounds are well-correlated with the periods of fermentation and storage in this study. The amounts of TMA, indole, sulphur-containing compounds, acetic acid, esters, and phenols increased constantly, while aldehydes decreased. The authors highlighted that the fermentation process favoured the delay of spoilage and generated new volatile compounds. Naturally fermented fish sauces are highly appreciated by consumers and are a very popular fish product around the world. Zeng *et al.* studied the chemical and microbial properties of Suan yu, a sanitary Chinese traditional low-salt fermented whole fish product.[82] They found that lactic acid bacteria, *Staphylococcus*, and yeast were the leading microorganisms in the six different Suan yu samples analysed, while *Enterobacteriaceae* and *Pseudomonas* were under the detection limit in the samples. Another study on the Suan yu sample was carried out by Gao *et al.*[83] These authors studied the autochthonous starter cultures (*Lactobacillus plantarum*, *Staphylococcus xylosus,* and *Saccharomyces cerevisiae*) on the volatile compounds of Suan yu. A small difference in the total number of volatiles between control and starter culture samples was detected, while the compound and proportion changed significantly. Among the detected volatiles, esters, higher alcohols, acids, carbonyls, and alkanes were identified as the principal volatile compounds in the Suan yu samples. The metabolism of microorganisms, including the hydrolysis and lipolysis of proteins, regulates the profile and amount of chemical substances during the fermentation period, which forms the sensory properties of different fish sauces in order.[84,85] Shrimp (*Penaeus vannamei*) is economically important but can be easily spoiled due to microbial action. Biopreservation is a preservation technique based on the addition of "useful" bacteria that are able to prevent the growth of detrimental microorganisms in food. Within these useful bacteria, two specific strains of lactic acid bacteria have formerly been chosen for a biopreservation strategy: *Lactococcus piscium*, for its ability to inhibit the sensory degeneration of seafood, and *Carnobacterium divergens*, for its ability to inhibit the growth of *Listeria monocytogenes*. The results were evaluated in a non-inoculated control batch. The two useful bacteria do not inhibit each other when co-inoculated in cooked and peeled shrimp samples. The inhibitory effect of *C. divergens* on *L. monocytogenes* in naturally contaminated shrimp has been verified. The use of a mixture of *L. piscium* and *C. divergens* is recommended for the biopreservation of seafood due to the elimination of the off-odour notes.[86] Wang *et al.* studied the mechanisms regulated by microbial metabolism in the flavour formation of traditional fish sauces collected from a local fish sauce processing factory in China.[87] In total, 56 volatiles were detected in the traditional fish sauce samples. Among these, 3-(methylthio)propanal had the highest average odour activity value, followed by 3-methylbutanal, 1-octen-3-ol, and 2-octenal. As a result of this study, five microbial genera, including *Halanaerobium*, *Halomonas*, *Tetragenococcus*, *Halococcus,* and *Candidatus Frackibacter,* found to be the main microbial flora responsible for volatile formation in the fish sauces.

Fish species are highly perishable because of their high water activity, neutral pH level, and presence of autolytic enzymes. Fish muscle is an important source of nonprotein nitrogen, and amino acids, nucleotides, and trimethylamine oxide (TMAO) serve as microbial substrates or electron acceptors. The rate of spoilage is highly temperature-dependent and can be inhibited by low storage temperatures. The main cause of spoilage in fish species is of microbial origin. Microbial spoilage of fish muscle involves the growth of bacteria, yeast, or fungi in high numbers and as products of their metabolism, giving rise to the sensory impressions perceived as spoilage. As a result of undesirable microbial action in the fish muscle, amines, sulphides, alcohols, aldehydes, ketones, and organic acids with unpleasant odours are formed. Although the microflora variation on freshly caught fish species and seafood products can be classified into similar microbial ecology groups, the microflora alterations are due to different toleration abilities of the microorganisms to the storage conditions.[88] The most important cause of deterioration in fish is known as bacterial activity. There are millions of microorganisms on the fish's skin, gills, stomach, and intestinal tract. In the post-mortem period, bacteria reach the fish meat through the gills, abdomen, and blood vessels, causing the formation of a fishy smell. Spoilage bacteria are contaminated as a result of the physical and chemical conditions in the fish or fishery products; however, spoilage clearly occurs in the growth of the microorganisms to high amounts ($>10^6$–10^7 cfu g^{-1}) and the growth and metabolism of microorganisms may be affected by the interaction between different groups. As a result of the activity of microorganisms on nonprotein nitrogen (free amino acids and nucleotides), volatile compounds such as TMA nitrogen, dimethylamine nitrogen, and ammonia, commonly known as total volatile basic nitrogen, are formed.[89] Nonprotein nitrogen is a readily available bacterial growth substrate. The spoilage of fresh fish is governed by the existence of TMAO, especially under conditions where oxygen is excluded. The undesirable bacteria use TMAO as the terminal electron acceptor in an anaerobic respiration process, resulting in off-odour notes due to the formation of TMA nitrogen.[90] Ethyl butyrate, methyl (methyl thio) methyl sulphide, carbon disulphide, methyl disulphide, *O*-2-methylpropyl hydroxylamine, *O*-3-methylbutyl hydroxylamine, 2-methyl-2-propanamine, and TMA have been reported as the principal volatiles detected in the headspace of prawns (*Penaeus plebejus*, *P. merguiensis*, *P. esculentus/semisulcatus*, and *Metapenaeus bennettae*) after storage on ice. These aroma compounds are mostly generated by bacterial (*Pseudomonas fragi* or *Shewanella putrefaciens*) degradation of bioactive components, mainly amino acids, and the sensory impression formed by the bacteria cultures was strongly related to the volatile pattern.[91] Each amino acid compound has a different kind of resistance to spoilage, so their impact on the overall aroma is quite variable. Among a number of compounds, TMAO is considered to be a well-known indicator of fish oxidation derived from the degradation of TMA.[88] Spoilage bacteria from gilt-head sea bream were identified by molecular techniques, and *Pseudomonas, Shewanella, Carnobacterium,* and *Lactobacillus* species were found to be abundant in sea bream kept at different temperatures and storage conditions.[92] In another study by the same authors in 2017, they investigated the microbial and non-microbial-originated volatile components on a model fish substrate with and without inoculation of spoilage bacteria from the sample.[79] Three types of spoilage bacteria (*Pseudomonas*, *Shewanella,* and *Carnobacterium/Lactobacillus* strains) were added to inoculate petri dishes with sterile fish juice agar and stored at 0 °C and 15 °C under air and a commercial modified atmosphere package (MAP CO_2: 60%, O_2: 10%, N_2: 30%). The authors highlighted that various C6–C10 alcohols and aldehydes were possibly of non-microbial origin. Volatiles such as 2- and 3-methyl-1-butanol, ethanol, 2- and 3-methylbutanal, 3-hydroxy-2-butanone, acetic acid, dimethyl sulphide, 3-methylbutyl acetate, and various ethyl esters were linked to microbial activity, while some of them increased during the storage period, making them potential chemical indices of spoilage.[79]

6.2.3 Effect of Processing on the Volatiles

6.2.3.1 Maillard Reaction

Fresh fish aromas are mostly defined by mild, green, and fishy-like odour notes. Although seafood has a very mild odour in its fresh form, they have a matrix rich in non-volatile aroma precursors, including amino acids, lipids, sugars, *etc.*[67] Especially processes such as cooking, drying, and fermentation have a substantial role in flavour formation from these precursors. Complex lipolytic and proteolytic reactions induced by various parameters, such as temperature and enzymes, are indeed a consequence of the progress of odour in processed seafood. A wide variety of changes during handling and processing that affect the development of aroma in fish consist of lipid oxidation, thermal degradation, and the Maillard reaction. These reactions are essential in the formation of complex processing flavours.[93]

The Maillard pathway, including Strecker degradation, is a thermally induced reaction between amino acids and reducing sugars, which leads to the formation of the characteristic heated flavour compounds. The compounds formed *via* the Maillard reaction can contribute to various aspects of seafood, such as odour, taste, and colour, especially in smoked fish.[94] From this point of view, the branched-chain essential amino acids (valine, isoleucine, and leucine), amino acids with sulphur (methionine and cystine/cysteine), and aromatic amino acids (phenylalanine and tyrosine) are the most significant flavour precursors as they are responsible for the formation of aroma- and taste-active compounds. During the early post-mortem process and storage, free amino acids initiate major changes and can be used as consistency parameters for processing and storage.[95] Additionally, the balance of desirable and undesirable seafood flavours depends on the marker compounds that affect the seafood quality. Thus, the investigation of these reactions is of utmost importance.

The realization of the Maillard reaction can be grouped into three stages.[96] In the first (early) stage, there is no browning or flavour generation; only the reduction reaction occurs between the carbonyl group of a reducing sugar and the free amino group of an amino acid or protein. After this, Amadori rearrangement takes place and produces 1-amino-l-deoxy-2-ketoses and simultaneously 2-amino-2-deoxyaldoses are formed by Heyns rearrangement. These compounds are substantial, non-volatile flavour precursors.[96,97] In the second (advanced) stage of the Maillard reaction, Amadori compounds produce degradation products, such as dicarbonyls (diacetyl, maltol, and pyruvaldehyde) and dehydration products such as furans. If the precursor is pentose, a furfural compound forms. On the other hand, hexoses are responsible for the production of 5-hydroxymethylfurfural (HMF).[98] Concurrently, amino acids are degraded to aldehydes with CO_2 loss. This decarboxylation and oxidative deamination of free amino acids is triggered by the dicarbonyls produced from Amadori rearrangement, and the produced compounds are aldehydes with one fewer carbon atom than the original amino acid. This whole aldehyde production process from protein oxidation is called Strecker degradation. The final stage of the Maillard reaction includes the formation of melanoidin pigments and heterocyclic volatile compounds.[92]

The compounds that arise from the Maillard pathway can be roughly classified into three groups: (1) sugar dehydration products, (2) amino acid degradation products, and (3) volatiles from further interactions.[99]

6.2.3.1.1 Sugar Dehydration Products. The first group comprises furans, pyrones, carbonyls, acids, and cyclopentenes. The formation pathway of furans was explained in the previous paragraph; they are produced in the advanced stage of the Maillard reaction. In addition, non-enzymatic reactions, such as furans, can be derived from the thermal degradation of sugars and oxidation of PUFA.[97] Furan-derived compounds are attributed with burnt, nutty, sweet, or caramel-like odour notes. Various furan compounds such as furfural, 5-methyl-2-furfural, 2-acetylfuran, furfuryl alcohol, and 2-pentyl furan were found in a diverse group of seafood, including smoked salmon,[100] grilled eel,[101] and European seabass.[68] Also, it was reported by Hwang *et al.* that the amount of furan compounds increased in seal protein hydrolysate with the addition of glucose in the medium, proving that they are sugar-degradation products.[102] The formation of furan compounds is highly dependent on the parameters of the Maillard reaction, such as temperature, time, pH, amino acid/sugar ratio, water content, and oxygen.[103] A study by Nie *et al.* (2013) investigated furan formation in a sugar–glycine model system during processing and the effects of reaction parameters, such as temperature, sugar type, and pH, on the amount of formed furans.[97] According to the results, the authors reported that acidic conditions resulted in a lesser formation of furan derivatives. Also, non-reducing sugars produced a higher amount of furans than sucrose. Thus, the formation of these compounds can be controlled with the careful selection of precursors and processing parameters.

6.2.3.1.2 Amino Acid Degradation Products. The second group of Maillard volatiles includes Strecker aldehydes and sulphur compounds formed by amino acid degradation.[99] Strecker degradation is one of the most crucial reactions for flavour development, and the reacting amino acids play an important role in the production of different volatiles. Actually, the type of amino acid is a deciding criterion for which volatiles will be formed. For example, alanine produces acetaldehyde, leucine forms 3-methylbutanal, and phenylalanine is responsible for the production of phenylacetaldehyde by Strecker degradation.[104] Also, a brief interaction with other amino acids and their respective volatiles leads to the formation of Glycine-formaldehyde, valine-isobutyraldehyde, cysteine-hydrogen sulphide, and methionine-methional.[104] All these aldehydes were common in many seafood products, including scallops, canned salmon, lobster tail, pastes of anchovy, herring, and shrimps.[104–106] Aldehydes are known to possess very low threshold values, and thus, their contribution to the overall aroma is predominant.[107] Within these mentioned aldehydes, phenylacetaldehyde was attributed with floral notes, 3-methylbutanal with herbaceous, pungent, meaty, and methional with cooked/boiled potato notes.[104]

Another important class of amino acid-degradation volatiles is sulphur compounds. These volatiles can be responsible for both the pleasant notes and the off-odours. Sulphur compounds are key odorants due to their low threshold values. They contribute a pleasant, mild odour at low amounts, and as the concentration increases, the odour characteristics turn into an oxidized, putrid aroma.[108] Thus, sulphur compounds play a critical role in the acceptance of foods. The formation of these compounds is generally due to the S-containing amino acids (methionine and cysteine) in seafood. Additionally, other S-containing precursors, such as thiamine (vitamin B1), and peptides, such as glutathione, also contribute to the formation of sulphur volatiles.[109] The configuration of S-volatiles can be divided into two groups: aliphatic and cyclic. Aliphatic sulphur compounds with the general name of polysulphides and another group of thiols are produced by microbial enzyme activity and are generally present in unprocessed, fresh, or stored seafood.[110] Polysulphides comprise compounds such as dimethyl of tetrasulphide, trisulphide, disulphide, and sulphide. The formation of these compounds is realized by the enzymatic degradation of cysteine. In addition, these compounds are highly reactive and have a tendency to react with thiols, resulting in the rearrangement and production of new compounds.[111] Within these compounds, dimethyl trisulphide and disulphide were found in a wide variety of seafood products. The green, vegetable-like odour notes in crayfish waste,[112] onion note in salted white herring,[113] and a rotten, sulphur odour in stored cods were attributed to dimethyl trisulphide,[114] while dimethyl disulphide gives rise to an onion or cabbage-like note in fermented fish sauce,[84] fresh and grilled eel,[101] stored cod fillets,[114] and stewed clam odour in clams and oysters.[63]

The second group of sulphur volatiles, *i.e.*, cyclic compounds, are mainly produced *via* the Maillard pathway in thermally processed sea products. The generation of these compounds takes place by the $-NH_2$ group and carbonyl reaction induced through applied heat. The representative groups of cyclic sulphur compounds in seafood are the derivatives of thiazoles, thiophenes, dithiazines, and thiolanes.[115] These compounds are responsible for the roasted, meaty, and shellfish odours in a variety of crustaceans.[116,117] The representative compounds of thiazoles include benzothiazole (sulphur, roasted odours), 2-acetylthiazole (meaty, roasted, popcorn odours), and 2-acetyl-2-thiazoline (hazelnut, grilled odours). Thiazoles are mainly generated by the cysteine/pyruvaldehyde interaction in the Maillard pathway.[118] The cooking process of fish and other seafood enhances the formation of these compounds, and, thus, they are commonly found more in cooked products.[119,120]

The thiophene compounds previously detected in processed fish are mainly thiophene, 2- and 3-methylthiophene, 2-ethylthiophene, 2- and 3-thiophenecarboxaldehyde, and 3-thiophene methanol. These compounds were found in cooked salmon and boiled fish from different species[114] and salt-fermented fish.[106] Thiophenes can also be generated from the further reaction of hydrogen sulphide with carbonyl compounds or furans. Another cyclic S-containing group, namely thiolanes, was also found in salt-fermented herring, anchovy,[101] and cooked salmon.[109]

6.2.3.1.3 Volatiles from Further Interactions. Last but not least, the third group of Maillard-derived compounds arises from further interactions between non-volatile precursors. Pyrazines are the featured N-containing compounds, as they are the most commonly found volatiles in this group. These compounds, especially their alkyl forms, are generated from the condensation of α-aminoketones in Strecker degradation and are generally responsible for the nutty or roasted odour note in thermally processed foods.[65,121] As the formation and distribution of pyrazines are directly related to the Maillard reaction, all parameters affecting the reaction, including heating time, temperature, and reactant ratio, play a crucial role in the presence of pyrazine derivatives.[121] Additionally, it was reported that the long duration and high intensity of applied heat promoted the formation of these compounds in processed seafood.[120] Also, Shye *et al.*[122] reported an interesting finding that the content of pyrazines increased with the concentration of added sugar in the medium. Hence, the limiting criterion of reaction rate can be defined as sugar concentration rather than free amino acids.

Similar to the previously mentioned Maillard-derived volatile groups, pyrazines also have very low thresholds, so they contribute greatly to the aroma, even if they are present in minor concentrations. Some of the pyrazine derivatives and their respective odour contributions to seafood are as follows: 2-ethylpyrazine with a nutty odour; 2,6-dimethyl pyrazine with a popcorn, roasted odour; trimethyl pyrazine with a roasty, earthy odour; and 2-acetylpyrazine with a nutty, popcorn odour notes.[65,121]

Pyrazines are followed by volatile groups of pyrroles, pyrrolines, pyridines, and oxazoles as process-derived compounds. Within these groups, 2-acetylpyrrole stands out as being present in a wide variety of food. This compound was reported to be found in cooked obscure puffer and roasted clam, contributing to nutty and bread odour notes.[120,123]

To sum up, the composite matrix of seafood, which is rich in precursors of volatiles, allows the formation of a wide variety of odorants during the handling and cooking processes. In order to obtain a demanded and preferred aroma in a fish or sea product, a comprehensive investigation of precursors should be carried out so that the formation of volatiles and their contribution to the overall aroma can be understood. As it was explained in detail in this section, heat-induced reactions, especially the Maillard reaction and thermal degradation, have a great and mostly desired effect on processed seafood. Also, the process-derived odorants and their respective odour notes in different kinds of seafood have been summarized in Table 6.1.

Table 6.1 Process-derived volatiles commonly detected in various seafoods.

Compound name	Odorant descriptors	Production pathway/ Precursor	Seafood matrix	Ref.
Sugar dehydration compounds				
Furfural	Sweet, caramel-like	Maillard reaction	Crayfish waste	81
	Bread, almond, sweet	Thermal decomposition of sugars	Fried grass carp	93
	Roasted, nutty	Wood smoke, Maillard reaction	Smoked salmon	69
5-HMF	Cardboard	Thermal decomposition of sugars	Fried grass carp	93
2-Acetyl-furan	Cooked, sweet	Maillard reaction	Smoked salmon	69
	Smoky, tobacco, narcotic	Maillard reaction	Fish and oyster sauce	53
2-Pentyl-furan	Sweet, spicy, green	—	Crayfish waste	81
	Green bean, butter	Lipid oxidation	Fried grass carp	93
	Fishy, grassy	Lipid oxidation	Cooked obscure puffer	89
Furfuryl alcohol	Cooked	Maillard reaction	Smoked salmon	69
	Fermented burnt sugar, sweet	—	Fish and oyster sauce	53
	Sour	—	Various fish species	88
Diacetyl	Butter	—	Smoked salmon	69
Amino acid degradation products				
Acetaldehyde	Sweet	—	Boiled cod and trout	94
	Sweet	Strecker degradation	Ripened anchovy	95
3-Methylbutanal	Malty	—	Boiled cod and trout	94
	Dark chocolate	Strecker degradation of leucine	Steamed *Coilia ectenes*	89
Phenylacetaldehyde	Sour	Strecker degradation of phenylalanine	Boiled squid, prawn and scallop	96

(continued)

Table 6.1 (*continued*)

Compound name	Odorant descriptors	Production pathway/ Precursor	Seafood matrix	Ref.
	Floral, honey-suckle		Skipjack tuna sauce	75
Methional	Boiled potato-like	—	Boiled cod and trout	94
	Cooked potato	—	Smoked salmon	69
	Potato	Strecker degradation of methio-nine	Boiled salmon	78
Hydrogen sulphide	Rotten egg	—	Lobster tail meat	97
	Sulphur	—	Roasted shrimps	98
	Rotten egg	Degradation of cysteine and/or methionine	Salted-dried white herring	82
Dimethyl sulphide	Sulphury	Degradation of methionine	Cooked mussels	99
	Seafood, sulphur, shrimp	—	Canned salmon	100
Dimethyl disulphide	Leek, onion, vegetal	—	Canned salmon	100
	Sulphury	Thermally generated from methional	Cooked mussels	99
Dimethyl trisulphide	Cabbage, sulphurous	Oxidation of methanethiol	Steamed *Coilia ectenes*	89
	Green vegetable-like	—	Crayfish waste	81
Diethyl disulphide	Sulphury, roasty	—	Cooked rainbow trout	34
Benzothiazole	Green, plastic, fruity	Strecker degradation	Smoked salmon	69
2-Acetylthiazole	Grilled hazelnut	Thermal degradation	Cooked mussels	99
3-Methylthiophene	Meaty, spicy	Strecker degradation	Canned salmon	100
Volatiles from further interactions				
2,5-Dimethyl-3-ethylpyrazine	Potato, roast	Maillard reaction	Fried grass carp	93

(*continued*)

Table 6.1 (*continued*)

Compound name	Odorant descriptors	Production pathway/ Precursor	Seafood matrix	Ref.
2-Methylpyrazine	Roasted nuts	Maillard reaction	Smoked salmon	69
2-Acetylpyrazine	Nutty, popcorn	—	Lobster tail meat	97
Pyrrolidin-2-one	Popcorn	Degradation of glutamic acid	Steamed and cooked red mullet	89
2-Acetylpyrrole	Nutty, walnut, bread	—	Cooked obscure puffer	89

6.3 Characteristics and Bioactive Compounds of Fish Protein Products

FPHs are currently considered high-added value products, a novel opportunity for the valourization of fish wastes. However, despite their remarkable health-promoting potential, consumer acceptance of their daily consumption highly depends on their organoleptic characteristics.

Wick *et al.* investigated the volatile constituents of FPCs for the first time. The researchers observed an aroma profile of FPCs dominated by amines, particularly TMA, ammonia, dimethylamine, ethylamine, diethylamine, and n-propylamine.[124] According to the results of the study, despite the presence of amines in high amounts, interestingly, the researchers determined a lack of fish-like odour and flavour in FPC samples. Similarly to these findings, Maga and Lorenz determined a significant loss in the odour intensity of FPCs when compared to the fish itself.[27] Gao *et al.* mentioned the off-flavour problems of bioactive peptides, which cause significant commercial concerns.[120] On the other hand, previously, Hevia *et al.* reported that some enzyme-solubilized FPC fractions possessed unpleasant, bitter, and acid-like odours.[125] To remove this kind of unpleasant odour, which may affect consumer preferences negatively, Fu *et al.* performed a study on removing the fishy flavour of silver carp protein by yeast fermentation.[29] Recently, Zhang *et al.* reported that the interactions between fish proteins and volatile flavour compounds play crucial roles in the deodorization of fish by-products such as FPHs and FPCs.[126]

Apart from sensorial properties, fish protein products, especially collagens, FPHs, *etc.* draw great interest due to their several functional bioactive properties, such as anti-skin ageing, anti-cardiovascular ailments, anti-hypertensive, anti-tumour, neuroprotective effects, high nutritional value, and remarkable amino acid profile.[127]

6.4 Conclusion

Fish and other seafood products are significant components of the human diet due to their high nutritional values and distinguishable and delicate flavour. Great progress in the comprehension of fish flavours has been achieved in recent years. These advances demonstrated that specific reactions of thermal degradation, the Maillard pathway, enzyme activity, and oxidation have a crucial role in the production of fish and seafood flavours. As fresh fish has a delicate aroma and can be masked by deteriorative odours, it is of the utmost importance to handle, process, and store the products in the most appropriate conditions. The information gathered until now for flavour alteration can be used in the generation of new aromas and the protection of the desirable flavour with external applications. Additionally, the use of wastes and by-products for the production of valourized seafood products, such as flavouring agents and fish protein hydrolysates, will be eased by this enhanced data collection.

Abbreviations

AP	Air packaging
FPC	Fish protein concentrate
FPH	Fish protein hydrolysate
HMF	5-Hydroxymethylfurfural
IL	Interleukin
MAP	Modified atmosphere packaging
PUFA	Poly-unsaturated fatty acid
ROS	Reactive oxygen species
TMA	Trimethylamine
TMAO	Trimethylamine oxide
TNF-α	Tumour necrosis factor-alpha
VP	Vacuum packaging

References

1. FAO (Food and Agriculture Organization of the United Nations), *FAOSTAT data*, 2017, Available from: http://www.fao.org/faostat/en/#data.
2. K. Elavarasan, V. Naveen Kumar and B. A. Shamasundar, *J. Food Process. Preserv*, 2014, **38**(3), 1207–1214.
3. M. Da Rocha, A. Alemán, G. C. Baccan, M. E. López-Caballero, C. Gómez-Guillén, P. Montero and C. Prentice, *J. Aquat. Food Prod. Technol.*, 2018, **27**(5), 592–608.

4. V. Mohr, in *Biochemical aspects of new protein food*, Pergamon, 1978, pp. 53–62.
5. E. Barzana and M. Garía-Garibay, in *Fisheries Processing*, Springer, Boston, MA, 1994, pp. 206–222.
6. R. Tahergorabi, K. E. Matak and J. Jaczynski, *J. Funct. Foods*, 2015, **18**, 746–756.
7. N. H. Herpandi, A. Rosma and W. A. Wan Nadiah, *Compr. Rev. Food Sci. Food Saf.*, 2011, **10**(4), 195–207.
8. A. Bougatef, N. Nedjar-Arroume, R. Ravallec-Plé, Y. Leroy, D. Guillochon, A. Barkia and M. Nasri, *Food Chem.*, 2008, **111**(2), 350–356.
9. S. Raghavan and H. G. Kristinsson, *Food Chem.*, 2009, **117**(4), 582–588.
10. S. Klomklao, S. Benjakul and H. Kishimura, *Int. J. Food Sci. Technol.*, 2013, **48**(7), 1483–1489.
11. P. Maturana, M. Martinez, M. E. Noguera, N. C. Santos, E. A. Disalvo, L. Semorile and A. Hollmann, *Colloids Surf., B*, 2017, **153**, 152–159.
12. S. Chakrabarti, F. Jahandideh and J. Wu, *Biomed. Res. Int.*, 2014, DOI: 10.1155/2014/608979.
13. B. Giménez, A. Alemán, P. Montero and M. C. Gómez-Guillén, *Food Chem.*, 2009, **114**(3), 976–983.
14. L. Picot, R. Ravallec, M. Fouchereau-Péron, L. Vandanjon, P. Jaouen, M. Chaplain-Derouiniot and P. Bourseau, *J. Sci. Food Agric.*, 2010, **90**(11), 1819–1826.
15. M. A. Siddik, M. R. Chaklader, M. J. Foysal, J. Howieson, R. Fotedar and S. K. Gupta, *Fish Shellfish Immunol.*, 2020, **97**, 465–473.
16. R. Vignesh, M. B. Haq, K. Devanathan and M. Srinivasan, *Arch. Appl. Sci. Res.*, 2011, **3**(5), 52–58.
17. I. Lassoued, L. Mora, R. Nasri, M. Aydi, F. Toldrá, M. C. Aristoy and M. Nasri, *J. Proteomics*, 2015, **128**, 458–468.
18. C. F. Chi, Z. H. Cao, B. Wang, F. Y. Hu, Z. R. Li and B. Zhang, *Molecules*, 2014, **19**(8), 11211–11230.
19. P. J. García-Moreno, A. Guadix, E. M. Guadix and C. Jacobsen, *Food Chem.*, 2016, **203**, 124–135.
20. C. B. Ahn, J. G. Kim and J. Y. Je, *Food Chem.*, 2014, **147**, 78–83.
21. I. Godinho, C. Pires, S. Pedro, B. Teixeira, R. Mendes, M. L. Nunes and I. Batista, *Appl. Biochem. Biotechnol.*, 2016, **178**(6), 1095–1112.
22. F. Subhan, H. Y. Kang, Y. Lim, M. Ikram, S. Y. Baek, S. Jin and S. Yoon, *Oxid. Med. Cell. Longevity*, 2017, DOI: 10.1155/2017/9703609.
23. A. V. Le Gouic, P. A. Harnedy and R. J. FitzGerald, in *Bioactive Molecules in Food*, ed. J. M. Mérillon and K. G. Ramawat, Springer International Publishing, 2018, pp. 1–35.
24. A. Kurniawan, A. Permadi and A. H. Purnomo, *J. Airaha*, 2020, **9**(1), 96–104.

25. H. D. Belitz, W. Grosch and P. Schieberle, Food Chemistry, Springer, 4th revised and extended edition, *Annu. Rev. Biochem.*, 2009, **79**, 655–681.
26. H. J. Weng, K. R. Cadwallader, H. H. Baek, L. R. D'Abramo and J. A. Sullivan, in *Flavor and Lipid Chemistry of Seafoods*, ACS Symposium Series, 1997, **674**, pp. 120–130.
27. J. A. Maga and K. Lorenz, *J. Milk Food Technol.*, 1972, **35**(3), 131–135.
28. S. Gu, W. Dai, Y. Chong, F. Lyu, X. Zhou and Y. Ding, *RSC Adv.*, 2020, **10**(19), 11292–11299.
29. X. J. Fu, S. Y. Xu and K. Jinmoon, *Food Sci. Biotechnol.*, 2009, **28**(1), 57–62.
30. D. B. Josephson, R. C. Lindsay and D. A. Stuiber, *J. Agric. Food Chem.*, 1984, **32**(6), 1344–1347.
31. H. H. F. Refsgaard, P. B. Brockhoff and B. Jensen, *J. Agric. Food Chem.*, 1998, **46**, 3473–3479.
32. Z. Bai, A. Pilote, P. K. Sarker, G. Vandenberg and J. Pawliszyn, *Anal. Chem.*, 2013, **85**(4), 2328–2332.
33. R. C. Lindsay, in *Seafoods: Chemistry, Processing Technology and Quality*, Springer, Boston, MA, 1994, pp. 75–84.
34. D. B. Josephson and R. C. Lindsay, in *Biogeneration of Aromas*, ed. T. H. Parliament and R. Croteau, ACS Symposium Series 317, American Chemical Society, Washington, DC, 1986, pp. 201–219.
35. G. Secci and G. Giuliana Parisi, *Ital. J. Anim. Sci.*, 2016, **15**(1), 124–136.
36. F. Shahidi and Y. Zhong, *Chem. Soc. Rev.*, 2010, **39**(11), 4067–4079.
37. J. Aalhus and M. Dugan, in *Encyclopedia of Meat Sciences*, Elsevier, United Kingdom, 2004.
38. M. Chaijan, S. Benjakul, W. Visessanguan and C. Faustman, *Food Chem.*, 2006, **99**, 83–91.
39. J. H. Sohn, Y. Taki, H. Ushio, T. Kohata, I. Shioya and T. Ohshima, *J. Food Sci.*, 2005, **70**, 490–496.
40. Q. Cai, Y. Wu, L. Li, Y. Wang, X. Yang and Y. Zhao, *J. Ocean Univ. China*, 2017, **16**(5), 855–862.
41. T. Aro, R. Tahvonen, L. Koskinen and H. Kallio, *Eur. Food Res. Technol.*, 2003, **216**, 483–488.
42. J. Cropotova, R. Mozuraityte, I. B. Standal and T. Rustad, *J. Aquat. Food Prod. Technol.*, 2019a, **28**, 505–518.
43. J. Cropotova, R. Mozuraityte, I. B. Standal and T. Rustad, *Food Control*, 2019b, **104**, 1–8.
44. Y. Wang, C. Li, Y. Zhao, L. Li, X. Yang, Y. Wu, S. Chen, J. Cen, S. Yang and D. Yang, *Food Chem.*, 2020, **323**, 126839.
45. C. P. Baron, I. V. Kjærsgård, F. Jessen and C. Jacobsen, , *J. Agric. Food Chem.*, 2007, **55**, 8118–8125.

46. M. N. Lund, M. Heinonen, C. P. Baron and M. Estévez, *Mol. Nutr. Food Res.*, 2011, **55**, 83–95.
47. E. R. Stadtman and R. L. Levine, *Amino Acids*, 2003, **25**, 207–218.
48. X. Wu, Z. Zhang, Z. He, Z. Wang, F. Qin, M. Zeng and J. Chen, *Molecules*, 2021, **26**(5), 1264.
49. S. Al-Dalali, C. Li and B. Xu, *Food Chem.*, 2022, **376**, 131881.
50. Y. L. Xiong, in *E. Antioxidants in Muscle Foods*, Wiley, New York, 2000, pp. 85–111.
51. I. Dalle-Donne, D. Giustarini, R. Colombo, R. Rossi and A. Milzani, *Trends Mol.Med.*, 2003, **9**, 169–176.
52. I. V. H. Kjaersgård, M. R. Nørrelykke, C. P. Baron and F. Jessen, *J. Agric. Food Chem.*, 2006, **54**, 9437–9446.
53. M. Pazos, R. Maestre, J. M. Gallardo and I. Medina, *Food Chem.*, 2013, **136**, 64–72.
54. M. Timm-Heinrich, S. Eymard, C. P. Baron, H. H. Nielsen and C. Jacobsen, *Food Chem.*, 2013, **136**, 1220–1230.
55. Y. Zhang, X. Ma and Z. Dai, *J. Food Process. Preserv.*, 2019, **43**(10), e14111.
56. A. Berardo, H. De Maere, D. A. Stavropoulou, T. Rysman, F. Leroy and S. De Smet, *Meat Sci.*, 2016, **121**, 359–364.
57. X. Feng and D. U. Ahn, *Radiat. Phys. Chem.*, 2016, **127**, 27–33.
58. A. Amici, R. L. Levine, L. Tsai and E. R. Stadtman, *J. Biol. Chem.*, 1989, **264**(6), 3341–3346.
59. F. Shahidi, X. Q. Han and J. Synowiecki, *Food Chem*, 1995, **53**(3), 285–293.
60. M. Muzaifa, N. Safriani and F. Zakaria, *Aquac. Aquar. Conserv. Legis.*, 2012, **5**(1), 36–39.
61. A. Gildberg, *Process Biochem.*, 1993, **28**, 1–15.
62. N. Haard, in *Safety and Quality Issues in Fish Processing*, ed. H. A. Bremner, Elsevier, 2002, pp. 220–253.
63. D. B. Josephson, in *Volatile Compounds in Foods and Beverages*, ed. H. Maarse and D. Marcel, New York, 1991, pp. 179–202.
64. S. Selli, C. Rannou, C. Prost, J. Robin and T. Serot, *J. Agric Food Chem.*, 2006, **54**(25), 9496–9502.
65. S. Selli, C. Prost and T. Serot, *Food Chem.*, 2009, **114**(1), 317–322.
66. J. Iglesias and I. Medina, *J. Chromatogr. A*, 2008, **1192**(1), 9–16.
67. R. C. Lindsay, *Food Rev. Int.*, 1990, **6**, 437–455.
68. F. Leduc, P. Tournayre, N. Kondjoyan, F. Mercier, P. Malle, O. Kol and G. Duflos, *Food Chem.*, 2012, **131**(4), 1304–1311.
69. J. Pino and S. E. Barzola-Miranda, *J. Raw Mater. Process. Foods*, 2020, **1**(2), 33–39.

70. S. Selli and G. G. Cayhan, *Microchem. J.*, 2009, **93**(2), 232–235.
71. G. Nieto, S. Bañón and M. D. Garrido, *Food Chem.*, 2011, **125**(4), 1147–1152.
72. EFSA Panel on Dietetic Products, Nutrition, and Allergies (NDA), *EFSA J.*, 2010, **8**(3), 1461.
73. C. Karahadian and R. C. Lindsay, *J. Am. Oil Chem. Soc.*, 1989, **66**(7), 953–960.
74. G. G. Beddows, M. Ismail and K. H. Steinkraus, *Int. J. Food Sci. Technol.*, 1976, **11**(4), 379–388.
75. H. R. Kim, H. J. Seo, D. S. Byun, M. S. Heu and J. H. Pyeun, *Fish. Sci.*, 2002, **68**, 1557–1562.
76. C. G. Beddows and A. G. Ardeshir, *Int. J. Food Sci. Technol.*, 1979, **14**(6), 603–612.
77. B. Kilinc, S. Cakli, S. Tolasa and T. Dincer, *Eur. Food Res. Technol.*, 2006, **222**(5), 604–613.
78. S. Klomklao, S. Benjakul, W. Visessanguan, H. Kishimura and B. K. Simpson, *Food Chem.*, 2006, **98**(1), 14–22.
79. F. F. Parlapani, A. Mallouchos, S. A. Haroutounian and I. S. Boziaris, *LWT- Food Sci. Technol.*, 2017, **78**, 54–62.
80. B. Nordvi, Ø. Langsrud, B. Egelandsdal, E. Slinde, G. Vogt, M. Gutierrez and E. Olsen, ,*J. Food Sci.*, 2007, **72**(6), S373–S380.
81. C. Li, J. Wu, Y. Li and Z. Dai, *Int. J. Food Sci. Technol.*, 2013, **48**(11), 2429–2437.
82. X. Zeng, W. Xia, Q. Jiang and F. Yang, *Food Control*, 2013, **30**(2), 590–595.
83. P. Gao, W. Wang, Q. Jiang, Y. Xu and W. Xia, *Int. J. Food Sci. Technol.*, 2016, **51**(7), 1630–1637.
84. A. Giri, K. Osako, A. Okamoto and T. Ohshima, *Food Res. Int.*, 2010, **43**(4), 1027–1040.
85. N. Udomsil, S. Chen, S. Rodtong and J. Yongsawatdigul, *Food Microbiol.*, 2016, **57**, 54–62.
86. T. Saraoui, J. Cornet, E. Guillouet, M. F. Pilet, F. Chevalier, J. J. Joffraud and F. Leroi, *Int. J. Food Microbiol.*, 2017, **241**, 69–77.
87. K. Wang, Y. Bao, H. Yang, Y. Chen Wang, D. J. M. Regenstein and P. Zhou, *Pol. J. Food Nutr. Sci.*, 2020, **70**, 301–312.
88. L. Gram and P. Dalgaard, *Curr. Opin. Biotechnol.*, 2002, **13**(3), 262–266.
89. K. Broekaert, B. Noseda, M. Heyndrickx, G. Vlaemynck and F. Devlieghere, in *Molecular Identification of the Dominant Microbiota and Their Spoilage Potential of Crangon Crangon and Raja Sp*, Ghent University, 2012, pp. 93–118.
90. P. Dalgaard, L. Gram and H. H. Huss, *Int. J. Food Microbiol.*, 1993, **19**, 283–294.

91. H. N. Chinivasagam, H. A. Bremner, A. F. Wood and S. M. Nottingham, *Int. J. Food Microbiol.*, 1998, **42**(1–2), 45–55.
92. F. F. Parlapani, K. A. Kormas and I. S. Boziaris, *J. Sci. Food Agric.*, 2015, **95**, 2386–2394.
93. H. H. Baek and K. R. Cadwallader, in *Flavor and Lipid Chemistry of Seafoods*, ed. F. Shahidi and K. R. Cadwallader, ACS Symposium Series 674, American Chemical Society, Washington, DC, 1997, pp. 85–94.
94. M. E. Bailey, in *Flavor of Meat, Meat Products and Seafoods*, ed. F. Shahidi, Blackie Academic & Professional, London, UK, 1998, pp. 267–289.
95. C. H. M. Chang, T. Ohshima and C. Koizumi, *J. Sci. Food Agric.*, 1992, **59**, 521–528.
96. J. E. Hodge, in *Chemistry and Physiology of Flavors*, ed. H. W. Schultz, E. A. Day and L. M. Libbey, AVI Publishing: Westport, CT, 1967, pp. 465–491.
97. S. Nie, J. Huang, J. Hu, Y. Zhang, S. Wang, C. Li and M. Xie, *Food Sci. Hum. Well.*, 2013, **2**(2), 87–92.
98. J. Xi, *Formation of flavor compounds by the reactionsof carbonyls and ammonium sulfide under mild conditions*, Rutgers The State University of New Jersey-NewBrunswick, 2005.
99. H. E. Nursten, *The Maillard Reaction: Chemistry, Biochemistry and Implications*, Royal Society of Chemistry, London, 2005.
100. V. Varlet, C. Knockaert, C. Prost and T. Serot, *J. Agric. Food Chem.*, 2006, **54**(9), 3391–3401.
101. X. H. Huang, X. Zheng, Z. H. Chen, Y. Y. Zhang, M. Du, X. P. Dong and B. W. Zhu, *Food Res. Int.*, 2019, **115**, 32–43.
102. C. F. Hwang, F. Shahidi, A. C. Onodenalore and C. T. Ho, in *Flavor and Lipid Chemistry of Seafoods*, ed. F. Shahidi and K. R. Cadwallader, American Chemical Society, ACS Symposium Series 674, 1997, ch. 8, pp. 76–84.
103. P. Labuza and W. M. Baisier, in *Physical Chemistry of Foods*, ed. H. Schwartzberg, Dekker, New York, 1992, pp. 595–640.
104. V. Varlet, C. Prost and T. Serot, *Food Chem.*, 2007, **105**(4), 1536–1556.
105. H. Y. Chung, I. K. S. Yung, W. C. J. Ma and J. S. Kim, *Food Res. Int.*, 2002, **35**, 43–53.
106. Y. J. Cha and K. R. Cadwallader, *J. Food Sci.*, 1995, **60**, 19–24.
107. S. Kesen, *J. Raw Mater. Process. Foods*, 2020, **1**, 13–21.
108. C. Alasalvar, P. C. Quantick and J. M. Grigor, in *Flavour and Lipid Chemistry of Seafoods*, ed. F. Shahidi and K. R. Cadwallader, American Chemical Society, ACS Symposium Series 674, Washington, DC, 1997, pp. 39–54.

109. L. Methven, M. Tsoukka, M. J. Oruna-Concha, J. K. Parker and D. S. Mottram, *J. Agric. Food Chem.*, 2007, **55**, 1427–1436.
110. C. Alasalvar, K. D. A. Taylor and F. Shahidi, *J. Agric. Food Chem.*, 2005, **53**, 2616–2622.
111. D. J. Rowe, *Perfum. Flavour*, 1998, **23**, 9–14.
112. U. Tanchotikul and T. C. Y. Hsieh, *J. Food Sci.*, 1989, **54**(6), 1515–1520.
113. H. Y. Chung, C. W. Yeung, J. S. Kim and F. Chen, *Food Chem.*, 2007, **104**(2), 842–851.
114. G. Olafsdottir, R. Jonsdottir, H. L. Lauzon, J. Luten and K. Kristbergsson, *J. Agric. Food Chem.*, 2005, **53**(26), 10140–10147.
115. G. MacLeod, in *Flavor of Meat and Meat Products*, Springer, Boston, MA, 1994, pp. 4–37.
116. Y. Okabe, Y. Inoue, Y. Kanda and T. Katsumata, *Eur. Food Res. Technol.*, 2019, **245**(1), 233–241.
117. D. Zhang, H. Ji and S. Liu and Gao, *J.Food Res. Int.*, 2020, **137**, 109517.
118. S. Kato, T. Kurata and M. Fujimaki, *Agric. Biol. Chem.*, 1973, **37**, 539–544.
119. K. Morita, K. Kubota and T. Aishima, *J. Sci. Food Agric.*, 2003, **83**(4), 289–297.
120. W. Wu, N. P. Tao and S. Q. Gu, *Eur. Food Res. Technol.*, 2014, **238**(2), 237–245.
121. T. Shibamoto and R. A. Bernhard, *Agric. Biol. Chem.*, 1977, **41**, 143–153.
122. S. C. Shye, B. S. Pan and C. M. Wu, *Zhongguo Nongye Huaxue Huizhi*, 1988, **26**, 545–551.
123. T. Kawai, Y. Ishida, H. Kakiuchi, N. Ikeda and T. Tsuneya, *Nippon Suisan Gakkaishi*, 1990, **56**, 795–802.
124. E. L. Wick, E. Underriner and E. Paneras, *J. Food Sci.*, 1967, **32**(4), 365–370.
125. P. Hevia, J. R. Whitaker and H. S. Olcott, *J. Agric. Food Chem.*, 1976, **24**(2), 383–385.
126. J. Zhang, Z. Chen, Z. Wang, L. Chen, Y. Ding, F. Lyu and X. Zhou, *Mini. Rev. Org. Chem.*, 2021, **18**(2), 259–269.
127. R. Gao, Q. Yu, Y. Shen, Q. Chu, G. Chen, S. Fen, M. Yang, L. Yuan, D. J. McClements and Q. Sun, *Trends Food Sci. Technol.*, 2021, **110**, 687–699.

CHAPTER 7

Soy Protein Flavours

JIAN LI*[a], XUEJIE LI[a], TAIJU DI[a] AND XUELI PANG[b]

[a]Beijing Advanced Innovation Center for Food Nutrition and Human Health, Beijing Technology and Business University, Beijing, China;
[b]Tobacco Research Institute of Chinese Academy of Agricultural Sciences, Qingdao, Shandong, China
*E-mail: lijian@th.btbu.edu.cn

7.1 Overview of SPI

7.1.1 Introduction of SPI

Soybeans [*Glycine max* (L.) Merr.] are one of the most widely grown and consumed legumes in the world. They originated in northern China and have been cultivated for more than 5000 years. As one of the most important food crops, soybeans are widely cultivated all over the world. Soybeans are rich in nutrients and contain abundant plant-derived proteins, isoflavones, and other bioactive phytochemicals, such as saponin and phytic acid.

Generally, processing will inhibit the activity of trypsin and chymotrypsin inhibitors, so the digestibility of soy protein isolates (SPI) is higher than that of soybean flour, and it can be easily absorbed by the body.[1] SPI is a high-quality protein with a relatively balanced amino acid profile. Except for methionine and cysteine, the content of other essential amino acids reached or exceeded the requirement of essential amino acids recommended by the World Health Organization (WHO).[2] Its protein digestibility-corrected amino acid fraction reaches a maximum of 1. In 1999, the United States Food and Drug Administration proposed that 'a daily intake of 25 g

Food Chemistry, Function and Analysis No. 41
Flavour and Consumer Perception of Food Proteins
Edited by Jing Zhao and Changqi Liu

Published by the Royal Society of Chemistry, www.rsc.org

of soy protein as part of an unsaturated fat and cholesterol diet can reduce the risk of heart disease.' In addition, soy protein has also been confirmed to have the effect of regulating blood fat and blood sugar.[3] In terms of functional properties, SPI has good solubility, gelation, emulsification, water absorption, water retention, foaming, and film-forming properties.[4] Studies have shown that SPI can replace milk protein and egg white as foaming agents.[5]

7.1.2 Applications and Limitations of SPI

Traditionally, soybeans were mainly processed into soybean oil or defatted soybean meal for animal feed. With the continuous growth of the world population and the increasing demand for protein, the market supply of proteins has been seriously challenged.[6] In addition, animal meat contains a higher content of cholesterol and saturated fatty acids that can cause a range of health problems, such as high blood pressure, heart disease, and obesity.[7] Moreover, animal meat products are ecologically burdensome and resource-intensive and are foods that raise ethical concerns, which prompted a search for high-quality plant-based alternatives to animal proteins.[8] Among plant proteins, soy protein has attracted much attention because of its readily available raw materials, low price, strong sustainability, and desirable nutritional properties.[9] Moreover, soy protein has some functional characteristics comparable to animal proteins and is one of the few plant proteins that can replace animal proteins.[10] For example, soy protein has ideal gelling properties, emulsifying properties, and water- and oil-holding capacity and is widely used in the production of high-protein beverages, infant formula, protein bars, breakfast cereals, bread, cakes, yoghurt, sausages, and meat products.

Although edible soy products are associated with various health benefits, soy protein is still not fully utilized as a functional ingredient in the food industry. The major limiting factor for the utilization of SPI is the typical grassy and beany off-flavours in SPI.[11] In production and practical applications, odours such as the beany odour unique to SPI are common in processed products, which negatively affect the flavour of food and consumers' acceptance of SPI-rich products.[12] Therefore, it is crucial to study and remove the off-flavour in SPI to promote the development and popularization of soybean products.

7.2 Off-flavours

The unpleasant flavours related to soybean products and SPI are described as beany, green, grassy, earthy, rancid, and muddy.[13] The composition of the off-flavour is very complex. It does not originate from one or a few specific substances; rather, it is a comprehensive effect resulting from the interaction between multiple flavour components and human senses of taste, smell, and touch. Lipid-derived odour components are the main reason for the quality decline in SPI during processing and storage. The main components of these flavours include non-volatile compounds, such as long-chain alcohols and esters, and volatile compounds, such as aldehydes, ketones, and short-chain alcohols.[14] Many studies have been conducted on the flavour of soybean protein and its products. The flavour compounds identified in soybean and its products and their identification methods have been summarized Table 7.1. These compounds can be classified as aldehydes, alcohols, ketones, acids, eaters, benzene and its derivatives, heterocyclic compounds, sulphur compounds, *etc.* Among them, the small molecule aldehydes and alcohols are dominant, which are derived from the oxidation of unsaturated fatty acids and are the main sources of the off-flavour in soybean protein.

7.2.1 Off-flavours in SPI

The headspace olfactory perception of SPI slurries was conducted by Boatright *et al.*, and the results showed that hexanal, methanethiol, 2-pentylfuran, and dimethyl trisulphide were the most important odour-active compounds in SPI. The flavour profile of SPI was simulated by combining these four odourants in water according to the concentration detected in SPI slurries.[35]

In addition to soy protein, there are also some studies on the flavour of soybean protein hydrolysates (SPHs). Cadwallader *et al.* researched the beef-like process flavour of extruded enzyme-hydrolyzed soybean protein by aroma extract dilution analysis. The results showed that 2-methyl-3-furanthiol, 2-methyl-3-(methylthio)furan, 2-methyl-3-(methyldithio)furan, bis(2-methylfuryl)disulphide, 2-ethyl-3,5-dimethylpyrazine, 2,6-diethylpyrazine, and 3,5-diethyl-2-methylpyrazine, which were detected with high flavour factors (FF), can be considered as the key flavour compounds.[44] Mottram *et al.* compared the aroma characteristics of acid-hydrolyzed and enzyme-hydrolyzed proteins from soy. Major differences were found between the two hydrolysates. The acidic hydrolyzed process led to the production of furans and furanones, pyrroles, and sulphur-containing compounds, while alcohols (including phenols) and pyrazines were mainly related to the enzyme-hydrolyzed process.[45]

7.2.2 Off-flavours in Soy Products

Lv *et al.* identified eight beany odour-active compounds (hexanal, (*E*)-2-hexenal, 1-octen-3-ol, hexanol, pentanol, acetic acid, benzaldehyde, and (*E*,*E*)2,4-decadienal), and four non-beany aroma-active compounds (3-methylbutanal, (*E*)-2-octenal, nonanal, and (*E*)-2-nonenal) in soymilk by dynamic headspace dilution analysis and gas chromatography-olfactometry-mass spectrometry (GC-O-MS).[24] Yang *et al.* analysed volatile compounds

Table 7.1 The summary of volatile compounds identified in soybean products.

No.	Compounds	Identification method[a]	Sources	No.	Compounds	Identification method[a]	Sources
1	(*E*)-2-Decenal	1, 5	15,16	78	Hexadecanoic acid	1	17
2	(*E*)-2-Hexenal	1, 3	15,18–22	79	Hexanoic acid	1	17,22
3	(*E*)-2-Heptenal	1	22,23	80	Octadecanoic acid	1	17
4	(*E*)-2-Nonenal	1, 2, 3, 4, 5, 8	16–19,21,24–28	81	Oleic acid	1	17
5	(*E*)-2-Octenal	1, 2, 3	15,17,18,21,26	82	Acetic acid	1	29
6	(*E*)-2-Undecenal	5	16	83	2,4-Bis(1,1-dimethylethyl)-phenol	1	17
7	(*E*,*E*)-2,4-Decadienal	1, 2, 4, 5, 7, 8	11,15,16,19,24–28	84	Phenol	1	17
8	(*E*,*E*)-2,4-Nonadienal	1, 2, 3, 5, 8	11,15,16,24,26,30,31	85	2-Methoxy-4-ethylphenol	1	29
9	(*Z*)-2-Heptenal	1,	17	86	2-Methoxy-4-vinylphenol	1	29
10	*cis*-4,5-Epoxy-(*E*)-2-decenal	2,	11	87	2,6-Dimethoxyphenol	1	29
11	*trans*-4,5-Epoxy-(*E*)-2-decenal	2,	11	88	4-Methylphenol	4	25
12	Acetaldehyde	5	32	89	2-Methoxy-4-vinylphenol	4	11,25
13	Propanal	3,	33	90	2-Methoxyphenol	4	25
14	Decanal	1,	15,17,22,34	91	Methanethiol	2	26,35–37
15	Pentanal	1, 3, 7	22,28,33	92	Maltol	1	17,29
16	Hexanal	1, 2, 3, 5, 6, 7, 8	11,15,16,18–24,26–36,38–41	93	Guaiacol	2, 3	21,26
17	Nonanal	1, 2, 5	15–18,22,26,34,38	94	Vinylguaiacol	2	26
18	Octanal	1, 3, 7	15,17,21,22,28,30	95	(*E*)-Isoeugenol	4	25
19	2-Hexenal	3	33	96	1,3,5-Benzenetriol	1	29

(continued)

Table 7.1 (*continued*)

No.	Compounds	Identification method[a]	Sources	No.	Compounds	Identification method[a]	Sources
20	2-Heptenal	3	33	97	2-Pentylfuran	1, 3, 5, 7, 8	15–17,20–24,28,29,32,34–36,38–40
21	2-Octanal	1,	22,34,38	98	Furfural	1	17
22	Heptanal	1, 7	22,28,38	99	Furfural	1	29
23	2,4-Decadienal	1	22,38	100	5-Methylfurfural	1	29
24	(*E*)-4,5-Epoxy-(*E*)-2-decenal	4	25	101	2-Ethylfuran	1, 3	21
25	(*Z*)-2-Nonenal	2, 4	25,26	102	2,3-(2*H*)-2-Methylbenzofuran	1	29
26	(*Z*)-4-Heptenal	2, 4	25,26	103	2,3,5-Trimethylpyrazine	1	29
27	3-Methylbutanal	1, 7	19	104	2,3-Diethyl-5-methylpyrazine	1	29
28	2-Methylbutanal	2, 3	21,26	105	2,5-Diethylpyrazine	1	29
29	2-Butyl-octenal	1	22	106	2,5-Dimethylpyrazine	1	29
30	Methional	2	11,26	107	2-Acetyl-2-thiazoline	2	26
31	(*E*,*Z*)-2,6-Nonadienal	2, 3	21,26	108	2-Acetyl-3-ethylpyrazine	1	29
32	(*E*,*Z*)-2,4-Decadienal	2, 5	11,16,26,31	109	2-Acetyl-3-methylpyrazine	1	29
33	Benzaldehyde	1, 3, 7	17,19,22,23,28–30,34,42	110	2-Acetylpyrazine	1	29
34	Benzeneacetaldehyde	1,	17	111	2-Acetylpyridine	1	29
35	4-Ethoxybenzaldehyde	5	16	112	2-Ethyl-5-methylpyrazine	1	29
36	2,2-Dideutero octadecanal	1	34	113	2-Isopropyl-3-methoxypyrazine	3	21

(*continued*)

Table 7.1 (*continued*)

No.	Compounds	Identification method[a]	Sources	No.	Compounds	Identification method[a]	Sources
37	Ethanol	3	33	114	2-Methoxy-3-(1-methylpropyl)-pyrazine	3	21
38	Decanol	1	23	115	2-Methoxy-3-isopropyl-(5 or 6)-methyl pyrazine	1	15
39	1-Butanol	1	17	116	2-Methyl-5-vinylpyrazine	1	29
40	1-Hexanol	1, 3, 5, 6, 7, 8	15–17,19–24,28,34,38,40,41,43	117	3-Ethyl-2,5-dimethylpyrazine	1	29
41	1-Pentanol	1, 3, 5, 6	16,19,22,23,33,41	118	3-Ethyl-2,5-dimethylpyrazine	3	21
42	1-Heptanol	1	23,29	119	3-Isobutyl-2-methoxypyrazine	3	21
43	1-Nonanol	1	15,19	120	D-Limonene	1	17
44	1-Octanol	1	15	121	Dodecamethylcyclohexasiloxane	1	17
45	1-Octen-3-ol	1, 2, 3, 5, 7, 8	11,15,16,18–20,22–24,26,28,33,34,38,42,43	122	Hexadecamethylcyclooctasiloxane	1	17
46	1-Tetradecanol	1	34	123	*N,N*-Dimethylformamide	1	17
47	2-Pentanol	1	23	124	Heptane	1	29
48	2-Hexanol	1	22	125	Pyrrole	1	29
49	Heptane-1,2,3-triol	1	38	126	2-Acetylpyrrole	1	29
50	2,3-Butadione	5	36	127	2,5-Diethyl-4-methylthiazole	1	29
51	2-Methoxyphenol	1	29	128	2-Acetylthiophene	1	29
52	2-Furanmethanol	1	29	129	2-Methylpyrazine	1	29
53	(*Z*)-6-Nonen-1-ol	4	25	130	2-Methylquinoxaline	1	29
54	1-Penten-3-ol	3	21	131	2-Acetyl-1-pyrroline	2, 4	25,26

(*continued*)

Table 7.1 (*continued*)

No.	Compounds	Identification method[a]	Sources	No.	Compounds	Identification method[a]	Sources
55	3-Methyl-1-butanol	1	29	132	2-Pentylpyridine	3, 4	25,30,31
56	1-Octen-3-one	1, 2, 3, 5, 8	11,16,21,24,26,30,31	133	3-Acetylpyrrole	4	25
57	6-Methyl-5-hepten-2-one	1	17	134	Dimethyl sulphide	2, 3	21,26
58	Dihydro-5-pentyl-2(3*H*)-furanone	1	15	135	Dimethyl trisulphide	2, 3, 5	21,26,30–32,35,36
59	2-Pentanone	1	29	136	2-Acetylthiazole	2	26
60	2-Heptanone	1, 3, 7	22,23,26,28–30,34,40,42	137	Pyridine	1	29
61	2-Octanone	3	30	138	Pentane	3	33
62	2-Nonanone	1	22	139	*n*-Butylbenzene	1	23
63	2-Decanone	3	30	140	Ethyl heptanoate	1	23
64	4-Methyl-2-heptanone	1	22	141	Ethyl hexanoate	1	23
65	Megastigmatrienone	1	29	142	Ethyl butyrate	1	23
66	1-Methyl-2-pyrrolidinone	1	29	143	2-Acetyl-1-pyrroline	2	11
67	Acetophenone	4	25	144	2-Isopropyl-3-methoxypyrazine	2	11
68	2,3-Butanedione	1, 2, 3	21,22,26	145	3-Hydroxy-2-methyl-4-pyranone	2	11
69	2,3-Pentanedione	1	22,23	146	4-Nonanolide	2	11
70	3-Hydroxy-2-butanone	1, 2	22,23,26	147	4-Hydroxy-2,5-dimethyl-3(2*H*)-furanone	2	11

(*continued*)

Table 7.1 (*continued*)

No.	Compounds	Identification method[a]	Sources	No.	Compounds	Identification method[a]	Sources
71	β-Damascenone	2	26	148	3-Hydroxy-4,5-dimethyl-2(5*H*)-furanone	2	11
72	Methyl isobutyl ketone	3	21	149	Levomenthol	—	40
73	1-Penten-3-one	3	21	150	Benzene, methyl(1-methyl ethyl)-	1	34
74	3-Octen-2-one	3	21,30	151	α-Terpinyl acetate	1	34
75	2′-Aminoacetophenone	2	11	152	1,4-Dione-2,5-cyclohexadiene	1	34
76	Acetophenone	2	31	153	Dimethyl trisulphide	7	28
77	Acetic acid	1	17,19	154	2-Pentylthiophene	7	28

[a]1: SPME-GC-MS (solid-phase microextraction-gas chromatography-mass spectrometry); 2: SAFE-GC-MS (solvent-assisted flavour evaporation-gas chromatography-mass spectrometry); 3: DHS-GC-MS (gas chromatography-mass spectrometry); 4: SBSE-GC-MS (stir bar sorptive extraction-gas chromatography-mass spectrometry); 5: SDE-GC-MS (gas chromatography-mass spectrometry); 6: TDS-GC-MS (thermal desorption-gas chromatography-mass spectrometry); 7: P&T-GC-MS (purge and trap-gas chromatography-mass spectrometry); 8: SPME-GC-FID (solvent-assisted flavour evaporation-gas chromatography-flame ionization detector).

across different production stages of fermented soybean tofu whey through headspace-gas chromatography-ion mobility spectrometry (HS-GC-IMS) combined with principal component analysis. The results showed there were 24 representative flavour compounds from all production stages of tofu whey. The concentrations of beany flavour compounds (hexanal, 1-octen-3-ol, (*E*)-2-hexenol, and heptanal) decreased after heating, whereas the aroma compounds increased. The fermented compounds, such as pentanal, 2-pentyfuran, 2,3-butanedione, and 2-heptanone, were produced after adding the coagulant.[46] Among these flavour compounds, some exhibit grassy and beany odours, while others elicit bitter, astringent, and pungent tastes. The combined effect of all these unpleasant odours results in an off-flavour in SPI. The reasons for the formation of off-flavours in SPI are extremely complicated, which mainly come from the oxidative degradation of lipids and the degradation of amino acids.[47]

7.2.3 The Production of Off-flavours in SPI

7.2.3.1 Autoxidation of Lipids

The fat content of soybeans is about 16–24%, and the C16-18 fatty acids account for a large proportion of it. Among them, oleic acid, linoleic acid, and linolenic acid account for about 20, 52, and 10%, respectively. These unsaturated fatty acids are prone to oxidation, which may generate various substances with off-flavours.[48] The autoxidation of lipids follows the mechanism of free radical reactions.[49] A small amount of lipid is catalysed by light, heat, or metal catalysts to decompose into unstable free radicals. In the presence of molecular oxygen, the free radicals react with oxygen to form peroxide radicals, which will then react with lipid molecules to form hydroperoxides and new free radicals. As the chain reaction proceeds, more lipid molecules are converted into hydroperoxides. Once hydroperoxide is formed, it immediately decomposes and then generates alkoxy radicals and hydroxyl radicals. Subsequently, the alkoxy groups decompose and generate aldehydes, ketones, alcohols, acids, alkanes, and furans.[50,51] Among them, aldehydes are the most unstable substances, which can be readily oxidized into acids, participate in polycondensation, and produce a variety of off-flavours.

7.2.3.2 Enzymatic Oxidation of Lipids

In addition to the autoxidation of PUFAs, off-flavour can also be formed by enzymatic oxidation during storage or processing. Once activated, lipoxygenases (LOXs) can directly catalyse the formation of hydroperoxides, which further decompose into small volatiles responsible for the beany flavour, such as alcohols, aldehydes, ketones, acids, and lactones.[52] Among PUFAs, linoleic acid can be oxidized to hexanal, nonanal, 1-hexanol, and 1-nonanol, while linolenic acid can be oxidized to (*E*)-2-hexenal, (*E*,Z)-2,6-nonadienal, (*E*,*E*)-2,6-nonadienal, (*E*,*E*)-2,4-nonadienal, (*E*)-3-hexen-1-ol,

(*Z*)-2-hexen-1-ol, (*Z*)-3-nonen-1-ol, and (*E*)-2-nonen-1-ol.[53,54] Most of these volatile compounds have been implicated as the source for the 'beany' flavour, and hexanal has been identified as one of the major volatiles responsible for the green odour of soybean.[55]

7.2.3.3 Amino Acid Degradation

Amino acid degradation is another important pathway for the development of off-flavour in SPI. In general, the degradation of amino acids produces pyrazines, benzene derivatives, and terpenes. Benzene derivatives can be formed from phenylalanine, tyrosine, and tryptophan. In addition, ammonia, hydrogen sulphide, and acetaldehyde are produced by the degradation of cysteine.[56] Moreover, proteolysis of SPI leads to a bitter taste. Hydrophobic amino acids such as valine, leucine, isoleucine, and tryptophan are important causes of the bitter taste.

7.2.3.4 The Hydrolysis of Soy Protein

Enzymatically hydrolyzed plant proteins, especially SPH, have been widely added to foods for their potential functional properties and physiological activities, such as antioxidative, antihypertensive, and immunomodulatory.[57,58] However, the SPH is often accompanied by the production of bitterness, which limits their utilization as food additives and ingredients and reduces their acceptance by consumers.[59] Studies have shown that the bitterness in SPH is related to low molecular mass peptides containing hydrophobic amino acids, such as tryptophan, leucine, isoleucine, and valine.[60] The molecular weight of bitter peptides is about 0.36–2.10 kDa[61], which is exposed from the inside of the protein to the surface after hydrolysis.[62]

7.2.3.5 Oleosins

Oleosins are the most abundant protein in plant seed oil bodies and play an important role in regulating oil body formation and lipid accumulation.[63] Among the formation mechanisms of the off-flavours in SPI, oleosins are worth mentioning here because of their close relationship with phospholipids (PLs), which are related to the formation of off-flavours in SPI.[64] In soybean, oleosins and PLs together form a monolayer membrane that surrounds the oil bodies. The structure of the membrane is such that the charged residues of oleosin and the polar heads of PLs are located on the outer surface of the oil body, while the hydrophobic domains of olein and the acyl chains of PLs are embedded in the triglyceride matrix.[48] During the pressing and *n*-hexane extraction of the oil, the oleosin-PLs membrane is separated from the oil body but remains in the defatted soybean meal. Because oleosins and soybean protein have the same isoelectric pH (4.5–5.0), the oleosin–PL complex coprecipitates along with soybean protein during the isoelectric point preparation of SPI. In addition, some free PLs and fatty acids can also combine with storage proteins to form oleosins during the pressing of the oil.[65] The PLs in oleosin–PL complexes are prone to autoxidation, which is the main cause of off-flavour generation.

7.2.3.6 *Phytochemicals*

In addition to some volatile compounds we have mentioned earlier, some non-volatile phytochemicals such as phenolic acids (170 ppm),[66] isoflavones (3 mg g^{-1}), and saponins (0.81 mg g^{-1})[67] are associated with the sour, bitter, and astringent flavour of the SPI. Most of these non-volatile components exist in a protein-binding state and are difficult to remove during the SPI manufacturing process.[48]

7.2.3.6.1 Phenolic Acids. In 1966, Arai *et al.* identified nine phenolic acids in hexane-defatted SPI and proposed that they may be related to the sour, bitter, and astringent flavour of SPI.[68] Soon, Rackies disputed the review on the basis that the phenolic acid content is low in SPI and pure phenolic compounds have almost no taste.[69] With the development of phenolic acid separation, purification, and identification, Maga and Lorenz concluded that a certain amount of phenolic acid was present in SPI and that it could contribute to astringency and bitterness.[70] In addition, there is a synergistic effect among the different phenolic acid combinations, which can lower the taste threshold.[70]

7.2.3.6.2 Isoflavones. Isoflavones are one of the phenolic compounds and account for 72% of the total phenolic content in the defatted soybean meal. Most of the isoflavones in raw soybeans exist in the hypocotyl and bind tightly to proteins during SPI production. It has been reported that the recovery rate of isoflavones in SPI is about 80–84%, and the isoflavone–protein complex can be stable for 2 years under normal temperature conditions.[71] Although the mechanism of action between isoflavones and soy proteins is unclear, it is worthy of affirmation that there is a nonspecific hydrophobic interaction in such interactions.[71] Isoflavones are available in the form of glycosides (usually 7-monoglycosides) and free aglycones, which contribute to the bitter and astringent taste.[47] In addition, isoflavones have a variety of aglycone isomers, which are more hydrophobic and tend to have a stronger bitter taste than their aglycone structures.[48]

7.2.3.6.3 Saponins. Saponins are widely distributed secondary metabolites in the plant kingdom that can protect against pathogens and herbivores in the plant defence system.[72] It is so named because its aqueous solution can form long-lasting foam, like soap.[73] Saponins are mainly divided into triterpenes and steroid glycosides, and their structures are very different due to the different number of attached sugar units at different positions in the molecule.[74] Saponins have been found in 12 kinds of legume crops, of which soybean seeds have the highest content (0.65%).[73] Soybean saponins are triterpene glycosides that can be classified into groups A and B based on the number of individual aglycones (soy saponins) and attached glycosyl groups.[74] Like isoflavones, saponins are also related to bitter, astringent, or other off-flavours of soy protein, but the saponins have good thermal stability and are not easy to lose during processing. It is reported that the concentration of total saponins in SPI was about 0.81 mg g^{-1}.[67]

7.2.3.7 Beany Flavour Released from Soy Protein

Proteins have a strong binding affinity for volatile compounds.[75] In the process of SPI separation, odour compounds are tightly bound to the protein body by covalent bonding, hydrogen bonding, or hydrophobic interaction.[76] As a result, the beany and other off-flavour molecules that bind to the SPI will be delivered to the food, causing unpleasant flavour characteristics. In addition, functional groups of ligands play an important role in the binding of flavour compounds to soy protein. The thermodynamic study of the adsorption of off-flavour compounds onto soy protein was conducted by Aspelund *et al.*[77] The results showed that the heat adsorption of aldehydes, ketones, and methyl esters was not significantly different from each other but was significantly different from that of hydrocarbons and alcohols. The dry soy protein matrix has the weakest adsorption effect on hydrocarbons and the strongest adsorption effect on alcohols. Gibb's free energy for these selected flavour compounds was negative, indicating that the interaction between these compounds and soybean dry protein was spontaneous.

7.2.4 Factors Affecting Off-flavours

Volatile compounds in SPI can be affected by various internal and external factors, including soybean varieties, protein structure, protein extraction method, pH, ionic strength, and processing conditions.[78] The various factors that influence the off-flavours in SPI are discussed in the following sections.

7.2.4.1 The Influence of Soybean Varieties

Different soybean varieties have distinct volatile compositions, and therefore the SPIs prepared from these varieties also vary in off-flavours. The differences in flavour profile among naturally bred soybeans are related to production places, sunlight, rainfall, soil quality, and temperature.[79] Therefore, even with similar processing methods, the flavour of soybean products will vary depending on the soybean varieties used. Kakumyan *et al.* compared the flavour profiles of 626 soybean varieties collected worldwide, and the selected four soybean varieties showed unique compositions of volatile compounds.[80] Through biochemical analysis, it was found that the uniqueness of the three of them was not related to the LOX enzyme, so there may be other factors affecting the flavour of soybean protein among soybean varieties.

In addition, with the development of genetic engineering, natural or artificial mutants for single, double, and triple LOX have been identified, and a series of new soybean varieties lacking LOX have been developed using these mutants.[81] These new varieties have a lower off-flavour content due to the removal of LOX. Hua *et al.* compared the flavour of protein yoghurt prepared with normal soy protein, 7S-lack soy protein, 11S-lack protein, and LOX-lack soy protein.[22] The LOX-lack soy yoghurt showed the lowest content of ethanol, 1-octen-3-ol, 2-octenal, and nonanal compared with other soy yoghurts. This indicates that the elimination of LOX hindered the oxygenation reaction of PUFAs and prevented the formation of off-flavours in soy protein.

7.2.4.2 The Influence of Protein Structure

The odour characteristic of SPI is also related to the structure of the protein. It has been reported that the unique subunit and secondary structure of soy protein can affect its solubility and beany flavour.[18] Oxidation-induced exposure to hydrophobic amino acids increased the binding affinity of SPI to aldehydes and alcohols. This is because soybean protein and volatile compounds are combined mainly through hydrophobic interaction in the solution.[82] However, as the degree of oxidation increases, the protein solubility decreases, and intermolecular aggregation occurs, resulting in a decrease in the flavour-binding sites.

The change in the protein conformation would significantly affect its ability to bind with flavour compounds, thus affecting the overall flavour formation. Factors such as pH, temperature, salt, high-pressure treatment, and oxidation can cause changes in protein conformation during processing, as discussed in Sections 7.2.4.3–7.2.4.6. Such changes can improve the functional properties of proteins to some extent and are inevitable in industrial production. The changes in protein conformation also alter their interaction with flavour substances, thus affecting the flavour quality of the products.[83] For example, chemical modification of a protein may cause protein unfolding and polymerization. Studies have shown that malondialdehyde (MDA) modification can change the binding of aroma compounds to SPI.[18] The results showed that the binding affinity between flavour compounds and SPI increased with an increase in MDA concentration in the range of 0–1 mM, which may be due to the exposure of hydrophobic amino acids hidden inside the protein. However, the binding affinity between flavour compounds and SPI decreased with the increase in MDA concentration at and above 0.25 mM, which was explained by the high degree of SPI denaturation and SPI aggregation, making the flavour-binding site inaccessible. The polarity and electrostatic charge of the protein may also change and result in different flavour-binding abilities. The degree of protein modification also affects the flavour adsorption of the protein. In general, as the degree of modification increases, the adsorption capacity weakens.[84]

7.2.4.3 The Influence of pH

The pH of the solution can influence the structure and degree of oxidative polymerization of volatile compounds. Meanwhile, pH can affect the charge distribution on the protein surface, protein conformation, and solubility, which can further affect the strength of the interaction between the volatile compounds and the protein. Guo *et al.* studied the binding of flavour compounds hexyl acetate (HxAc), heptyl acetate (HpAc), linalyl formate (LiFo), linalyl acetate (LiAc), geraniol, linalool, limonene, and myrcene to SPI in pH 3.0, 6.0, and 9.0 aqueous solutions. Similar to the interactions between BLG and flavour compounds, the binding capacity of HxAc, HpAc, LiFo, LiAc, geraniol, linalool, and SPI increased with the increase in pH. The increase in headspace concentration and 'salting-out' effect were also observed in the interactions between limonene, myrcene, and SPI.[85]

7.2.4.4 The Influence of Temperature

Temperature is an important factor that affects the binding between proteins and flavour compounds. The change in temperature may cause the protein to unfold or aggregate.[86] A change in temperature may also cause a change in the protein-binding sites and binding constants, which will affect the flavour adsorption capacity of the proteins.[87] The temperature change may also impact the rheological properties of food and, thus, affect the release of flavour substances. Heat treatment in industrial production usually does not destroy the primary structure of proteins; rather, it causes changes in the secondary, tertiary, and quaternary structures of proteins.[88] The protein structure unfolds when heated, exposing the hydrophobic groups embedded in the interior and enhancing the flavour adsorption capacity of the protein. Studies have shown that the protein surface hydrophobicity increased with temperature from 20 °C to 70 °C, while it decreased when the temperature reached 70 °C, which will result in lower binding sites between flavour compounds and proteins.[88] This phenomenon is related to the denature and aggregation of proteins at high temperatures. Chen *et al.* studied the effect of preheated treatment on the binding of flavour compounds with SPI in an aqueous model. The results showed that the binding ability of HxAc, HpAc, and preheated SPI decreased with the increase in temperature (80, 90, and 100 °C). However, the binding ability of LiFo, LiAc geraniol, linalool, and preheated SPI increased with increasing temperature (80, 90, and 100 °C).[89]

7.2.4.5 The Influence of Salt

The effect of salt on the binding between proteins and flavour compounds is achieved by changing the ionic strength of the matrix. High ionic strength may cause protein denaturation, aggregation, and precipitation, which then change the binding capacity of the proteins. Damodaran and Kinsella studied the effect of anions on the binding of 2-nonanone to bovine serum albumin and found that the binding capacity of 2-nonanone to bovine serum albumin increased with increasing salt concentrations (NaCl, KCl, $CaCl_2$, and $MgCl_2$).[90] Anions can stabilize or destabilize protein structures due to their ability to bind to macromolecular dipole groups and water molecules, which, in turn, affects the hydrophobic interactions within the protein. However, there are few studies about the interaction between flavour compounds and soybean proteins affected by salt.

7.2.4.6 Other Influencing Factors

High-pressure treatment can denature proteins and affect the interactions between proteins and flavour compounds. Protein oxidation may cause changes in protein structure, thereby reducing the binding affinity for flavour compounds.[91] Temperature, pressure, and pH value during processing will affect not only the binding strength between protein and flavour but also the activity of LOX, then affecting the formation of off-flavours.[92,93]

7.3 Methods for Removal of the Off-flavours

The off-flavour of SPI limits its use in the food industry. To improve product quality and expand the application of SPI in food processing, many researchers have attempted to remove the off-flavour compounds from soy proteins. In the past few years, a variety of technologies have been developed to remove the off-flavour in proteins. The most commonly used is the physical method, including thermal, high pressure, ultrasound, salting-out, β-cyclodextrin, irradiation, *etc*. Solvent extraction, biotic elicitors, and LOX null mutants are the commonly used methods of chemical, biological, and genetic engineering.

7.3.1 The Cultivation of Soybean Varieties Without LOXs

In 1932, Andre and Hou first discovered that soybean protein products produced a beany flavour due to the enzymatic reaction of PUFAs, such as linoleic acid and linolenic acid, and LOX played a major role in this reaction. Soy seeds are especially rich in LOX, which accounts for about 1% of the total protein and is mainly found in the epidermis and cotyledons.[94] Theorell *et al.* obtained the LOX crystals from soy protein for the first time in 1974. Most of the information mainly comes from research on soybean LOX isoenzymes.[95] At least six LOX (LOX1-6) isoenzymes have been found in soybean plants, of which LOX1, LOX2, and LOX3 are found in dormant soybean seeds and are the products of specific genes.[96] The remaining isoenzymes only exist in plant tissues and are considered to be part of the plant defence mechanism.[97] LOX can specifically catalyse PUFAs containing *cis, cis*-(1*Z*,4*Z*)-pentadiene structure to generate hydroperoxides, which are subsequently cleaved by hydroperoxide lyase to produce C_6 aldehydes, which are responsible for the beany, grassy, or green odour in SPI. In 1985, Matoba *et al.* investigated the flavour between normal soybean and seeds lacking LOX1, LOX2, and LOX3 and found LOX2 was related to the formation of hexanal.[98] Furuta *et al.* also reported that the lower contents of 1,3-diethyl-2-thibarbituric acid and *n*-hexanal were identified in soybeans without LOX.[99] The absence of LOX also contributes to the flavour improvement of soy products, such as soymilk and tofu. Kobayashi reported that the soymilk made from soybeans without LOX2, 3 and LOX1, 2, 3 had lower contents of off-flavour compounds, including hexanal, 2-pentylfuran, 2-heptanal, 1-octen-3-ol, 2-hexenal, 2,4-decadienal, and 1-pentanol.[16] Wilson found that tofu made from soybeans without LOX-2 had less beany flavour than normal tofu.[100] Besides, some studies showed that different types of LOX isoenzymes had different beany flavour removal effects, and the soybeans lacking LOX1, LOX2, and LOX3 still contained enzymes that convert linolenic acid to hydroperoxides.[101] The contents of beany flavour compounds can be significantly reduced in LOX1- or LOX2-absent soybeans, while only the absence of LOX3 will increase the content of hexanal. However, cultivating soybean varieties without LOX is not without disadvantages, as LOX is related to maintaining the protease inhibitor content, which is vital for plants for protection against insects and pests.[102] In addition, LOX is involved in the synthesis of several important secondary metabolites, such as jasmonic acid (JA), which play an important role in some physiological processes.[103] Hence, the complete removal of LOX may adversely affect the plant-defence mechanism.

7.3.2 Thermal Treatment

LOX in soybeans is greatly affected by temperature, so thermal treatment can reduce the off-flavour contents by inactivating enzyme activity. The thermal treatment methods used to remove off-flavour mainly include blanching, hot grinding, ultrahigh temperature (UHT) treatment, superheated steam treatment, and far-infrared heating. The blanching process is mainly used in the production of soymilk, which mainly includes three processes: soaking the soybean seeds in hot water (>80 °C) for more than 5 min, grinding and extracting with hot water, removing okara, and cooking. In addition to inactivating LOX and eliminating the off-flavour of soybean, the blanching process also changes the gel properties and stability of soymilk.[104] Lv *et al.* reported that the LOX activity decreased with the extension of blanching time, and the contents of five beany flavour compounds (*n*-hexanal, 1-octen-3-ol, (*E*)-2-hexenal, *n*-hexanol, and (*E*,*E*)-2,4-decadienal) and three non-beany flavour compounds were significantly decreased. However, the decrease in non-beany flavour compounds' content was less than that in beany flavour compounds, which helped in reaching a balance between beany and non-beany flavour compounds.[19] Zhang *et al.* compared the flavour of soymilk prepared by three grinding methods (ambient, cold, and hot grinding) and three heating processes (stove cooking, one-phase UHT, and two-phase UHT). The results showed that hot-grinding and two-phase UHT treatments have the best effect on removing beany flavour compounds.[24] Although blanching and hot grinding can largely remove the off-flavour of soy protein, the PUFAs in the residue will continue to oxidize in the presence of water and oxygen because the LOX enzyme cannot be completely inactivated under these conditions.[105] The superheated steam can provide heat beyond the corresponding saturation temperature at a given pressure to dry the product and avoid further oxidation of the residue. Chong *et al.* treated the soymilk with superheated steam at 9.3 min and 119 °C, and the results showed that the superheated steam treatment could reduce the off-flavour content in soymilk by inactivating the LOX.[92] Far-infrared heating is a technique to reduce the content of beany flavour compounds by denaturing the LOX. Compared to traditional heat treatment methods, far-infrared heating can quickly transfer heat into soybean seeds through the strong absorption of infrared by water, so the heating time of the material is short and the degree of denaturation is small.[106]

7.3.3 High-pressure Treatment

High pressure has been applied to inactivate microorganisms and enzymes and unfold proteins without destroying most food quality parameters, such as colour, flavour, and nutrient content.[107] Cornelly *et al.* studied the effect of high-pressure processing (HPP) on the activity of LOX in whole soybeans and soymilk, and the results showed that completely inactivated LOX needed either very high pressure (800 MPa) or a combined pressure/temperature treatment (60 °C/600 MPa).[108] Although the combination of high-pressure and high-temperature processing to completely inactivate LOX can reduce the beany flavour content, this processing method may cause other undesirable chemical changes that may result in the destruction of amino acids and vitamins, browning reactions, and the formation of cooked flavour.[109]

As an emerging technology, ultra-HPP has been studied as an alternative to traditional high-temperature and HPP technology to improve food quality.[110] Ultra-high-pressure homogenization (UHPH) has been commonly used in soymilk production to improve the elaboration and conservation of soymilk. In addition, the UHPH treatment also plays an important role in the overall sensory quality of soymilk. Ferragut *et al.* have characterized the volatile profile of soymilk treated with the UHPH treatment.[105] Compared with the base and pasteurization products, UHPH-treated soymilk (300 MPa, 80 °C) had low levels of furan compounds and good sensorial acceptance.

7.3.4 Irradiation Treatment

In the traditional heating process, heat is transferred from the heating medium to the surface of the sample through thermal convection and then transferred to the interior through thermal conduction. When the sample interior reaches the set temperature, the sample surface has been subjected to high temperatures for a long time, which can lead to deterioration of food quality.[111] Radio frequency (RF) heating is a new type of heating technology that can simultaneously heat the inside and outside of food. The frequency range of electromagnetic energy for RF heating is 1 MHz–300 MHz, which can generate heat rapidly inside the food by the displacement of molecular friction and space charge displacement.[112] In 2018, Jiang *et al.* proposed to process soybean RF heating to rapidly inactivate LOX and eliminate beany flavour compounds while maintaining the quality and functional properties of soybean products. The results showed that the inactivation rate of LOX was 94.30% after RF treatment for 210 s while the inactivation rate was 3.08% after conventional treatment (110 °C) for the same duration. The contents of hexanal decreased from 521 to 116 $\mu g L^{-1}$, and that of 1-hexanol decreased from 271 to 6.99 $\mu g\ L^{-1}$ after RF-heating treatment.[113]

Compared with thermal treatment, irradiation, as a non-heat treatment method, has certain advantages in maintaining the colour, flavour, nutrients, and texture of food.[114] The wavelength of gamma rays is very short, similar to ultraviolet rays and microwaves, and does not induce neutrons, so there will be no 'meltdown' and chain reactions, and the irradiated food will not be radioactive.[115] Broetto *et al.* irradiated soybeans with 2.5, 5, and 10 kGy gamma irradiation to evaluate the effect of gamma irradiation on the chemical constitution and the activity of LOX in different soybean cultivars. The results showed that the gamma irradiation induced a small increase in protein and lipid content, and the activity of LOX decreased in all tested samples with the increase in irradiation dose.[116]

7.3.5 Biotic Elicitor Treatments

The inherent antioxidant potential of soybean seeds has been demonstrated to reduce the beany flavour, and the effect is mediated by secondary metabolites such as anthocyanins, tocopherols, and phenolics.[117] Therefore, it is a new strategy to reduce the beany flavour of soybeans by eliciting the production of desirable secondary metabolites.[103] Biotic elicitors are chemical molecules that can promote the biosynthesis of specific compounds and induce a series of physiological changes.[118] Kumari *et al.* used the phytohormones methyl jasmonate (MJ) and salicylic acid (SA) as biotic elicitors on the different growth stages of soybeans (bloom, pod development, and seed development). The results showed that 0.01 mM SA could reduce the LOX and hydroperoxide lyase activity by 2.3- and 2.4-fold in high off-flavour soybean cultivars, and SA is the more effective biotic elicitor in reducing beany flavour than MJ.[103] In another study, chitosan (CHT) and JA biotics as elicitors were applied to the germination process of soybean. The results showed that CHT and JA treatment can significantly reduce the amount of thiobarbituric acid, which is an indicator of lipid peroxidation responsible for off-flavour generation.[64]

7.3.6 Ultrasound Treatment

As a green, physical, and non-thermal technology, ultrasound has been widely used in the fields of processing and agriculture to improve the quality and safety of foods.[119] Ultrasound propagates in the food matrix and induces the compression and decompression of particles, with the effects of dynamic agitation, cavitation, shear stresses, and turbulence, thereby changing the physical and chemical properties of the matrix.[120,121] The chemical and mechanical effects induced by ultrasound can change the hydrophobicity of proteins, which will affect their flavour-binding ability as flavour compounds interact with proteins mainly through hydrophobic interactions.[119] Arora *et al.* studied the effect of ultrasound on the removal of off-flavour in SPI by β-cyclodextrin embedding. The results showed that ultrasonic treatment at 50 °C for 5 min, followed by treatment with phospholipase A2 and β-cyclodextrin, can remove 99% of PLs in SPI, which is related to the beany flavour. This is because ultrasound can weaken the interaction between soy protein and phospholipids, which is beneficial for the formation of PL–cyclodextrin inclusion complexes.[122]

7.3.7 Solvent Extraction Treatment

There are two main reasons organic solvents can remove the off-flavour of soybean protein. On the one hand, organic solvents can dissolve lipids, polyphenols, and other non-volatile off-flavour precursors; on the other hand, organic solvents can extract volatile off-flavour compounds. However, different solvents have different effects on the removal of off-flavour in soybean protein. Ethanol, methanol, aqueous acetone, isopropanol, and hexane/alcohol azeotropes have proved to be effective in removing PLs and improving the flavour of SPI.[123] Compared to hexane, the methanol/chloroform mixtures have a higher extraction rate of lipids and thus can sufficiently remove off-flavours in SPI.[124] But methanol/chloroform are not food-grade and cannot be used in the food industry. Besides, proteins are severely denatured, and many functions are destroyed after being exposed to organic solvents.[124] Wang *et al.* studied the effect of ethanol or isopropanol washing on the removal of beany, earthy, and astringent flavours from pea protein. The results showed that higher concentrations of ethanol and isopropanol (50 and 80%) were more effective in removing flavour compounds (the total peak area decreased by 82 to 94%) with some modifications to the chemical composition and functionalities of pea protein.[125] Previous studies have compared the effects of azeotropic mixtures such as hexane–ethanol, hexane–methanol, and hexane–isopropanol on removing the off-flavour of soybean protein and on the function of proteins, and the results showed that the effect of hexane–ethanol azeotropic-mixed solvent is the most ideal.[106]

7.3.8 β-Cyclodextrin Treatment

To improve the sensory quality of SPI, it is necessary to remove the off-flavour precursors, such as free fatty acids and PLs. An ideal method is to remove the precursors and compounds related to off-flavour while causing minimal damage to the functional properties of SPI.[48] It is well known that cyclodextrin, especially β-cyclodextrin, can form molecular inclusion complexes with lipophilic molecules by overcoming the hydrophobic interactions of the protein–PL complex or transferring PLs from the air–water interface to the water face.[126] According to the United States Food and Drug Administration, 2% of β-cyclodextrin can be added to food products, and β-cyclodextrin-mediated removal or reduction of off-flavours has shown considerable promise in food applications.[127] However, the binding affinity of long-chain fatty acids and PLs to oleosins is much higher than that of short-chain carbonyl compounds. There are two C18 acyl chains in PLs with a free energy charge of −13.6 to −16.3 kcalmol^{-1} at 25 °C, so higher energy is required to separate the protein–PL complexes to achieve the combination of β-cyclodextrin and PLs. Besides, a polar head and two acyl chains are attached to the glycerol backbone in PLs, and this steric structure will hinder the binding of β-cyclodextrin and PLs.[48] Therefore, to achieve the combination of β-cyclodextrin and PLs, it is necessary to destroy the interaction among the PLs–protein complex. Arora *et al.* studied the removal effect of PLs and off-flavours in SPI by the treatment of sonication, β-cyclodextrin, and phospholipase A2. The results showed that 99% of the PLs and free fatty acids in soybean protein could be removed by this method, which contributes to improving the flavour of SPI.[122] In this treatment, ultrasound and phospholipase A2 disrupt the interaction between PLs and soy protein, which facilitates the formation of inclusion complexes between β-cyclodextrin and PLs.

7.3.9 Salting-out Treatment

As mentioned in Section 7.2.3.5, large amounts of PLs and free fatty acids bind to oleosins, which are closely related to the off-flavour in SPI. In this regard, an ideal method was proposed for removing the off-flavour in SPI by selectively precipitating these oleosins. Samoto found that the oleosins precipitated in 30–40% saturated ammonium sulphate solutions, while the major storage proteins, such as 7S and 11S, remained unaffected.[128] And there were approximately 30% oleosins in the precipitate of 30–40% ammonium sulphate solution, which was much higher than the expected oleosin content (7.2%) of SPI.[129] This may be due to the salting-out of some denatured components of 7S and 11S storage proteins bound to oleosins. However, it is impractical to remove oleosins by salting-out ammonium sulphate in large-scale production. And there were 0.9% lipid residues in SPI after salting out by ammonium sulphate, which is still above the threshold of aldehydes and ketones odour production.[128] In this regard, Samoto *et al.*[130] proposed a two-step pH-solubility method to precipitate the oleosins in SPI. First, the preliminary extraction of defatted soybean flour at pH 8.0 and the 11S storage protein were precipitated from the extracted protein by adjusting pH to 5.8 with 1 mM sodium sulphate. The precipitated 11S protein was removed by centrifugation, and the pH of the supernatant was adjusted to 5.0 at 55 °C for 15 min. The oleosins and 7S protein were precipitated after adjusting the pH to 5.5 and 4.5, respectively. Although the two-step pH-solubility method is an improvement on ammonium sulphate salting-out, the process results in a lower yield of oleosin-free SPI, and there was no report about the content of residual lipid and off-flavour in the treated SPI.

7.4 Conclusion

As one of the most popular plant proteins, the off-flavours of soybean, both taste and odour, restrict its application in food industry production. It is imperative to get a method for eliminating the off-flavour while retaining the functionality of soybean protein. At present, there are a variety of methods for reducing the off-flavour of soybean protein, which can be divided into two categories. The first is to inhibit the LOX activity, and the second is to eliminate or cover up the off-flavour caused by the oxidation of PUFAs.[131] However, traditional off-flavour removal methods, such as organic solvent reagent extraction and heat treatment, cause soybean protein denaturation while removing the off-flavour. In recent years, some gentler methods for off-flavour removal have been proposed. Compared to direct heating treatment, microwave heating and RF heating can effectively reduce the off-flavour in soybean protein and avoid the functional destruction of soy protein.

On the other hand, the binding affinity between off-flavour and soybean protein is an important factor affecting consumers' acceptance of soy protein products. Therefore, the structural modification of soybean protein may also provide some new perspectives for the removal of off-flavour. For example, treatments such as adding organic solvent reagents, extrusion, shear, and enzymatic treatment will change the spatial structure of soy protein, resulting in a change in the binding affinity of flavour compounds.

To reduce the off-flavour of soy protein, multidisciplinary approaches involving agronomy, genetic engineering, process treatments, and off-flavour perception and/or suppression are needed to solve the off-flavour issue.[48] And whether these off-flavour removal strategies can be applied in commercial production needs to be studied further. The ease of integration within commercial operations and affordability will be the ultimate measures of the success of those treatments.[48]

Abbreviations

CHT	Chitosan
FF	flavour factor
GC-O-MS	Gas chromatography-olfactometry-mass spectrometry
HpAc	Heptyl acetate
HPP	High-pressure processing
HS-GC-IMS	Headspace-gas chromatography-ion mobility spectrometry
HxAc	Hexyl acetate
JA	Jasmonic acid
LiAc	Linalyl acetate
LiFo	Linalyl formate
LOXs	Lipoxygenases
MDA	Malondialdehyde
MJ	Methyl jasmonate
PLs	Phospholipids
PUFAs	Polyunsaturated fatty acids
RF	Radio frequency
SA	Salicylic acid
SPH	Soybean protein hydrolysates
SPI	Soy protein isolates
UHPH	Ultra-high-pressure homogenization
UHT	Ultrahigh temperature
WHO	World Health Organization

References

1. M. Lokuruka, *Afr. J. Food, Agric., Nutr. Dev.*, 2010, **10**, 2439–2459.
2. J. E. Tang, D. R. Moore, G. W. Kujbida, M. A. Tarnopolsky and S. M. Phillips, *J. Appl. Physiol.*, 2009, **107**, 987–992.

3. Y. Yang, T. A. Churchward-Venne, N. A. Burd, L. Breen, M. A. Tarnopolsky and S. M. Phillips, *Nutr. Metab.*, 2012, **9**, 57.
4. M. Ogara, M. D. D. Layo, A. M. Pilosof and R. A. Macchi, *J. Am. Oil Chem. Soc.*, 1992, **69**, 184–187.
5. J. Li, X. Yang, M. S. Swallah, H. Fu, L. Ji, X. Meng, H. Yu and B. Lyu, *Int. J. Food Sci. Technol.*, 2022, **57**, 188–200.
6. D. Chéreau, P. Videcoq, C. Ruffieux, L. Pichon, J.-C. Motte, S. Belaid, J. Ventureira and M. Lopez, *OCL*, 2016, **23**, D406.
7. M. A. Asgar, A. Fazilah, N. Huda, R. Bhat and A. A. Karim, *Compr. Rev. Food. Sci. F*, 2010, **9**, 513–529.
8. X. J. Li and J. Li, *Cereal Food World*, 2020, **65**, 0040.
9. K. Inouye, M. Shiihara, T. Uno and T. Takita, *J. Agric. Food. Chem.*, 2002, **50**, 1652–1658.
10. C. Hall, C. Hillen and J. Garden Robinson, *Cereal Chem.*, 2017, **94**, 11–31.
11. S. Kaneko, K. Kumazawa and O. Nishimura, *J. Agric. Food. Chem.*, 2011, **59**, 12204–12209.
12. Z. Ma, J. I. Boye, B. K. Simpson, S. O. Prasher, D. Monpetit and L. Malcolmson, *Food Res. Int.*, 2011, **44**, 2534–2544.
13. L. Bott and E. Chambers IV, *J. Sens. Stud.*, 2006, **21**, 308–321.
14. D. Zhu and S. Damodaran, *Food Chem.*, 2018, **264**, 319–325.
15. C. M. Zhang, Y. F. Hua, X. F. Li, X. Z. Kong and Y. M. Chen, *Food Chem.*, 2020, **333**, 127469.
16. K. b. y. Akio, Y. Tsuda, N. Hirata, K. Kubota and K. Kitamura, *J. Agric. Food Chem.*, 1995, **43**, 2449–2452.
17. X. Zhao, L. Sun, X. Zhang, H. Liu and Y. Zhu, *Colloids Surf., B*, 2020, **196**, 111345.
18. J. Wang, M. M. Zhao, C. Y. Qiu and W. Z. Sun, *Food Res. Int.*, 2018, **105**, 150–158.
19. Y. C. Lv, H. L. Song, X. Li, L. A. Wu and S. T. Guo, *J. Food. Sci.*, 2011, **76**, S20–S25.
20. X. F. Li, X. Liu, Y. F. Hua, Y. M. Chen, X. Z. Kong and C. M. Zhang, *RSC Adv.*, 2019, **9**, 2906–2918.
21. K. T. Kaczmarska, M. V. Chandra-Hioe, D. Frank and J. Arcot, *LWT-Food Sci. Technol.*, 2018, **87**, 225–233.
22. Y. P. Zhou, X. F. Li, Y. F. Hua, X. Z. Kong, C. M. Zhang, Y. M. Chen and S. D. Wang, *LWT-Food Sci. Technol.*, 2019, **115**, 108431.
23. A. Achouri, J. I. Boye and Y. Zamani, *Food Chem.*, 2006, **99**, 759–766.
24. Y. Zhang, S. T. Guo, Z. S. Liu and S. K. C. Chang, *J. Agric. Food Chem.*, 2012, **60**, 7457–7462.

25. A. K. Nedele, S. Gross, M. Rigling and Y. Y. Zhang, *Food Chem.*, 2021, **334**, 127591.
26. P. R. Lozano, M. Drake, D. Benitez and K. R. Cadwallader, *J. Agric. Food Chem.*, 2007, **55**, 3018–3026.
27. S. Yuan and S. K. C. Chang, *J. Agric. Food Chem.*, 2007, **55**, 426–431.
28. M. Solina, P. Baumgartner, R. L. Johnson and F. B. Whitfield, *Food Chem.*, 2005, **90**, 861–873.
29. J. S. Cai, Y. Y. Zhu, R. H. Ma, K. Thakur, J. G. Zhang and Z. J. Wei, *Food Chem.*, 2021, **340**, 127880.
30. Q. Lei and W. L. Boatright, *J. Food Sci.*, 2001, **66**, 1306–1310.
31. W. L. Boatright and Q. Lei, *J. Food Sci.*, 1999, **64**, 667–670.
32. W. L. Boatright, *J. Am. Oil Chem. Soc.*, 2002, **79**, 317–323.
33. S. Min, Y. Yu, S. Yoo and S. S. Martin, *J. Food Sci.*, 2005, **70**, C1–C7.
34. N. N. Wu, L. J. Wang, X. Q. Yang, S. W. Yin, Z. Teng and E. L. Zheng, *J. Am. Oil Chem. Soc.*, 2011, **88**, 1621–1631.
35. E. C. Ang and W. L. Boatright, *J. Food Sci.*, 2010, **68**, 388–393.
36. W. L. Boatright and Q. Lei, *J. Food. Sci.*, 2000, **65**, 819–821.
37. Q. Lei and W. L. Boatright, *J. Food. Sci.*, 2010, **68**, 1568–1572.
38. W. B. Navicha, Y. F. Hua, K. Masamba, X. Z. Kong and C. M. Zhang, *J. Food. Meas. Charact.*, 2018, **12**, 1695–1706.
39. E. J. Lee, H. Kim, J. Y. Lee, K. Ramachandraiah and G. P. Hong, *Foods*, 2020, **9**, 818.
40. P. Phuhongsung, M. Zhang and B. Bhandari, *Food Res. Int.*, 2020, **137**.
41. S. H. Yoo and Y. H. Chang, *Prev. Nutr. Food Sci.*, 2016, **21**, 338–347.
42. E. J. Lee, H. Kim, J. Y. Lee, K. Ramachandraiah and G. P. Hong, *Foods*, 2020, **9**, 818.
43. S. H. Yuan and S. K. C. Chang, *J. Food Sci.*, 2010, **72**, S481–S486.
44. H. H. Baek, C. J. Kim, B. H. Ahn, H. S. Nam and K. R. Cadwallader, *J. Agric. Food Chem.*, 2001, **49**, 790–793.
45. M. D. Aaslyng, J. S. Elmore and D. S. Mottram, *J. Agric. Food Chem.*, 1998, **46**, 5225–5231.
46. Y. Yang, B. Wang, Y. Fu, Y. G. Shi, F. L. Chen, H. N. Guan, L. L. Liu, C. Y. Zhang, P. Y. Zhu, Y. Liu and N. Zhang, *Food Chem.*, 2021, **346**, 128880.
47. G. MacLeod, J. Ames and N. L. Betz, *Crit. Rev. Food Sci. Nutr.*, 1988, **27**, 219–400.
48. S. Damodaran and A. Arora, *Annu. Rev. Food Sci. Technol.*, 2013, **4**, 327–346.
49. E. Frankel, *Prog. Lipid Res.*, 1984, **23**, 197–221.
50. E. N. Frankel, W. E. Neff and E. Selke, *Lipids*, 1981, **16**, 279–285.
51. J. Lee, S. Min, E. Lee and D. Min, *J. Food Sci.*, 2003, **68**, 1933–1937.

52. Z.-q. Jiang, M. Pulkkinen, Y.-j. Wang, A.-M. Lampi, F. L. Stoddard, H. Salovaara, V. Piironen and T. Sontag-Strohm, *LWT-Food Sci.Technol.*, 2016, **68**, 295–305.
53. H. Jelen, in *Food flavors: Chemical, sensory and technological properties*, CRC Press, Boca Raton, FL, 2011.
54. M. Xu, Z. Jin, Y. Lan, J. Rao and B. Chen, *Food Chem.*, 2019, **280**, 83–95.
55. D. Kaneko, T. Igarashi and K. Aoyama, *J. Agric. Food Chem.*, 2014, **62**, 1658–1663.
56. A. Adams, V. Polizzi, M. van Boekel and N. De Kimpe, *J. Agric. Food Chem.*, 2008, **56**, 2147–2153.
57. B. P. Singh and S. Vij, *LWT-Food Sci. Technol.*, 2018, **91**, 303–307.
58. R. Liu, L. G. Wang, Y. Liu, T. Wu and M. Zhang, *Food Hydrocolloids*, 2018, **81**, 39–47.
59. X. H. Tong, Z. T. Lian, L. M. Miao, B. K. Qi, S. Zhang, Y. Li, H. Wang and L. Z. Jiang, *LWT-Food Sci. Technol.*, 2020, **134**, 110151.
60. J. P. Wu and R. E. Aluko, *J. Pept. Sci.*, 2007, **13**, 63–69.
61. M. R. Kim, S. Y. Choi and C. H. Lee, *J. Microbiol. Biotechn.*, 1999, **9**, 509–513.
62. X. D. Sun, *Int. J. Food Sci. Technol.*, 2011, **46**, 2447–2459.
63. W. X. Liu, H. L. Liu and L. Q. Qu, *Theor. Appl. Genet.*, 2013, **126**, 2289–2297.
64. S. Kumari, L. J. Memba, A. Dahuja, T. Vinutha, S. Saha and A. Sachdev, *Food Chem.*, 2016, **205**, 264–271.
65. M. Ohtsuru, M. Kito and Y. Takeuchi, *Journal of the Agricultural Chemical Society of Japan*, 1976, **40**, 2261–2266.
66. F. Sosulski, *J. Am. Oil Chem. Soc.*, 1979, **56**, 711–715.
67. P. A. Ireland, S. Z. Dziedzic and M. W. Kearsley, *J. Sci. Food Agric.*, 1986, **37**, 694–698.
68. S. Arai, M. Noguchi, M. Yamashita, H. Kato and M. Fujimaki, *Agric. Biol. Chem.*, 1970, **34**, 1569–1573.
69. J. J. Rackis, D. J. Honig, D. J. Sessa and F. R. Steggerda, *J. Agric. Food Chem.*, 1970, **18**, 977–982.
70. J. A. Maga and K. Lorenz, *J. Sci. Food Agric.*, 1974, **25**, 797–802.
71. D. A. Rickert, L. A. Johnson and P. A. Murphy, *J. Agric. Food Chem.*, 2004, **52**, 1726–1734.
72. J. M. Augustin, V. Kuzina, S. B. Andersen and S. Bak, *Phytochemistry*, 2011, **72**, 435–457.
73. Z. Kerem, H. German-Shashoua and O. Yarden, *J. Sci. Food. Agric.*, 2005, **85**, 406–412.
74. H. Barakat, V. Reim and S. Rohn, *Food. Res. Int.*, 2015, **76**, 142–149.
75. K. Wang and S. D. Arntfield, *Flavour Fragr. J.*, 2017, **32**, 92–101.

76. Y. Lan, M. Xu, J.-B. Ohm, B. Chen and J. Rao, *Food Chem.*, 2020, **278**, 665–673.
77. T. G. Aspelund and L. A. Wilson, *J. Agric. Food Chem.*, 1983, **31**, 539–545.
78. Y. Zhu, Y. Wang, J. Li, F. Li, C. Teng and X. Li, *J. Agric Food. Chem.*, 2017, **65**, 4728–4735.
79. R. S. J. Keast and J. J. Lau, *J. Food Sci.*, 2006, **71**, S567–S572.
80. P. Kakumyan, M. Kato, M. Hajika and K. Matsui, *Biosci. Biotechnol. Biochem.*, 2009, **73**, 1844–1848.
81. J. Wang, H. Kuang, Z. Zhang, Y. Yang, L. Yan, M. Zhang, S. Song and Y. Guan, *Crop J.*, 2020, **8**, 432–439.
82. Q. X. Zhou and K. R. Cadwallader, *J. Agric. Food Chem.*, 2006, **54**, 1838–1843.
83. S. Damodaran and J. E. Kinsella, *J. Agric. Food Chem.*, 1981, **29**, 1253–1257.
84. I. Suppavorasatit and K. R. Cadwallader, *J. Agric. Food Chem.*, 2012, **60**, 7817–7823.
85. J. Guo, Z. Y. He, S. Y. Wu, M. M. Zeng and J. Chen, *J. Food Biochem.*, 2019, **43**, e12817.
86. Y. J. He, C. Y. Zhou, C. B. Li and G. H. Zhou, *Food Chem.*, 2021, **346**, 128976.
87. J. Kuehn, T. Considine and H. Singh, *J. Agric. Food Chem.*, 2008, **56**, 10218–10224.
88. K. Wang and S. D. Arntfield, *Food Hydrocolloids*, 2015, **43**, 410–417.
89. J. Guo, Z. Y. Hen, S. F. Wu, M. M. Zeng and J. Chen, *Food Chem.*, 2019, **290**, 16–23.
90. S. Damodaran and J. E. Kinsella, *J. Biol. Chem.*, 1980, **255**, 8503–8508.
91. F. Zhou, M. Zhao, G. Su and W. Sun, *J. Agric. Food Chem.*, 2014, **62**, 9544–9552.
92. W.-K. Chong, S.-Y. Mah, A. M. Easa and T.-C. Tan, *J. Food Sci. Technol.*, 2019, **56**, 4371–4379.
93. R. Wang, X. Zhou and Z. X. Chen, *Food Chem.*, 2008, **106**, 603–611.
94. A. V. Torres-Penaranda, C. Reitmeier, L. A. Wilson, W. R. Fe Hr and J. M. Narvel, *J. Food Sci.*, 1998, **63**, 1084–1087.
95. K. Thiele, *FEBS Letters*, 1999, **449**, 7–11.
96. D. R. Iassonova, L. A. Johnson, E. G. Hammond and S. E. Beattie, *J. Am. Oil Chem. Soc.*, 2009, **86**, 59–64.
97. S. A. Boué, B. Y. Shih, C. H. Carter-Wientjes and T. E. Cleveland, *J. Agric. Food Chem.*, 2005, **53**, 4778–4783.
98. T. Matoba, H. Hidaka, H. Narita, K. Kitamura, N. Kaizuma and M. Kito, *J. Agric. Food Chem.*, 1985, **33**, 852–855.

99. S. Furuta, Y. Nishiba, M. Hajika, K. Igita and I. Suda, *J. Agric. Food Chem.*, 1996, **44**, 236–239.
100. L. Wilson, in *Comparison of Lipoxygenase-Null and Lipoxygenase-Containing Soybeans for Foods*, Lipoxygenase and Lipoxygenase Pathway Enzymes, 1996.
101. J. M. King, I. K. Svendsen, W. R. Fehr, J. M. Narvel and P. J. White, *J. Am. Oil Chem. Soc.*, 1998, **75**, 1121–1126.
102. W. Carvalho, M. Oliveira, E. Barros and M. A. Moreira, *Plant Physiol. Biochem.*, 1999, **37**, 497–501.
103. S. Kumari, A. Dahuja, T. Vinutha, S. K. Lal, A. Kar and R. D. Rai, *J. Agric. Food Chem.*, 2015, **63**, 700–706.
104. X. Y. Peng and S. T. Guo, *Food Hydrocolloids*, 2015, **43**, 58–65.
105. F. H. Poliseli-Scopel, J. J. Gallardo-Chacón, B. Juan, B. Guamis and V. Ferragut, *Food Chem.*, 2013, **141**, 2541–2548.
106. H. J. Li, H. M. Zhou, K. X. Zhu and W. Peng, *Food Research and Development*, 2012, **33**, 215–220.
107. W. Kowalczyk, C. Hartmann, C. Luscher, M. Pohl, A. Delgado and D. Knorr, *Innov. Food Sci. Emerg.*, 2005, **6**, 318–326.
108. C. van der Ven, A. M. Matser and R. W. van den Berg, *J. Agric. Food Chem.*, 2005, **53**, 1087–1092.
109. K. C. Kwok and K. Niranjan, *Int. J. Food Sci. Technol.*, 2010, **30**, 263–295.
110. F. H. Poliseli-Scopel, M. Hernández-Herrero, B. Guamis and V. Ferragut, *LWT-Food Sci. Technol.*, 2012, **46**, 42–48.
111. N. P. Brunton, J. G. Lyng, W. Q. Li, D. A. Cronin, D. Morgan and B. McKenna, *Food Res. Int.*, 2005, **38**, 337–344.
112. L. Manzocco, M. Anese and M. C. Nicoli, *Food Res. Int.*, 2008, **41**, 1044–1049.
113. Y. M. Jiang, S. K. Wang, F. He, Q. L. Fan, Y. L. Ma, W. X. Yan, Y. L. Tang, R. J. Yang and W. Zhao, *Int. J. Food Sci. Technol.*, 2018, **53**, 2738–2747.
114. K. Zhong, X. S. Hu, G. H. Zhao, F. Chen and X. J. Liao, *Food Chem.*, 2005, **92**, 473–479.
115. O. B. Wood and C. M. Bruhn, *J. Am. Diet Assoc.*, 2000, **100**, 246–253.
116. E. A. de Barros, F. Broetto, D. F. Bressan, M. M. P. Sartori and V. E. Costa, *Radiat. Phys. Chem.*, 2014, **98**, 29–32.
117. A. Dahuja and T. R. Madaan, *Plant Foods Hum. Nutr.*, 2003, **58**, 1–8.
118. R. Radman, T. Saez, C. Bucke and T. Keshavarz, *Biotechnol. Appl. Biochem.*, 2003, **37**, 91–102.
119. D. Kang, W. Zhang, J. M. Lorenzo and X. Chen, *Crit. Rev. Food Sci. Nutr.*, 2020, **61**, 1914–1933.
120. O. A. Higuera-Barraza, C. L. Del Toro-Sanchez, S. Ruiz-Cruz and E. Márquez-Ríos, *Ultrason. Sonochem.*, 2016, **31**, 558–562.

121. X. Chen, Y. Zou, D. Y. Wang, G. Y. Xiong and W. M. Xu, *Food Chem.*, 2020, **331**, 127369.
122. A. Arora and S. Damodaran, *Food Chem.*, 2011, **127**, 1007–1013.
123. L. A. Johnson and E. W. Lusas, *J. Am. Oil Chem. Soc.*, 1983, **60**, 229–242.
124. W. Roland, L. Pouvreau, J. Curran, F. Velde and P. Kok, *Cereal Chem.*, 2017, **94**, 58–65.
125. Y. Wang, B. Guldiken, M. Tulbek, J. D. House and M. Nickerson, *Food Res. Int.*, 2020, **132**, 109085.
126. A. Arora and S. Damodaran, *J. Am. Oil Chem. Soc.*, 2011, **88**, 213–222.
127. N. L. Schnell, *Enzyme Microb. Tech.*, 2021, **145**, 10974.
128. M. Samoto, C. Miyazaki, J. Kanamori, T. Akasaka and Y. Kawamura, *Journal of the Agricultural Chemical Society of Japan*, 1998, **62**, 935–940.
129. J. T. C. Tzen, Y. Cao, P. Laurent, C. Ratnayake and A. H. C. Huang, *Plant Physiol.*, 1993, **101**, 267–276.
130. M. Samoto, M. Maebuchi, C. Miyazaki, H. Kugitani, M. Kohno, M. Hirotsuka and M. Kito, *Food Chem.*, 2007, **102**, 317–322.
131. B. Wang, Q. Zhang, N. Zhang, K. H. Bak, O. P. Soladoye, R. E. Aluko, Y. Fu and Y. H. Zhang, *Trends Food Sci. Technol.*, 2021, **112**, 336–347.

CHAPTER 8

Pulse Protein Flavour

MINWEI XU* AND BINGCAN CHEN

Department of Plant Sciences, North Dakota State University, Fargo, North Dakota, USA
*E-mail: Minwei.xu@ndsu.edu

8.1 Introduction

Pulses are the edible seeds of plants in the legume family (Fabaceae). Dry common beans (*Phaseolus vulgaris*), dry broad beans (*Vicia faba*), dry peas (*Pisum sativum*), chickpeas (*Cicer arietinum*), cowpeas (*Vigna unguiculata*), pigeon peas (*Cajanus cajan*), lentils (*Lens culinaris*), Bambara beans (*Vigna subterranean*), lupins (*Lupinus* spp.), and several minor dry beans are characterized as pulses by the United Nations' Food and Agriculture Organization (FAO). To differentiate pulses from other vegetable crops that are harvested still green, the term 'pulses' is limited to the crops harvested as dry grains.[1] Pulses are prepared for consumption in many ways, some examples being whole seeds and split pulses.[2] Pulses can be traced back to 11 000 years ago in the Fertile Crescent, which was home to some of the earliest civilizations.[1] Nowadays, pulses are grown in 173 different countries in areas including Asia, the Middle East, Europe, North America, South America, Australia, and Africa. As of 2020, the world's largest producers of pulses were India, Canada, Russia, Brazil, Nigeria, China, the USA, Ethiopia, Australia, and Turkey.[3] The proximate composition of pulses is characterized as high in carbohydrate and protein and low in fat. The protein content of pulses generally falls between 15% and 30%. Carbohydrate is the major component of pulses, comprising about 60–70%. The fat

Food Chemistry, Function and Analysis No. 41
Flavour and Consumer Perception of Food Proteins
Edited by Jing Zhao and Changqi Liu

Published by the Royal Society of Chemistry, www.rsc.org

content in pulses is between 0.4% and 4%, with an average value of 1.5%.[4] Pulses are good sources of dietary fibre, resistant starch, protein, and vitamins.[4] In addition to their nutrient composition, the health benefits of pulses have been well documented by Hall *et al.*[4] Pulses contain health-promoting antioxidants, such as carotenoids and phenolic compounds. Pulse products are gluten-free and have a low glycaemic index. However, the antinutrients present in pulse seeds pose a major concern. Patterson *et al.*[5] summarized the antinutritional factors in pulses, which include phytate, enzyme inhibitors (trypsin inhibitors, chymotrypsin inhibitors, and α-amylase inhibitors), polyphenolics (including tannins), lectins, and saponins.

Pulses are considered 'poor man's meat' due to their high protein content and affordability. Among 11 pulses, lupin has the highest crude protein content, ranging from 32% to 44%.[4] Pea, kidney bean (*P. vulgaris*), and pinto bean (*P. vulgaris*) have 14–31%, 17–27%, and 18–25% protein content, respectively. In general, the protein content of pulses is higher than that of wheat or corn. The high protein content, in combination with cholesterol-free, low allergenicity, and non-genetically modified organism states, has made pulses a major plant-derived contender. A more well-balanced and health-conscious diet also diverts the trajectory of plant protein-based foodstuffs and challenges animal-derived protein consumption.[6] Pulses have been incorporated into various food products, such as baby foods, plant-based milk, bean curd, meat products, baked goods, and extruded products.[7] One of the most popular plant protein-based products that have been introduced recently is meat analogues, anticipated to be a complement to animal proteins.[7–10] While soybean and wheat gluten are the most common ingredients for meat analogues, pulse proteins are hypoallergenic substitutes.

Nevertheless, pulse proteins remain underexploited or dampened in food applications. There are three main impediments: the low protein digestibility corrected amino acid score; the lack of two sulphur-contained amino acids, cysteine and methionine; and the bitter taste and beany flavour.[4,10] The beany flavour and extent of bitterness in pulse proteins have led to different preferences between consumers from Asian and Western countries. While consumers from Asian countries are familiar with the beany flavour, many consumers from Western countries do not accept foods that taste bitter and have a green/beany flavour. The overall consensus is that the bitterness and green/beany flavour of pulse proteins are not ideal for food products such as meat analogues and non-dairy milk. For this reason, the beany flavour in pulse flour and derivate products has prevented these products from being exploited to their full potential for human consumption.[11]

This chapter reviews the current literature on the flavour of pulse proteins. Section 8.2 discusses the basic protein structure that would affect the flavour of pulse protein; Section 8.3 compares the main protein extraction methods: dry extraction and wet extraction; and Section 8.4 lists the categories and sources of major flavour-related compounds in pulse protein and the interaction between flavour compounds and proteins. Section 8.5 presents the effect of processing technologies on pulse protein flavours.

8.2 Pulse Protein Profiles

8.2.1 Amino Acid Profile of Pulse Proteins

Natural proteins contain up to 21 different primary amino acids (see Table 8.1), linked together *via* amide bonds. Histidine, threonine, lysine, methionine, valine, isoleucine, leucine, and phenylalanine are essential amino acids.

Pulse proteins have a well-balanced amino acid profile with a high level of lysine.[19] Consuming an adequate amount of lysine is especially important for child development, as dietary lysine deficiencies may increase the probability of growth retardation. Pea proteins have great potential as a source of lysine, as they have a higher lysine content than soybeans.[15] In comparison to soy protein isolates, the ratio of essential to total amino acids (E : TN) was higher in pea protein isolates, which met the minimum E : TN ratio (0.36) suggested by FAO/WHO/UNU. It was also observed that most pulse protein isolates fulfil the requirements for amino acids recommended by FAO/WHO/UNU.[15]

Pulse proteins have a low content of sulphur-containing amino acids, such as methionine and cysteine. As shown in Table 8.1, most pulses have a low methionine and cysteine content, which is less than 3.2 g per 100 g and 1.1 g per 100 g, respectively. The only exception is the methionine content in pinto beans, which is about 7 g per 100 g. Compared to pulse proteins, cereal proteins have low levels of lysine, leucine, and phenylalanine but are relatively high in sulphur-containing amino acids. In this perspective, cereal protein is a complement to pulse proteins.[13]

The twenty-first new proteinogenic amino acid, which has been recognized as a natural amino acid, is selenocysteine. Selenocysteine has a structure similar to that of cysteine but with an atom of selenium (Se) taking the place of the usual sulphur and forming a selenic group. The function of selenocysteine is more related to enzymes than storage proteins. Selenocysteine is mainly found in animal products, while the concentration of selenocysteine in pulses is very low, which is 1.8 ng g^{-1} and 0.9 ng g^{-1} (dry basis) in chickpea and cowpea, respectively.[18] Selenocysteine is one of the major sources of Se in pulse foods. The bio-accessibility of Se (ranging from 12% to 29%) is higher than that of iron (ranging from 4% to 10%) from pulses but is somewhat comparable to the bio-accessibility of zinc from pulses, which ranges from 27% to 56%.[18,20] One of the possible reasons is that selenocysteine can be destroyed due to heat processing in pulses, similar to cysteine. The content of selenocysteine in various

Table 8.1 Effect of pulse germination on the amino acids profile.[a,4,12–18]

	Taste	Chickpea	Lentil	Dry pea	Cowpea	Lupin	Pigeon pea	Pinto bean	Kidney bean	Black bean	Navy bean
Alanine	Sweet	5.4	5.1	5.1	4.3	2.8	6.3	4.7	4.2	NR	4.7
Arginine	Bitter, sharp	7.9	6.4	7.1	7.3	10.9	3.2	6.6	6.1	6.7	8.2
Aspartic acid	Sour, umami	11.9	11.0	11.7	11.9	8.4	10.5	11.1	12.4	NR	13.6
Cysteine	Sulphurous	—[b]	—[b]	—[b]	1.0	0.6	NR	0	—[c]	1	1.1
Glutamic acid	Umami, meaty taste	16.0	14.7	16.7	18.5	26.1	24.7	15.3	17.8	NR	10.6
Glycine	Sweet, pleasant	5.9	5.7	6.4	4.1	3.7	4.0	4.1	4.1	NR	4.9
Histidine	Bitter, salty, sour	6.0	4.7	5.4	2.8	3.1	3.4	2.0	2.6	2.9	3.2
Isoleucine	Bitter, salty, sour	3.3	3.3	3.5	4.3	3.2	4.1	4.8	4.3	4.1	5.0
Leucine	Bitter, salty, sour	7.5	7.2	7.8	7.5	7.4	8.6	8.9	8.0	7.7	9.1
Lysine	Bitter, salty	3.5	7.5	6.6	6.6	7.6	5.9	7.8	6.7	6.7	7.2
Methionine	Bitter	0.6	0.1	0.2	1.2	0.3	3.2	7.0	2.3	2.0	1.0
Phenylalanine	Bitter, metallic, obnoxious	5.3	4.3	4.5	5.6	3.3	9.7	6.0	8.3	5.1	6.4
Proline	Sweet, minerally, salty	4.1	3.9	4.1	3.6	4.3	5.4	4.8	4.0	NR	NR
Serine	Sweet, food-like	6.1	5.9	5.8	5.6	6.0	4.8	6.3	5.8	NR	6.9
Threonine	Sweet, moderately fatty	3.2	3.4	3.4	3.7	4.3	2.6	4.6	4.0	3.1	5.1

(continued)

Table 8.1 (*continued*)

	Taste	Chickpea	Lentil	Dry pea	Cowpea	Lupin	Pigeon pea	Pinto bean	Kidney bean	Black bean	Navy bean
Tryptophan	Bitter, minerally, stale	NR	NR	NR	1.1	0.3	NR	NR	1.2	NR	NR
Tyrosine	Flat	2.1	2.3	2.5	3.2	4.3	0.4	5.3	—[d]	NR	3.9
Valine	Bitter, flat	3.9	4.4	4.4	4.9	3.5	5.7	5.8	5.1	4.9	6.1
Selenocysteine	NR	1.8[e]	NR	NR	0.9[e]	NR	NR	NR	NR	NR	NR

[a]All the units are g per 100 g protein, except selenocysteine, which is ng g^{-1} grains (dry basis). NR, not reported. Taste is adapted from Schiffman and Dackis.[12] Glutamic acid indicated both glutamic acid and glutamine; aspartic acid indicated both aspartic acid and asparanin. Chickpea, lentil, and pea adapted from Xu *et al.*;[13] cow pea, lupin, and black bean adapted from Hall *et al.*,[4] pigeon pea adapted from Oshodi *et al.*;[14] pinto bean adapted from Tan *et al.*;[15] kidney bean adapted from Wu *et al.*;[16] navy bean adapted from Jafari *et al.*[17]

[b]Cysteine and lysine are reported together as lysine.

[c]Cysteine and methionine are reported together as methionine

[d]Tyrosine and phenylalanine are reported together as phenylalanine.

[e]Data of selenocysteine is from Khanam and Platel.[18]

pulses has not been widely studied yet. Although the taste of selenocysteine is seldom mentioned in the literature, the flavour of Se–methylselenocysteine has been reported as garlic-like. The content, function, and taste of selenocysteine in pulse seeds need more study.

8.2.2 The Composition of Pulse Proteins

Pulse seed proteins are empirically classified based on their solubility and can be divided into four major classes: globulin, albumin, prolamin, and glutelin. Globulin and albumin are major storage proteins in pulse seeds. Globulins are soluble in salt solutions, while albumins are soluble in water and are regarded as metabolic and enzymatic proteins with cytosolic functions. Prolamins and glutelins are alcohol-soluble and dilute acid- or alkaline-soluble proteins, respectively.[4] Prolamins and glutelins are two groups of plant storage proteins present in small amounts in pulse seeds. The sum of prolamins and glutelins is usually less than 10%.[19] More than 80% of the pulse proteins are globulins and albumins (see Table 8.2).

Globulins are responsible for 44–80% of pulse storage proteins (see Table 8.2). Globulins can be further divided into 11S and 7S based on their sedimentation coefficients. The names of 11S and 7S proteins vary in different pulses. For example, legumin and α-conglutin are 11S proteins, while vicilin, convicilin, β-conglutin, γ-conglutin, and phaseolin are 7S proteins (see Table 8.2). Legumins (Mw ~ 300–400 kDa) are hexamers. The six monomers interact non-covalently and assemble into two trimeric intermediates. Each monomer has a molecular weight of about 63 kDa and consists of an acidic subunit of 40 kDa and a basic subunit of 22–23 kDa.[24] Legumins have been found in dry peas, lentils, chickpeas, and navy beans (*P. vulgaris*).[23,25,27] However, a legumin-like protein is named α-conglutin (11S) in lupin.[24] Vicilin has been reported in many pulses, such as adzuki bean (*Vigna angularis*), chickpea, lentil, and dry pea. Hall *et al.*[4] reported that vicilin in adzuki bean contains three subunits with molecular weights of 29 kDa, 33 kDa, and 60 kDa, respectively. Convicilin (7S) is reported as another globulin in pulses such as chickpeas, lentils, and peas. However, it is more widely accepted that convicilin is a subunit of vicilin. Two subunits of convicilin are identified with molecular weights of 77 kDa and 72 kDa. β-Conglutin (7S) is a globulin in lupin. Each of its monomers is formed by three subunits with molecular weights of 17–20 kDa, 25–46 kDa, and 53–64 kDa.[28] γ-Conglutin (7S) in lupin is a tetrameric protein of about 50 kDa. It is composed of two subunits with a molecular weight between 17 kDa and 29 kDa.[28] Phaseolin (7S) is a trimer protein composed of three subunits of 46 kDa, 49 kDa, and 50 kDa. Phaseolin is mainly found in common beans, such as navy bean.[15]

Albumin is the second major group of pulse storage proteins. Compared to the globulins of beans, pea albumins contain more of the essential amino acids tryptophan, lysine, threonine, and methionine.[29] The 2S albumins commonly consist of two subunits, which are linked

Table 8.2 Protein content and proportion of protein classes in pulses.[a,4,19,21–26]

		Globulin			Albumin				Prolamin	Glutelin
Pulses	Content	Protein	Svedberg unit	Mw kDa	Content	Protein	Svedberg unit	Mw kDa	Content	Content
Chickpea	56%	Legumin	11S	320–400	12%	Albumin	2S	20	3%	18.%
		Vicilin	7S	145–190						
Lentil	51%	Legumin	11S	340–360	17%	Albumin	2S	NR	4%	11%
		Convicilin	7S	290						
		Vicilin	7S	175–180						
Lupin	44–60%	α-Conglutin	11S	20–22, 42–52[b]	9–22%	δ-Conglutin	2S	4–9	NR	NR
		β-Conglutin	7S	17–20, 25–46, 53–64[b]						
		γ-Conglutin	7S	17–29[b]						
Pea	55–65%	Legumin	11S	320–410	18–25%	Albumin	2S	68	NR	NR
		Convicilin	8S	180–210						
		Vicilin	7S	150						
Navy bean	46–81%	Legumin	11S	320–400	11–31%	Albumin	2S	NR	<1%	<1%
		Phaseolin	7S	120–150						

[a]NR, not reported. Mw, molecular weight. Chickpea adapted from Chang *et al.*;[21] lentil adapted from Ghumman *et al.*[22] and Swanson;[23] lupin adapted from Bennetau-Pelissero[24] and Hall *et al.*;[4] pea adapted from Lu *et al.*;[19] navy bean adapted from Zhang and Romero[25] and Yeboah *et al.*[26]

[b]Mw of protein subunits.

by disulphide linkages. Similar to globulin, albumin subunits vary with different pulses. The albumin (2S) of chickpeas is about 20 kDa. It is composed of two polypeptides of 10 kDa and 12 kDa.[21] When it comes to lupin, the albumin is named δ-conglutin (2S), which is a monomeric protein consisting of two subunits of around 4 kDa and 9 kDa, respectively.[30] Pea seeds contain two major albumins, PA1 and PA2. PA1 consists of two proteins, PA1a, with a molecular weight of about 6 kDa and PA1b with a molecular weight of about 4 kDa.[30] The molecular weight of PA2 is about 26 kDa.[31]

Overall, the content range of these proteins varies across different references. Take the lentil proteins as an example. Sulieman *et al.*[32] reported that the albumin and globulin content in lentils is 56–64% and 26–29%, respectively. However, Hall *et al.*[4] reported albumin and globulin content in lentils as 17% and 51%, respectively. Such a huge difference could be due to the extraction method of proteins, which will be discussed in Section 8.3.

Proteins from pulse seeds are promising ingredients for formulating new food products. No publications have reported the aroma/odour of protein molecules because the polymers are non-volatile. However, the volatiles attached to the proteins are likely responsible for the aroma/odour of proteins. Specific interactions between volatile compounds and pulse proteins may lead to a disturbance of their balance and, therefore, generate different flavours. During food processing and storage, the flavour of food products depends on a fine balance between the release and retention of aroma compounds, which, in turn, is controlled by the reversibility of the binding of the aroma compounds. Semenova *et al.*[33] studied the role of the faba bean (*Vicia faba*) globulin state in the binding and release of an aroma compound in an aqueous medium. Denatured globulin, either by acidic or heat treatment, would result in the loss of the protein's capacity for binding hexyl acetate. In other words, the interaction of pulse protein and volatile compounds is critical to flavours.

8.3 Preparation of Pulse Protein Extracts

One of the most valuable ingredients extracted from pulses is their protein fraction. There are mainly three types of pulse protein-enriched products: pulse flours, pulse protein concentrates, and pulse protein isolates. Generally, pulse flour is the flour of dehulled pulse seeds. The protein content of pulse flour is normally lower than 50%.[34] Pulse protein concentrates and isolates are two primary commodities available on the market. Although the boundary between pulse protein concentrates and isolates varies between pulses, pulse protein concentrates generally refer to a pulse protein product with a content range from 50% to 80%, while pulse protein isolates refer to products with a protein content larger than 80%.[35,36] Generally, pulse flour is prepared by the dry milling of dehulled pulses.

Pulse protein concentrate is usually prepared by either the dry separation method or the wet processing method, while pulse protein isolate is produced by the wet processing method.[7]

8.3.1 Dry Processing

Dry milling is the process of separating the anatomical parts of the pulses. Milling generally involves the recovery of the main tissue, the cotyledon (85% of the pulse seed mass), and the concomitant removal of the testa and embryo.[37] In addition, milling generally reduces the particle size of starch and protein in the cotyledon.

Dry milling normally includes the processes of pulse cleaning, grading, tempering, conditioning, and milling (see Figure 8.1).[38] Dry milling is used to separate the testa and embryo from pulse flours. The pulse flours normally contain a protein content lower than 50%. Starch is the other major component of pulse flours. On the other hand, the testa and embryo are relatively rich in nutritional components such as dietary fibre, vitamins, minerals, and fat. Thus, while milling increases the palatability of pulse flour, it decreases the nutritional value of the obtained pulse flour.

Air classification is a technological method of producing protein concentrates by removing starch from pulse flours according to their size and density (see Figure 8.1).[39] Pulse flours, including proteins and starches, can be reduced into particles of different sizes by milling. Air classification of the pulse flour is performed in an air classifier with a spiral air stream. A fine fraction containing around 75% of the protein and a coarse fraction containing most of the starch granules are separated based on their density. However, some starch granules remain embedded in the protein matrix. By repeating air classification, the separation of starch and protein can be improved. Reichert *et al.*[40] extracted pea protein from four varieties of

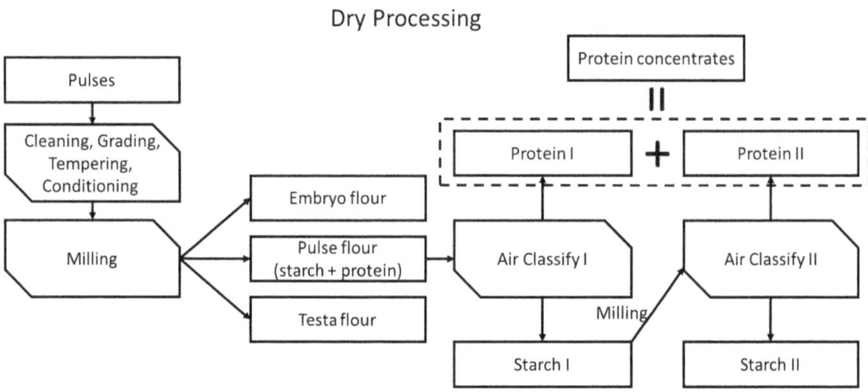

Figure 8.1 Extraction of pulse flour and pulse protein concentrate by dry processing.

field pea seeds with an Alpine air classifier. Pea flour was separated into starch I and protein I. Thereafter, starch I was fed again into the Alpine air classifier to produce starch II and protein II. The yield was improved by the combination of protein I and protein II.

8.3.2 Wet Processing

Wet processing is a promising method for extracting pulse protein concentrates or isolates. Generally, wet processing contains three steps: extraction, separation, and drying. First, a water solution, a water/alcohol solution, an alkaline solution, an acid solution, or a salt solution could be employed for protein extraction. Thereafter, isoelectric precipitation, salt out, diafiltration, or ultrafiltration is used to separate the pulse protein. Finally, the pulse protein is collected by the drying process, including spray drying or freeze drying (see Figure 8.2). The specific extraction, separation, and collection methods can be combined to obtain different pulse protein products.

Pulse protein concentrates can be extracted with direct water or water/alcohol extraction from pulse flours. Cai *et al.*[41] reported protein concentrate extracted from chickpea, faba bean, common bean, lentil, pea, and mung bean (*Vigna radiata*) with the water extraction process. Two hundred grams of each seed was extracted twice with 500 mL of water. The protein content of chickpea, faba bean, and smooth pea protein concentrates was 50.5%, 65.5%, and 67%, respectively.

Water extraction/ultrafiltration has been developed for high-quality pea-protein isolate.[42] After wet milling, soluble proteins were extracted from pea flours at pH 7.3. The supernatants were ultrafiltered with hollow fibre to remove oligosaccharides. Finally, the pH was adjusted to 8.3 with a sodium hydroxide solution. The protein isolates were dried with a wheel rotary spray drier. The purity of protein isolates obtained was 89 and 92% on the pilot and industry scales, respectively.

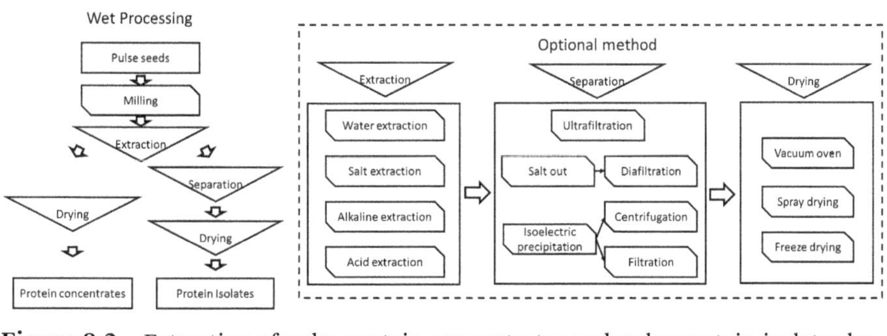

Figure 8.2 Extraction of pulse protein concentrates and pulse protein isolates by wet processing.

Alkaline extraction/isoelectric precipitation is a method based on the solubilization of proteins in an alkaline solution followed by an isoelectric precipitation process (see Figure 8.2). The method takes advantage of the solubility of pulse storage proteins, which is high at pH 8–10 while low at pH 4–5.[43,44] Briefly, pulse flours are dispersed in an alkaline solution at a flour : water ratio ranging from 1 : 5 to 1 : 15. After stirring for 1 h, the supernatant is collected, and the pH is adjusted to the isoelectric points, normally pH 4.5. Precipitates are collected, dried, and milled into powder to make pulse protein isolates into powder.[43] Boye *et al.*[7] reported that the purity of protein isolates extracted from chickpea, pea, great northern bean (*P. vulgaris*), lentil, lima bean (*P. lunatus*), and mung bean was 90.5%, 90.1%, 91.2%, 89.3%, 91.2%, and 88.3%, respectively.

Acid extraction is a method similar to alkaline extraction except for the initial pH value of the solution. Alli *et al.*[45] obtained bipyramidal crystalline proteins of kidney beans by acid extraction/isoelectric precipitation. The protein was extracted from kidney bean flour with a citric acid solution (pH 4.0) at room temperature. The precipitation was achieved by keeping the temperature at 4 °C for 18 h. The freeze-dried protein isolates contained 95.7% protein and 2.6% total carbohydrate.

Salt extraction/diafiltration is a method developed based on the solubility of pulse protein isolates in salt solutions. In this process, an appropriate salt solution with the desired ionic strength is employed to extract the protein. Then, protein precipitation is induced by diluting the salt solution. After centrifugation or filtration, the pellets are dried and stored. Marquez *et al.*[46] extracted 10 g of soluble bean proteins with a 100-mL buffer solution (0.1 M phosphate and 0.5 M NaCl, pH 7.6). After stirring and centrifugation, the supernatant was dialysed against distilled water, freeze-dried, and stored. The protein content was reported to range from 74.7% to 84.2%.

8.3.3 Impact of Extraction Method on the Protein Flavours

Pulse protein flavours have been widely reported due to their specific off-flavours, which impede the popularization of pulse protein products. The protein extraction process affects the flavour of pulse protein.[11] However, only a few articles have studied the effect of the extraction process on the pulse protein flavours.

Dry processes are physical methods for producing pulse flours or pulse protein concentrates from pulse seeds. The volatiles are attached to not only pulse proteins but also the remaining starches. Volatiles, physically trapped by proteins and starches, are supposed to evaporate during dry milling and air classification.

Wet processes, on the other hand, involve both chemical reactions and physical interactions. The flavour profile evolves during the wet extraction process. Water-soluble volatiles are supposed to be removed, while volatiles, such as esters, alcohols, and carboxylic acids, would react with other chemicals in the alkaline or acid environment. Meanwhile, new volatiles could be developed by these chemical reactions.

Murat *et al.*[11] studied the pea protein flavour with water extraction and isoelectric precipitation. The flavour of pea protein powders is different from that of pea flour. Some compounds disappear (*e.g.*, 1-octanol, 5-pentyldihydro-2(3*H*)-furanone, 1-hexanol, and 1-nonanol) while some appear (*e.g.*, 3-methylthiopropanal, benzeneacetaldehyde, (*E*)-2-octenal, (*E*,*E*)-3,5-octadien-2-one, and 4-ethyl-benzaldehyde). Only a few odour-active compounds are common in both pea flour and pea protein isolates (*e.g.*, hexanal and hexanoic acid).

Xu *et al.*[43] studied volatile component changes of germinated chickpea, lentil, and yellow pea flours over the alkaline extraction/isoelectric precipitation process characterized by headspace-solid phase microextraction (HS-SPME) coupled with gas chromatography mass spectrometry and olfactory (GC–MS/O). After extraction, 77 out of 124 volatile compounds were stripped from the initial pulse flours. In total, 55% of aliphatic alcohols, 19% of aliphatic aldehydes, 33% of aliphatic esters, 62% of aliphatic ketones, 58% of benzene derivatives, 60% of carboxylic acids, 40% of furanoids, 72% of hydrocarbons, 17% of others, 33% of pyrazines, 33% of terpenes, and 50% of unknown components were eliminated from the pulse flours by the extraction procedure. Four possible reasons were proposed by Xu *et al.* for this:[43] first, water-soluble volatiles can be easily removed by water. Second, saponification may occur to a lesser extent during extraction so that esters in pulse flours can be removed by water. Third, carboxylic acids, such as acetic acid and butanoic acid, may be neutralized by alkalinity. Furthermore, volatile compounds might also diffuse into a porous system or be physically trapped by carbohydrates and cyclic oligosaccharides.

Although a significant amount of odorous volatiles disappeared, a minor proportion of the volatiles could stem from pulse flour and/or newly formed volatiles during wet processing. Xu *et al.*[43] identified 19 new volatiles, mainly including 3 aliphatic alcohols, 4 aliphatic aldehydes, 2 aliphatic esters, 1 aliphatic ketone, 1 benzene derivative, and 1 furanoid.

8.4 Volatile Profiles of Pulse Proteins

Aroma/odour is the key aspect of food flavour. It is very difficult to identify the flavour of a food product without aroma/odour. Therefore, it is reasonable that there has been so much focus on aroma/odour. Yet, as noted by Reineccius,[47] flavour perception is slowly being recognized as a multifaceted sense, depending heavily upon other sensory attributes such as taste, texture, and appearance. The flavour scientists strive to broaden the definition of flavour to include other senses besides aroma/odour. On that note, this section will focus on volatile compounds related to aroma/odour for two reasons. First, aroma/odour is much more important than other flavour perceptions; second, less work has been reported on the analysis of non-volatile substances.[47]

8.4.1 Major Flavour Compounds in Pulse Proteins

Pulse proteins are described as floral, green, peanut, caramel, grass, and beany.[11,48–52] The principal volatile compounds bound to pulse proteins that have been studied are variable but mainly involve hydrocarbons (alkanes), carboxylic acids, esters, lactones, aliphatic alcohols, aldehydes, and ketones.

8.4.1.1 Aliphatic Alcohols

Aliphatic alcohols are one of the major volatile compounds in pulse proteins. Hexanol, heptanol, and 2-octen-1-ol are related to grass and green flavours. These volatiles have been reported in not only pulse proteins but also pulse seeds. Xu *et al.*[48] reported a strong green flavour in raw pulse seeds, including chickpeas, lentils, and yellow peas. The hexanol level increased following the germination of pulse seeds. Mushroom flavour is another typical flavour attributed to 1-octen-3-ol. Vara-ubol *et al.*[49] reported that 1-octen-3-ol had mushroom flavour at 1 and 10 mg L^{-1} while being beany at 100 and 1000 mg L^{-1}. The flavour of 1-octen-3-ol changed depending on the concentration levels that were present. This could explain why some bean-related volatile compounds have been reported in other flavours, such as mushrooms. 1-Nonanol is another common aliphatic alcohol in protein isolates. The odour of 1-nonanol is described as rose and orange. Sharan *et al.*[50] reported 1-nonanol in fava bean seeds identified by GC–MS/O. 1-Nonanol is not only present in pulse proteins, such as chickpeas, lentils, dry peas, and fava beans, but has also been reported in hemp protein isolates.[53]

8.4.1.2 Aliphatic Aldehyde

Aliphatic aldehydes include important volatile compounds related to green and grassy odours. Aliphatic aldehydes are also responsible for other flavours, such as floral, sweet, and citrus. Octanal, nonanal, and decanal are reported as having a citrus odour, while heptanal is reported as being sweet. These minor volatile components of pulse proteins play an important role in understanding pulse protein flavours.

Hexanal is one of the most common volatiles in plants, including fruits, cereals, grains, and oilseeds.[54] Hexanal is considered one of the key aromas in pulse proteins. The odour of hexanal can be described as grassy and green. Hexanal is mainly decomposed by linoleic acid, which is rich in pulses.[55] Guldiken *et al.*[56] identified 20 aldehydes in lentil protein isolates. Interestingly, 78% of the total peak area of aldehyde was hexanal. In addition, 2,6-nonadienal is very important in pulse protein flavour. The odour of 2,6-nonadienal is characterized as cucumber-like.[57] 2,6-Nonadienal has been reported in chickpeas, lentil protein isolates,[43] and lupin seeds.[52] 2,6-Nonadienal has been considered an oxidation product of lipids.

8.4.1.3 Aliphatic Ketones

Aliphatic ketones are another source of the aroma/odour of pulse proteins. Twelve ketones have been identified from chickpea, lentil, and yellow pea seeds.[48] However, only three of them have been found in pulse protein isolates.[43] One of the major ketones in pulse proteins is 3,5-octadien-2-one. It has isomers such as (*E,E*)-3,5-octadien-2-one, (*E,Z*)-3,5-octadien-2-one, (*Z,Z*)-3,5-octadien-2-one, and (*Z,E*)-3,5-octadien-2-one. It is hard to distinguish these isomers based on the flavour description, retention time, and molecular weight. Therefore, 3,5-octadien-2-one stands for one or more isomers in this section. Xu *et al.*[43] reported that 3,5-octadien-2-one has a spicy, earthy, green pepper aroma in chickpea, lentil, and dry pea protein isolates with the HS-SPME–GC–MS/O method. Murat *et al.*[11] employed solvent-assisted flavour evaporation coupled with the GC-O method and reported 3,5-octadien-2-one with an odour of mushrooms, dill, and broccoli.

In addition, 3-octen-2-one has been found in pulses, either in flours or in protein isolates. Oomah *et al.*[58] detected 3-octen-2-one in the flour of fava beans in 13 different low-tannin genotypes. 3-Octen-2-one has also been reported in the protein isolates of chickpeas, lentils, dry peas, and fava bean.[43,50]

8.4.1.4 Benzene Derivatives

Compounds containing benzene rings were called aromatic compounds because of the sweet or pleasant odours found in many benzene derivatives. Although scientists have already realized that not all aromatic compounds have a sweet odour and not all compounds with a sweet odour are aromatic, benzene derivatives are still related to the pleasant flavour in pulse proteins. One or more of acetophenone, benzaldehyde, eugenol, maltol, methyleugenol, phenol, and phenylethyl alcohol have been discovered in pulse proteins such as chickpeas, lentils, dry peas, and fava beans. The flavours are described as pleasant, such as sweet, caramel, floral, clove, and almond (see Table 8.3).

Phenylethyl alcohol is a typical floral aroma compound. Phenylethyl alcohol is mainly formed as a metabolite from the degradation of amino acids *via* keto acids. The bioconversion of L-phenylalanine and L-leucine into phenylethyl alcohol and 3-methyl-1-butanol is a key step in the formation of odourants.[52,59] Phenylethyl alcohol has been reported in chickpeas, lentils, dry peas, and adzuki beans for its floral aroma.[48,60] Germination could increase the intensity of phenylethyl alcohol in pulse seeds.[48]

Table 8.3 Pulse protein-related volatile compounds.[43,50]

Compound name	Aroma description	Protein source
Aliphatic alcohols		
2-Nonen-1-ol	Melon, mint, medicine	CP, LE, YP
2-Octen-1-ol	Cucumber, grass, green	CP, LE, YP, FB
2-Penten-1-ol	NR	LE, YP
1-Hexanol	Lemmon, grass	CP, LE, YP, FB
1-Nonanol	Rose–orange odour	CP, LE, YP
1-Octen-3-ol	Mushroom	CP, LE, YP, FB
2,4,7,9-Tetramethyl-5-decyn-4,7-diol	NR	LE
2-Ethyl-1-hexanol	Mild, oily, sweet, slightly floral	LE, YP, FB
3-Octanol	Citrus, nut, moss, herbal, earthy	LE, FB
Aliphatic aldehydes		
2-Decenal	NR	CP
2-Hexenal	NR	LE
2,6-Nonadienal	Green cucumber skin	CP, LE
7-Tetradecenal	NR	YP
7,10-Hexadecadienal	NR	CP, LE, YP
9-Hexadecenal	NR	CP, LE, YP
Decanal	Citrus, soap, orange peel	CP, LE, YP, FB
Heptadecanal	NR	CP, LE, YP
Heptanal	Sweet	CP, LE, YP, FB
Hexadecanal	NR	LE
Hexanal	Green, grass	CP, LE, YP, FB
Nonanal	Plastic, citrus peel-like	CP, LE, YP, FB
Octanal	Green, citrus	CP, LE, YP
Pentadecanal	NR	CP, LE, YP
Tetradecanal	NR	CP, LE, YP
Tridecanal	Unpleasant, grass, pungent	CP, LE, YP
Undecanal	NR	CP, LE
Aliphatic esters		
2,2,4-Trimethyl-1,3-pentanediol diisobutyrate	NR	CP, LE, YP
3-Hydroxy-2,2,4-trimethylpentyl isobutyrate	NR	CP, LE, YP
Dimethyl glutarate	NR	LE, YP
Aliphatic ketones		
3,5-Octadien-2-one	Spicy, earthy, green pepper	CP, LE, YP, FB
3-Octen-2-one	Beany, metal, mushroom	CP, LE, YP, FB
6,10,14-Trimethyl-2-pentadecanone	NR	LE
Benzene derivatives		
Acetophenone	Sweet, almond, pungent,	CP, LE, YP, FB

(continued)

Table 8.3 (*continued*)

Compound name	Aroma description	Protein source
	chemical, flower, bitter, must	
Benzeneacetaldehyde	Floral, honey, sweet	LE
Benzaldehyde	Almond, sweet, woody	CP, LE, YP, FB
Eugenol	Pleasant, clove-like	CP, LE
Maltol	Caramel	CP, LE, YP
Methyleugenol	Clove-like, fresh	CP
Phenol	Peppery, woody	CP, LE, YP
Phenylethyl Alcohol	Floral, rose, honey	CP, LE, FB
Carboxylic acids		
2-Methylhexanoic acid	Acidic, sour, dairy fatty with cooked lard-like nuances	YP
Hexanoic acid	Sickening, sweaty, rancid, sour, sharp, pungent	CP, LE, FB
Furanoids		
1-(2,4-Dimethyl-furan-3-yl)-ethanone	NR	LE
5-Pentyldihydro-2(3*H*)-furanone	Peach, coconut	CP, LE, YP
Furfural	Soy sauce-like, penetrating odour	CP, LE, YP
Hydrocarbons		
Tetradecane	NR	LE, YP
4(1*H*)-Pteridinone, 5,6,7,8-tetrahydro-6-methyl-	NR	YP
Pyrazines		
2-Methoxy-3-(1-methylethyl) pyrazine	Pea pod, earthy	CP, LE, YP
2-Methoxy-3-(1-methylpropyl) pyrazine	Green, earthy, a little spicy	LE, YP
Others		
1,3,2-Dioxaborolane, 4,5-dimethyl-2-phenyl-	NR	CP
1-(2-Butoxyethoxy)ethanol	NR	CP, LE
Geraniol	Honey, tea, floral	CP, LE, YP
Pyrrole	NR	LE

Protein from chickpea, lentil, and yellow pea adapted from Xu *et al.*,[43] protein from fava bean adapted from Sharan *et al.*[50]

CP, Chickpea; FB, Fava bean; LE, Lentil; NR, Not reported; YP, Yellow pea.

8.4.1.5 Furanoids

Furonoids refer to compounds having a structure based on furan. Furanoids have not been widely reported in pulse protein-related flavours. Xu *et al.*[48] tentatively identified furfural in the pulse protein-related flavours. The furfural response of pulse protein samples was under the detection limitation of GC–MS, while a strong soy-sauce-like flavour was smelled by panellists through GC–O at the same retention time. Researchers noticed the olfactory threshold of furfural is low, which is only 14.1 μg L^{-1} in red wine.[61] This explains why the nose is more sensitive than MS to furfural. In addition, furfural was further confirmed by its calculated retention index (RI) value (1490) which is close to the reference RI value (1503).[48]

5-Pentyldihydro-2(3*H*)-furanone is an important flavour compound with a peach flavour. The flavour is strong in chickpea, lentil, and dry pea flours and improves significantly following pulse germination.[48] Murat *et al.*[11] have also identified 5-pentyldihydro-2(3*H*)-furanone from pea proteins and pea flours.

8.4.1.6 Pyrazines

Pyrazines are heterocyclic aromatic organic compounds with the chemical formula $C_4H_4N_2$. Three different 3-alkyl-2-methoxypyrazines have been identified as aroma constituents inherent in green peas, namely, 3-isopropyl-2-methoxypyrazine, 3-*sec*-butyl-2-methoxypyrazine, and 3-isobutyl-2-methoxypyrazine.[62] Normally, the intensity of pyrazine in pulse proteins or pulse flours is low. The signal is hard to detect by GC–MS. However, the threshold of alkyl pyrazines is extremely low, *e.g.*, the 2-methoxy-3-isopropyl pyrazine has a threshold of 0.002 μg L^{-1} in water.[57] Therefore, pyrazines are of great importance in pulse proteins, which could be one of the major sources of beany, earthy, and pea pod flavours. Pyrazine can be formed during the Maillard reaction and fermentation.[47,63] The pyrazine originally attached to pulse proteins is also worth studying. However, the detection limit of current methods needs to be improved to test the tiny amount of pyrazine in pulse proteins.

8.4.1.7 Others

In addition to the volatiles mentioned above, other volatiles such as esters, terpenes, carboxylic acids, and hydrocarbons also contribute to the flavour of pulse proteins. The volatiles have low odour values (odour concentration/odour threshold) due to either a low amount in the pulse proteins, such as esters, carboxylic acids, and terpenes, or a high odour threshold, such as hydrocarbons.

Moreover, some volatiles exist in small amounts but have a strong odour that can be identified by olfactory rather than MS, such as furfural and 2-methoxy-3-isopropyl pyrazine.[43] Xu *et al.*[48] identified these compounds based on the flavour description and retention time of authentic standards, which need skilled panellists. However, eight compounds with low thresholds have not been identified due to the panellists' lack of experience. The odours are of great importance in pulse protein flavour, such as photinia-like, toasted soybean, sweaty, sweet, plastic, and stinky odours. A new detection method needs to be developed to deal with compounds with low threshold but strong flavours.

8.4.2 Flavour Sources of Pulse Protein

8.4.2.1 Lipids

Similar to other vegetables, fatty acids, amino acids, and carbohydrates are sources of pulse protein flavours.[47] Lipid is not a major component in pulse seeds. The lipid content in pulses is between 0.4% and 4%, with an average value of 1.5%.[4] Most pulses contain less than 3% (w/w) of total lipid content, except chickpeas and lupins.[4] Nevertheless, lipid degradation is still one of the major sources of flavour compounds. In pulse seeds, about 80% of the fatty acids are C18-fatty acids, which include stearic acid, oleic acid, linoleic acid, and linolenic acid.

Lipid degradation includes two major types, hydrolysation and oxidation, which are catalysed by lipases and lipoxygenases, respectively (see Figure 8.3). Lipases directly attack the fatty acid ester bonds of triglycerides, producing free fatty acids. The flavours developed by lipase action depend on the composition of the fat. The release of short-chain fatty acids, such as butyric, caproic, and capric, causes particularly disagreeable odours and flavours, whereas the long-chain fatty acids (C12 and above) produce candle-like or, at alkaline pH, soapy flavours.[64]

Oxidative deterioration of free and esterified unsaturated fatty acids is primarily responsible for the formation of odours in pulses. Lipoxygenase attacks linoleic and linolenic acids to yield hydroperoxides. However, the lipoxygenases are unique in pulses, which result in different volatile profiles from the pulse lipids. Although one fatty acid would result in different volatile compounds with a different group of enzymes, there are still some common volatiles that can be summarized in the lipoxygenase pathway. Oleic acids would be oxidized and result in hexanal and hexanol with a green and grassy odour. Linoleic acids can be oxidized and result in 3-hexenal, 1-octen-3-one, and 1-octen-3-ol with the flavour attributes of natural green, metal, and mushroom, respectively. Linolenic acids can be oxidized into 2,4-nonadienal, 2-hexenal, and pentyl furan with the flavours of green, cucumber, and beany.[65] In fact, some volatiles can be derived from different fatty acids. Take hexenal as an example. It could be derived from either linoleic acid or linolenic acid.[66] Murat *et al.*[11] summarized the source of different volatiles from different fatty acids in various pulses. In green pea, 1-penten-3-ol and 1-hexanol came from fatty acid breakdown. 1-Pentanol has been identified in peas and originated from fatty acid breakdown. 1-Octen-3-ol, a beany compound, was found in green peas and reported to come from fatty acids. 1-Octanol has been found in many studies in peas. The presence of aliphatic saturated and unsaturated alcohols suggests enzymatic oxidation of pulse lipids.

As summarized by Hall *et al.*,[4] most of the pulses are rich in linoleic acid, ranging from 21% to 57%, while being low in steric acids (<5%). Specifically, kidney beans, great northern beans, and pinto beans are high in linolenic acid, which is 46%, 43%, and 43%, respectively. Chickpea is

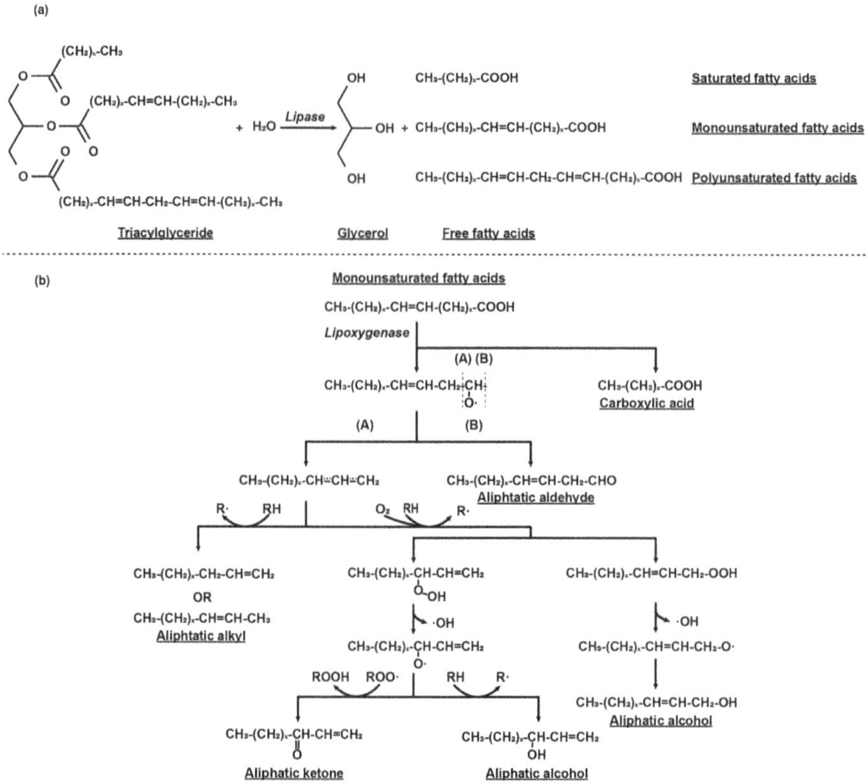

Figure 8.3 Volatiles in pulse seeds originate from lipids with the synergistic effect of lipase (a) and lipoxygenase (b). The monounsaturated fatty acid is employed as an example of the oxidative deterioration of lipids. RH, R•, ROO•, •OH, and O• represent lipid, lipid radical, lipid hydroperoxyl radical, hydroxyl radical, and oxyl radical, respectively.

high in linoleic acid (57%), while lupin is high in oleic acid (52%). The difference in lipid composition would result in different volatile profiles. Xu *et al.*[48] analysed the volatile profiles of chickpea, lentil, and dry pea with principal component analysis and hierarchical clustering analysis. The volatile profiles of chickpeas are totally different from those of lentils and dry peas, which have similar fatty acid profiles.[4] Khrisanapant *et al.*[67] studied 11 types of pulses using HS–SPME–GC–MS and GC–flame ionization detector. Volatile aldehydes, alcohols, ketones, esters, terpenes, and hydrocarbons were the chemical groups identified across all the pulses. Different fatty acid profiles and lipoxygenases were considered the major factors contributing to the volatile differences.

Overall, most of the aliphatic alcohols, ketones, aldehydes, and alkyls originate from lipids (see Figure 8.3). Esters and carboxylic acids could be derived from the reaction of these compounds.

8.4.2.2 Proteins

Protein is one of the major components of pulse seeds. Generally, the protein contents of pulse seeds are in the range of 18% to 30%. The breakdown of proteins is the source of many flavour compounds in food. Proteins are degraded into peptides and amino acids by proteases and peptidases, respectively.

Peptides and amino acids are not volatiles but could provide sweet, bitter, salty, umami, and sour tastes. As shown in Table 8.1, five amino acids are responsible for sweet taste, nine amino acids are responsible for bitterness, and two amino acids are responsible for umami taste. In pulses, most of the amino acids are in the blocks of storage proteins rather than in the free forms. Processing technologies such as cooking, fermentation, or germination could hydrolyse part of the proteins into free amino acids, therefore, increasing flavour, *e.g.*, umami taste from glutamic acid and aspartic acid. Fermentation, germination, and heating are the three possible processes to release the umami taste. For example, soy sauce is full of umami taste. A similar sauce can be produced by fermenting steamed pea seeds.[68] Pigeon peas and lupins are relatively high in glutamic and aspartic acids, while pinto beans and navy beans are relatively low in glutamic and aspartic acids (see Table 8.1).

Nine amino acids are related to the bitter taste (see Table 8.1). Although the bitterness of pulse proteins is mainly due to saponin and isoflavones, Roland *et al.*[62] reported that the release of the nine amino acids during pulse processing could also contribute to the bitterness of pulse products.

In addition, amino acids can be converted into volatiles through the Ehrlich–Neubauer pathway or Maillard reaction.[65] The Ehrlich–Neubauer pathway is employed to demonstrate the generation of volatiles originating from amino acids (see Figure 8.4). It leads to the transformation of amino acids into α-keto acids, followed by the decarboxylation step leading to the production of branched chains or aryl aldehydes, alcohols, and acids. Pulse proteins would be infected by microorganisms if not stored as required, which may result in volatiles converted from amino acids through the Ehrlich–Neubauer pathway. Although the Ehrlich–Neubauer pathway has been reported in microorganisms rather than in plant metabolism, the aryl compounds are still considered to be related to aromatic amino acids such as phenylalanine and tyrosine. Benzene derivatives, such as phenylacetaldehyde, benzaldehyde, acetaldehyde, and phenylethyl alcohol, are the oxidation products of α-keto acid, which could originate from phenylalanine or tyrosine (see Table 8.3).[65] Most of the benzene derivatives are considered to have pleasant flavours, such as floral, almond, and clove (see Table 8.3).

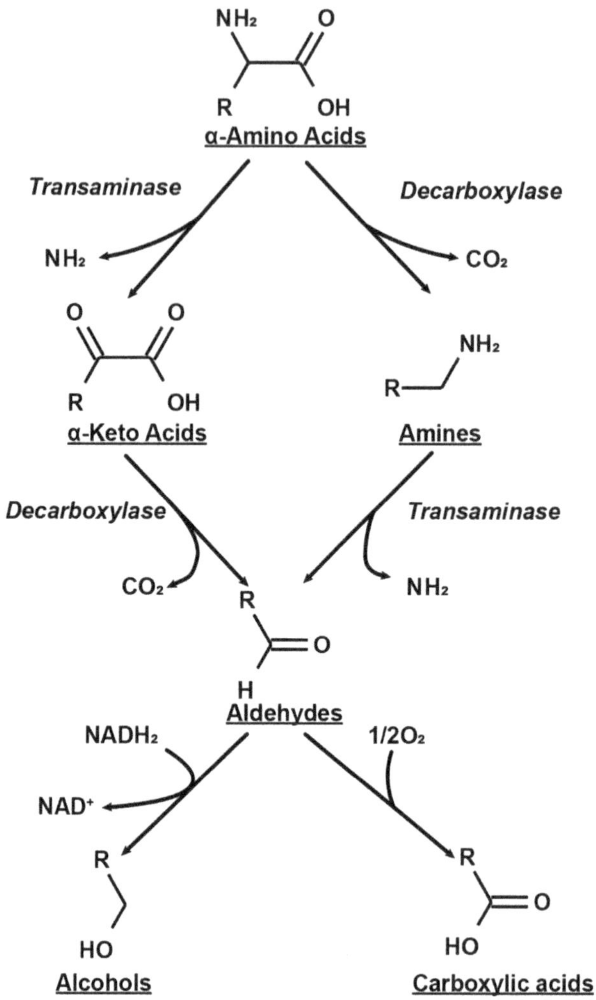

Figure 8.4 General scheme of the Ehrlich–Neubauer pathway. Reproduced from ref. 65 with permission from Taylor and Francis, Copyright 2011.

The Maillard reaction is another important pathway for converting amino acids into volatiles.[69] The Maillard reaction, including Strecker degradation, is a series of reactions between amino acids and reducing sugar. Alkyl pyrazines are derived through Strecker degradation of amino acids. Some of the alkyl pyrazines have extremely low olfactory thresholds, such as 2-methoxy-3-(1-methylpropyl) pyrazine (0.002 $\mu g\ L^{-1}$; water), 2-ethyl-3,5-dimethylpyrazine (0.04 $\mu g\ L^{-1}$; water), and 2,3-diethyl-5-methylpyrazine (0.09 $\mu g\ L^{-1}$; water).[65] These compounds are present in extremely low concentrations but are the main compounds that contribute to the perceived green pea aroma in pulses owing to their low threshold.[48,62] A methoxypyrazine was identified in both pea flour and pea protein isolates, as reported by Murat *et al.*[11] Xu *et al.*[43] found that the protein isolates extracted from germinated dry pea have an increased earthy odour following the increased concentration of 2-methoxy-3-(1-methylpropyl)-pyrazine. Overall, benzene derivatives and pyrazines may have originated from pulse proteins.

8.4.2.3 *Carbohydrates*

If one considers plant metabolism, it is clearly established that plants metabolize carbohydrates into other plant needs, *e.g.*, lipids and amino acids. Therefore, one may state that nearly all plant flavours come indirectly from carbohydrate metabolism, as all of the other flavour precursors come from carbohydrate metabolism. However, a few flavour constituents come directly from carbohydrate metabolism.

A few flavour compounds can also directly originate from carbohydrates, such as furanoids, saccharides, and glucosinolates.[47] Take furanoid as an example. It is one of the important volatiles that originated in carbohydrates. The enolization reaction of hexoses and pentoses gives rise to 1,2-endiol and 2,3-enediol, respectively. Those enediols are important intermediates of caramel aroma development, such as the formation of furfural and 5-pentyldihydro-2(3*H*)-furanone, which have been identified in the protein isolates extracted from germinated chickpea, lentil, and yellow pea.[43,48]

8.4.3 The Interaction of Flavour Compounds and Proteins

Pulse protein contains both pleasant and unpleasant flavours. It is desirable to keep the pleasant flavour while removing the unpleasant flavour when using pulse proteins as food ingredients. Pulse proteins are complex systems due to their diverse chemical compositions and structures. Proteins are known to interact with volatile compounds, directly affecting their release, removal, masking, and perception of flavours. Understanding the strength and nature of the interactions between protein ingredients and volatiles is therefore becoming particularly important to improve the overall sensory profile and quality of protein ingredients.

The affinity of volatile compounds for proteins depends on the molecular structure of both the volatile compounds and the proteins. Therefore, different volatiles could have different binding affinities to proteins due to the amino acid profile and conformation of the protein. The multiple functional groups of volatile compounds could make the protein–flavour interaction more complicated. Therefore, the flavour profile is very hard to predict and control in the real food system. However, the principle of interaction between pulse protein and volatile compounds still needs to be reviewed for a better understanding of pulse protein flavours.

In general, the interaction between pulse proteins and volatile compounds can be categorized into physical trappings and molecular interactions. Physical trappings involve the adsorption of flavour compounds onto the inner or outer surface of pulse protein. Molecular interactions can be classified as covalent chemical bonds and non-covalent physicochemical bonding, including van der Waals forces, hydrogen bonds, hydrophobic interactions, and ionic bonds.[70] Interactions between pulse proteins and flavours based on covalent bonding can be defined as irreversible interactions, while those based on physical trapping and non-covalent physicochemical bonding could be defined as reversible interactions.[47]

8.4.3.1 Irreversible Interaction

The release of volatiles is one of the key prerequisites for perceiving flavours. Irreversible interaction, however, would result in low efficiency in releasing volatiles from pulse proteins. Therefore, it is important to discriminate between irreversible and reversible interactions between volatiles and non-volatiles. Proteins with a high content of sulphur-containing amino acids, such as cysteine and methionine, would potentially have more covalent bonds between volatiles and proteins. Commonly, pulse proteins have low cysteine and methionine contents, which are less than 1 and 2%, respectively, in most pulse seeds (see Table 8.1). Pinto beans have a relatively high methionine content, which is about 7% of total amino acids. Compared to cereal and meat proteins, pulses are low in sulphur-containing amino acids. As a result, volatiles in pulses are easier to remove by extraction compared to those in cereals and meat.

8.4.3.2 Reversible Interaction

Reversible interaction between volatiles and proteins includes physical trappings, van der Waals forces, hydrogen bonds, hydrophobic interactions, and ionic bonds.

The physical trap is flavour compounds trapped in a sealed cavity. Flavour compounds will be released once the shell or wall of the cavity has been destroyed. Hydrolysis of pulse proteins could break the sealed cavity and result in the release of volatiles.

Van der Waals force is a distance-dependent interaction between atoms or molecules. Unlike ionic or covalent bonds, these attractions do not result from a chemical electronic bond; they are comparatively weak and, therefore, more susceptible to disturbance. The van der Waals force quickly vanishes at longer distances between interacting molecules. Thus, even physical mixing could affect the van der Waals force.

A hydrogen bond is primarily an electrostatic force of attraction between a hydrogen (H) atom and an electronegative atom or group, such as nitrogen (N) and oxygen (O). H, N, and O are three of the most common elements made up of protein and volatiles. Hydrogen bonds are widely spread between proteins and volatiles. However, it is a weaker force compared to ionic bonding.

The hydrophobic effect is the observed tendency of non-polar or amphiphile substances to aggregate in an aqueous solution and exclude water molecules. Most flavours bind reversibly with protein *via* hydrophobic interactions with its non-polar interior region due to the presence of a non-polar (aliphatic) segment in the flavour compounds. In other words, any factors that modify the conformation of proteins, thereby exposing or disrupting the interior hydrophobic area, would alter the protein flavour-binding affinities.

Ionic bonding is a type of chemical bonding that involves the electrostatic attraction between oppositely charged ions or between two atoms with sharply different electronegativities. It is the primary interaction occurring in ionic compounds. The amino group ($-NH_2$) and carboxyl group (-COOH) are the two main protein groups. The pH value could significantly affect the flavour-binding capacity of proteins. The $-NH_3$ works at a low pH value (pH < 7) in the form of NH_4^+ for ionic-binding volatiles, while -COOH works at a high pH value (pH > 7) in the form of $-COO^-$. The amount of $-NH_2$ and -COOH groups exposed to the environment could affect the ionic-bonding capacity. Hydrolysed proteins could have more $-NH_2$ and -COOH groups than native proteins exposed to the environment. In addition, arginine, lysine, asparagine, and glutamine contain two or more $-NH_2$ groups, while aspartic acid and glutamic acid contain two -COOH groups. These amino acids contribute to the ionic-bonding capacity of proteins.

Overall, these five forces together interact between proteins and volatiles. Changes in temperature, pH, water activity, and ionic strength can affect the relationship between pulse protein and volatiles. Multiple processes have been developed to improve the balance between proteins and volatiles.

8.5 Processing Technologies

With an understanding of the basic interactions between flavour compounds and pulse proteins, physical, chemical, and biological methods have been developed to improve the flavour of pulse proteins. However, the odour detection limit of our nose is about 10^{-19} moles, which surpasses the most sensitive analytical instrument in the modern world.[47] It seems hard to remove all the undesirable flavours with currently available technologies. The strategies for mitigating off-flavours normally contain both reducing and masking off-flavours.

8.5.1 Soaking

Soaking, a non-thermal process, is effective in reducing the content of non-nutritional factors and improving the composition and digestibility of pulse seeds. Soaking is also beneficial for flavour removal. Water-soluble volatiles could be released from pulse seeds during the soaking process.

Pulse proteins are considered to have a compact structure, with starch granules in native pulse seeds. The transfer efficiency of volatile compounds is limited at the beginning of soaking. Water-soluble volatiles bound to pulse protein with weak forces, such as the Van der Waals force and hydrogen bond, could be released into the water when the seeds have absorbed enough water.[62]

Soaking alone is not considered an effective odour-removal strategy. Soaking has always been employed as a pre-conditioning method for other processes such as boiling, germination, and fermentation. For example, pre-soaking pulses before boiling is a strategy to obtain a soft texture and reduce the processing time.[62]

Nevertheless, the soaking process could also form new flavour compounds. With the increase in water activity, enzymes or microbes are activated to convert precursors into flavour compounds. During the soaking process of soybeans, 1-octen-3-ol is very likely formed during the soaking phase *via* a biological (enzyme)-controlled mechanism.[71] Therefore, keeping a low temperature (<10 °C) is a method for reducing the activities of enzymes or microbes.

8.5.2 Boiling

Boiling could be the oldest heat treatment for pulse seeds. Boiling includes two functional steps: soaking and heating. Boiling could mitigate the flavour of pulse protein based on four different mechanisms. First, soaking is a step to release water-soluble volatiles from pulse proteins into water. Second, boiling could evaporate some of the volatiles out of the water. Low-boiling-point chemicals such as hexanal, heptanal, 1-octen-3-ol, 3,5-octadien-2-one, and 2-pentylfuran get decreased in adzuki beans after boiling.[60] Third, high temperatures would result in chemical reactions that could remove odours. The boiling of yellow pea significantly decreased the total sulphur compounds, which could covalently bond to proteins. Fourth, part of the proteins could be denatured and expose the interior hydrophobic groups. The volatiles with a hydrophilic or covalent bond to the proteins are anticipated to be released from the denatured protein molecules. Ma *et al.*[72] boiled navy beans, red kidney beans, green lentils, and yellow peas with four different pre-boiling treatments. Pre-boiling significantly reduced the total volatiles by 61.75%. Specifically, pre-boiling coupled with freeze-drying removed most of the ketones and aldehydes.

However, high temperatures could generate new flavours. New odours could be either beany-related flavour or beany flavour-masker. Benzene derivatives can be developed by boiling.[60] Ethylbenzene, 1,3-dimethylbenzene, *p*-xylene, and styrene have been increased from not detected to 42, 40, 98, and 261 μg kg^{-1}, respectively, in the boiled adzuki beans. Increased benzene derivatives can be considered an improvement due to their pleasant flavour.[60]

8.5.3 Roasting

The roasting process, sometimes denoted as parching, is used to prepare some specific recipes from pulses. The basis of this method is the exposition of usually whole seed to dry heat at a temperature ranging from 200 °C to 250 °C. Traditional roasting of pulses is used primarily to reduce non-nutritional factors, enhance flavour, and extend product shelflife.

Roasting has positive effects on pulse flavours. The roasting process can develop a lot of pleasant flavours, which could mitigate undesirable pulse flavour. The roasting process could develop pleasant flavours from alcohols and aldehydes such as phenylethyl alcohol, benzaldehyde, and benzene acetaldehyde.[60] These compounds are related to almond, floral, and clove-like flavours. In addition, the high temperature could result in the denaturation of pulse proteins. Enzymes, such as lipoxygenase, are responsible for the degradation of lipids. Roasting could denature these enzymes to prevent the generation of undesirable flavours from lipid oxidation.[11]

On the other hand, roasting has negative effects on pulse flavours. Nitrogenous flavour compounds develop during the roasting process. These compounds generally have a beany, roasted nut flavour and an earthy and sharp taste, such as 2-ethyl-5-methyl-pyrazine, 2-ethyl-6-methyl-pyrazine, 2-methyl-6-propyl-pyrazine, and 2-methyl-5-propyl-pyrazine.[72] The Maillard reaction, specifically Strecker degradation, could be the major pathway for generating pyrazine with the precursor of amino acids. The parameters of the roasting process, such as moisture, temperature, and time, need to be controlled for developing the aroma/odour of pulse seeds.

8.5.4 Extrusion

The extrusion process is an automated technology characterized by high temperature, high pressure, and short processing time.[73] Extrusion technology has been applied to plant proteins since the 1960s. Following the current hot topic of plant protein, the extrusion process has been used to produce meat analogues with pulse proteins.[74] A typical extrusion system consists of a mixer, bin feeder, preconditioner, extrusion barrel configuration, die, and knife.[75] Extrusion cooking is a complex process that includes hydration, forming, cooking, unfolding, and alignment functions. Low-moisture meat analogues (15–30% moisture content) and high-moisture meat analogues (>50% moisture content) have been developed as two different extrusion products.

Extrusion processing can induce the denaturation and unfolding of pulse proteins. The disulphide bond could be detached at a temperature higher than 80 °C. In addition, the denaturation process may expose side chains and increase chain flexibility, digestibility, and protein–starch cross-linking. Volatiles bound to pulse proteins with physical trappings, van der Waals forces, hydrogen bonds, hydrophobic interactions, ionic bonds, and even covalent bonds may be released during the extrusion process. In this sense, the volatiles would be released with the water vapour when low-moisture meat analogues puff off from the die. On the other hand, high-moisture meat analogues would form a gel with a high water content. High-moisture meat analogues could have a higher capability of entrapping air bubbles or liquid phases. Therefore, released volatiles could be retained by the gel.

Due to the complicated process and multiple factors, different researchers have drawn varied conclusions about the effect of the extrusion process on flavour attributes of final products. Simons *et al.*[76] extruded pinto, navy, and black beans into low-moisture meat analogues. The volatiles of these products exceeded the minimum flavour value for acceptability. It is speculated that increased the extrusion temperature from 80 °C to 120 °C at zone 5 could result in the evaporation of volatiles, although new volatiles would be generated. Martin *et al.*[77] extruded mixtures of lentil, lupin, and faba bean protein isolates into low-moisture meat analogues. The flavours of the products were described as green/grassy, flour-like, savoury-like, and earthy. Asgar *et al.*[78] indicated that one of the primary constraints of meat analogues in the current market is the substantial off-flavour issue. Therefore, de-flavoured pulse protein products could be a better ingredient for producing meat analogues.

8.5.5 Germination

Germination is the process of improving the nutritional value and reducing anti-nutritional factors in pulse seeds. Basically, germination starts with seed hydration. Water is absorbed by dry seeds through a concentration gradient and diffusion. In turn, the respiratory activity increases, and gibberellic acids are synthesized in the germ. With the increased mobility, which is attributed to water, gibberellic acids move from the embryo to the aleurone layer as a molecular signal. Enzymes such as lipase and lipoxygenase are synthesized with the signal and secreted into the endosperm. Lipids are degraded and used as an energy source for sprouting and the new flavours.[79]

Aliphatic aldehydes, alcohols, and ketones are closely related to the degradation of pulse lipids. Increased activity of lipoxygenase could result in increased volatiles during pulse germination.[43] Xu *et al.*[48] have studied the dynamic flavour change of chickpea, lentil, and yellow pea during pulse germination. In total, 124 volatiles have been recorded during pulse germination. Most of the volatiles increased during the long period of germination. However, after one or two days of germination, the volatiles in the pulse seeds did not develop well. Despite the increase in beany flavour, carriers of beany flavour might be loosened so that the binding effect of beany flavour may be weakened, which would result in the easy removal of beany flavour with post-processing. Interestingly, some volatiles decreased in the first days of chickpea germination. For example, 2,4-nonadienol and hexanal, two of the major contributors of beany flavour, decreased by 72% and 70%, respectively, after two days of chickpea germination.

The germination of pulse seeds could also mask the undesirable flavour of pulse seeds. Kaczmarska *et al.*[52] studied the flavour of germinated lupin with GC-MS/O. In untreated lupin seeds, beany, green odours associated with 1-penten-3-ol, hexanal, 1-hexanol, (*E*)-2-nonenal, and (*E*)-2-octenal were the strongest. After three days of germination, meaty (2-methylbutanal) and sulphur-like odours (dimethyl sulphide and dimethyl trisulphide) significantly increased ($P < 0.001$) and became predominant odours in the germinated lupin. Moreover, sweet odour notes (2,3-butanedione, methyl isobutyl ketone, (*E*,*Z*)-2,6-nonadienal, and guaiacol) increased after the germination of lupin. The beany and green characteristics of lupin have not been significantly changed.

Courtney *et al.*[80] made cookies with germinated pinto bean flour. After two days of germination, the pinto bean was dried, milled, and mixed with other ingredients. The cookies were then baked at 190 °C for 7 min, after which they were removed from the oven and allowed to cool for 2 min on the baking tray. The sensory evaluation showed that no significant ($P > 0.05$) difference could be observed in terms of flavour acceptability.

8.5.6 Fermentation

Fermentation is the desirable process of biochemical modification of the primary food matrix brought about by microorganisms and their enzymes. Fermentation is a process that provides us with a variety of unique flavours to enjoy, such as soy sauce, cheese, yoghurt, bread, beer, and wine.[81] Microorganisms can produce a large number of different classes of flavour compounds. Esters, acids, carbonyls, alcohols, terpenes, lactones, pyrazines, and sulphur compounds can be formed *via* reactions such as oxidation, reduction, isomerization, esterification, and hydrolysis during fermentation. The flavour change of the fermentation process is mutually dependent on the different varieties of microorganisms and pulses.

Fermentation could reduce the off-flavours of pulse seeds. Yeast can secrete enzymes such as alcohol dehydrogenase and aldehyde dehydrogenase, which would remove the beany-related volatiles. The genera *Saccharo-*

myces and *Kluyveromyces* have been employed to decrease the green beany flavour in soybeans by the degradation of aldehydes and alcohols.[82]

Fermentation can produce pleasant flavours to mask the off-flavours. Ben-harb *et al.*[83] screened 55 strains from different microbial species to improve the aroma of pea and pea/milk emulsions. Two yeast strains, *Candida catenulata* and *Geotrichum candidum,* were able to colonize pea-based products and positively influence the release of volatile compounds by generating a roasted/grilled aroma for 100% pea emulsion and a fruity, lactic aroma for pea/milk emulsion.

However, fermentation could also increase beany-related flavours by some microorganisms. Kaczmarska *et al.*[52] fermented lupin flour with *Lactobacillus delbrueckii* subsp. *bulgaricus* and *Streptococcus thermophilus* for 20 h at 30 °C. A significant increase in the mushroom odour and earthy characteristics was attributed to the increased volatiles such as penten-3-one, octen-3-one, and 3-octen-2-one. The flavour of beany and green odours also increased due to the increased concentration of hexanal and (*E*)-2-octenal.

Fermented pulses have also been developed into baking products. Natural fermentation of pigeon pea has been developed to improve the sensory of pigeon pea flours.[84] With the pasta cooking process, fortified pasta with 5% and 10% fermented pigeon pea flour had an acceptability score similar to control pasta.

In summary, the relationship between fermentation and off-flavours of pulses cannot be simply summarized as positive or negative. The species of microorganisms, varieties of pulse seeds, and fermentation conditions, such as time and temperature, are the factors that could affect the flavour of final products.

8.5.7 Others

Many other processes are also available for improving the flavour of pulse proteins, such as chemical modification, spray drying, and freeze-drying.[47] These methods changed the flavours of pulse proteins through chemical and physical reactions involving covalent bonds, physical traps, van der Waals forces, hydrogen bonds, hydrophobic interactions, and ionic bonds.

In addition to these methods, genetic modification is always neglected by food scientists. A breeding programme could be an effective solution to eventually solving the flavour problem. In fact, flavour phenotyping is expensive, subject to environmental variation, not amenable to high-throughput assays, and always beyond the means of most breeding programmes. Fortunately, plant breeders have realized the importance of new traits, such as flavour and functionality of crops, which are equally important as yield and disease resistance. The close collaboration between breeders and food scientists will fill the gaps and set up a new paradigm for the future food system.

8.6 Conclusion and Future Trends

Although pulse proteins are of increasing importance in the global plant-based food market, there is still a strong need to improve their aroma/odour. This chapter shows the composition, source, and interaction of aroma/odour compounds with proteins. The flavour of pulse proteins is closely related to the composition of pulse seeds. Normally, the pulse odour is considered a combination of grassy, green, earthy, and mushroom flavours. It is hard to remove all the volatiles because of the interaction between proteins and volatiles. Therefore, future research on improving pulse protein flavour should focus on the combination of removing and masking aroma/odour.

Abbreviations

–COOH	Carboxyl group
–NH2	Amino group
·OH	Hydroxyl radical
C	Carbon
CP	Chickpea
E	*Trans*
E/TN	Essential to total amino acids
FAO	Food and agriculture organization
FB	Fava bean
GC	Gas chromatography
H	Hydrogen
HS	Headspace
kDa	kilo Dalton
LE	Lentil
MS	Mass spectrometry
Mw	Molecular weight
N	Nitrogen
NR	Not reported
O·	Oxyl radical
O	Olfactory
PA1	Pea protein 1
PA2	Pea protein 2
PA1a	Pea protein 1a
PA1b	Pea protein 1b
R·	Lipid radical
RH	Lipid
RI	Retention index
ROO·	Lipid hydroperoxyl radical
Se	Selenium
SPME	Solid phase microextraction

UNU	United Nations University
WHO	World Health Organization
YP	Yellow pea
Z	*Cis*

References

1. USA Pulses, *What are pulses*, Available from: https://pulses.org/what-are-pulses.
2. R. N. Tharanathan and S. Mahadevamma, *Trends Food Sci. Technol.*, 2003, **14**, 507–518.
3. FAO, *FAOSTAT Domain: Crops and livestock products*, Available from: https://www.fao.org/faostat/en/#data/QCL.
4. C. Hall, C. Hillen and J. G. Robinson, *Cereal Chem.*, 2017, **94**, 11–31.
5. C. A. Patterson, J. Curran and T. Der, *Cereal Chem.*, 2017, **94**, 2–10.
6. A. V. C. Schneider, *Br. J. Nutr.*, 2002, **88**, 243–250.
7. J. Boye, F. Zare and A. Pletch, *Food Res. Int.*, 2010, **43**, 414–431.
8. T. Kim, *J. Chem. Inf. Model*, 2019, **9**, 1689–1699.
9. O. P. Malav, S. Talukder, P. Gokulakrishnan and S. Chand, *Crit. Rev. Food Sci. Nutr.*, 2015, **55**, 1241–1245.
10. P. Kumar, M. K. Chatli, N. Mehta, P. Singh, O. P. Malav and A. K. Verma, *Crit. Rev. Food Sci. Nutr.*, 2017, **57**, 923–932.
11. C. Murat,M. H. Bard, C. Dhalleineand N. Cayot, *Food Res. Int.*, 2013, **53**, 31–41.
12. S. S. Schiffman and C. Dackis, *Percept. Psychophys.*, 1975, **17**, 140–146.
13. M. Xu, PhD thesis, *Improvement of the Physicochemical Attributes and Antioxidant Profiles from Pulse Seeds through Germination*, North Dakota State University, 2019, pp. 75–78.
14. A. A. Oshodi, O. Olaofe and G. M. Hall, *Int. J. Food Sci. Nutr.*, 1993, **43**, 187–191.
15. E.-S. Tan, N. Ying-Yuan and C.-Y. Gan, *Food Chem.*, 2014, **152**, 447–455.
16. W. Wu, W. P. Williams, M. E. Kunkel, J. C. Acton, Y. Huang, F. B. Wardlaw and L. W. Grimes, *J. Agric. Food Chem.*, 1996, **44**, 1296–1301.
17. M. Jafari, A. R. Rajabzadeh, S. Tabtabaei, F. Marsolais and R. L. Legge, *Food Chem.*, 2016, **208**, 35–41.
18. A. Khanam and K. Platel, *Food Chem.*, 2016, **194**, 1293–1299.
19. Z. X. Lu, J. F. He, Y. C. Zhang and D. J. Bing, *Crit. Rev. Food Sci. Nutr.*, 2019, **8398**, 1–13.
20. S. Hemalatha, K. Platel and K. Srinivasan, *Food Chem.*, 2007, **102**, 1328–1336.

21. Y. W. Chang, I. Alli, A. T. Molina, Y. Konishi and J. I. Boye, *Food Bioprocess. Technol.*, 2012, **5**, 618–625.
22. A. Ghumman, A. Kaur and N. Singh, *LWT-Food Sci. Technol.*, 2016, **65**, 137–144.
23. B. G. Swanson, *J. Am. Oil Chem. Soc.*, 1990, **67**, 276–280.
24. C. Bennetau-Pelissero, in *Bioactive Molecules in Food*, ed. J.-M. Mérillon and K. G. Ramawat, Springer International Publishing AG, Cham, Switzerland, 2018, pp. 1–43.
25. Y. Zhang and H. M. Romero, *Int. J. Biol. Macromol.*, 2020, **162**, 1516–1525.
26. F. K. Yeboah, I. Alli, B. K. Simpson, Y. Konishi and B. F. Gibbs, *Food Chem.*, 1999, **67**, 105–112.
27. Y. Ladjal-Ettoumi, H. Boudries, M. Chibane and A. Romero, *Food Biophys.*, 2016, **11**, 43–51.
28. K. Chilomer, K. Zaleska, D. Ciesiołka, P. Gulewicz, A. Frankiewicz and K. Gulewicz, *Acta. Soc. Bot. Pol.*, 2010, **79**, 11–20.
29. Y. Liu, H. Su and H. I. Song, *Food Anal. Methods*, 2018, **11**, 343–354.
30. P. R. Shewry and M. J. Pandya, *Seed Proteins*, 1999, 563–586.
31. K. Jaouani, I. Karmous, M. Ostrowski, E. E. Ferjani, A. Jakubowska and A. Chaoui, *J. Plant Physiol.*, 2018, **226**, 64–76.
32. M. A. Sulieman, A. B. Hassan, G. A. Osman, M. M. El Tyeb, E. A. El Khalil, A. H. El Tinay and E. E. Babiker, *Pakistan J. Nutr.*, 2008, **7**, 801–805.
33. M. G. Semenova, A. S. Antipova, T. A. Misharina and R. V. Golovnya, *Food Hydrocolloids*, 2002, **16**, 557–564.
34. A. S. Sandberg, in *Functional Foods: Concept to Product*, ed. M. Saarela, Woodhead Publishing Limited, Padstow, 2nd edn, 2011, pp. 358–382.
35. A. Mohanan, M. T. Nickerson and S. Ghosh, *Food Chem.*, 2020, **316**, 126282.
36. S. K. Johnson, J. Clements, C. B. J. Villarino and R. Coorey, in *Gluten-Free Ancient Grains: Cereals, Pseudocereals, and Legumes: Sustainable, Nutritious, and Health-Promoting Foods for the 21st Century*, ed. J. Awika and J. Taylor, Elsevier, Cambridge, 1st edn, 2017, pp. 179–221.
37. A. Patel, H. C. Devraja, P. Sharma and R. R. B. Singh, *Food Technology II*, Available from: http://agrimoon.com/food-technology-ii-pdf-book-free-download/.
38. P. R. Shewry, in *Principles of cereal science and technology*, AACC International Press, St. Paul, 3rd edn, 2010, **51.**
39. M. Shapiro and V. Galperin, *Chem. Eng. Process. Process Intensif.*, 2005, **44**, 279–285.
40. R. D. Reichert, *J. Food Sci.*, 1982, **47**, 1263–1267.

41. R. Cai, B. Klamczynska and B. K. Baik, *J. Agric. Food Chem.*, 2001, **49**, 3068–3073.
42. M. Fredrikson, P. Biot, M. L. Alminger, N. G. Carlsson and A. S. Sandberg, *J. Agric. Food Chem.*, 2001, **49**, 1208–1212.
43. M. Xu, Z. Jin, Z. Gu, J. Rao and B. Chen, *Food Chem.*, 2020, **314**, 126184.
44. Y. Lan, B. Chen and J. Rao, *Food Hydrocolloids*, 2018, **80**, 245–253.
45. I. Alli, B. F. Gibbs, M. K. Okoniewska, Y. Konishi and F. Dumas, *J. Agric. Food Chem.*, 1993, **41**, 1830–1834.
46. U. M. Lanfer Marquez, R. M. C. Barros and F. M. Lajolo, *Food Chem.*, 1996, **55**, 179–184.
47. G. Reineccius, in *Flavor Chemistry and Technology*, CRC Press, Boca Raton, 2nd edn, 2006.
48. M. Xu, Z. Jin, Y. Lan, J. Rao and B. Chen, *Food Chem.*, 2019, **280**, 83–95.
49. S. Vara-ubol, E. C. Ivz, D. H. Chambers and J. Hall, *J. Sens. Stud.*, 2004, **19**, 15–26.
50. S. Sharan, G. Zanghelini, J. Zotzel, D. Bonerz, J. Aschoff, A. Saint-Eve and M.-N. Maillard, *Compr. Rev. Food Sci. Food Saf.*, 2020, 401–428.
51. Y. Lan, M. Xu, J. Ohm, B. Chen and J. Rao, *Food Chem.*, 2019, **278**, 665–673.
52. K. T. Kaczmarska, M. V. Chandra-Hioe, D. Frank and J. Arcot, *LWT-Food Sci. Technol.*, 2018, **87**, 225–233.
53. P. Shen, Z. Gao, M. Xu, J. Rao and B. Chen, *Food Hydrocolloids*, 2020, **108**, 106065.
54. G. Reineccius, in *Source Book of Flavors*, CBS Publishers & Distributors, Maryland, 2nd edn, 1994.
55. R. Amarowicz and R. B. Pegg, *Eur. J. Lipid Sci. Technol.*, 2008, **110**, 865–878.
56. B. Guldiken, R. Green and M. T. Nickerson, *Eur. Food Res. Technol.*, 2021, 593–604.
57. G. A. Burdock, in *Fenaroli's handbook of flavor ingredients*, CRC Press, Boca Raton, 6th edn, 2010.
58. B. D. Oomah, M. Razafindrainibe and J. C. Drover, *J. Sci. Food Agric.*, 2014, **94**, 473–481.
59. A. Paraskevopoulou, A. Chrysanthou and M. Koutidou, *Food Res. Int.*, 2012, **48**, 568–577.
60. S. Bi, A. Wang, F. Lao, Q. Shen, X. Liao and P. Zhang, *Food Chem.*, 2021, **339**, 127878.
61. S. Chen, A. Wang, M. Qian, Z. Li and Y. Xu, *J. Agric. Food Chem.*, 2021, **67**, 4876–4884.
62. P. M. T. Roland, L. Pouvreau, J. Curran, F. Van De Velde and P. M. T. De Kok, *Cereal Chem.*, 2017, **94**, 58–65.

63. L. Bott and E. Chambers IV, *J. Sens. Stud.*, 2006, **21**, 308–321.
64. D. J. Sessa and J. J. Rackis, *J. Am. Oil Chem. Soc.*, 1977, **54**, 468–473.
65. H. Jelen, in *Food flavors: chemical, sensory and technological properties*, CRC press, 2011.
66. S. Azarnia, J. I. Boye, T. Warkentin, L. Malcolmson, H. Sabik and A. S. Bellido, *Food Chem.*, 2011, **124**, 326–335.
67. P. Khrisanapant, B. Kebede and S. Y. Leong, *Foods*, 2019, **8**, 1–19.
68. P. Hajeb and S. Jinap, , *Crit. Rev. Food Sci. Nutr.*, 2015, **55**, 778–791.
69. Ä. Juraj, P. Cviková, H. Peter, K. Miroslava, K. Simona, H. Lukáš, B. Marek, T. Lenka, O. Bučko and T. Jana, *J. Microbiol. Biotechnol. Food Sci.*, 2017, 209–220.
70. K. Wang and S. D. Arntfield, *Flavour Fragr. J.*, 2017, **32**, 92–101.
71. C. Zhang, Y. Hua, X. Li, X. Kong and Y. Chen, *Food Chem.*, 2020, **333**, 127469.
72. Z. Ma, J. I. Boye, S. Azarnia, B. K. Simpson, Z. Ma, J. I. Boye, S. Azarnia, B. K. Simpson, Z. Ma, J. I. Boye, S. Azarnia and B. K. Simpson, *Int. J. Food Prop.*, 2016, **19**, 2251–2271.
73. R. Osen, S. Toelstede, F. Wild, P. Eisner and U. Schweiggert-Weisz, *J. Food Eng.*, 2014, **127**, 67–74.
74. R. Osen, S. Toelstede, P. Eisner and U. Schweiggert-Weisz, *Int. J. Food Sci. Technol.*, 2015, **50**, 1390–1396.
75. S. Vatansever, M. C. Tulbek and M. N. Riaz, *Cereal food*, 2020, **65**, 1–11.
76. C. W. Simons, C. H. Iii, M. Tulbek, M. Mendis, T. Heck and S. Ogunyemi, *J. Sci. Food Agric.*, 2014, 1–5.
77. A. Martin, V. Schmidt, R. Osen, J. Bez, E. Ortner and S. Mittermaier, *J. Sci. Food Agric.*, 2020, 1–11.
78. M. A. Asgar, A. Fazilah, N. Huda, R. Bhat and A. A. Karim, *Compr. Rev. Food Sci. Food Saf.*, 2010, **9**, 513–529.
79. M. Xu, J. Rao and B. Chen, *Crit. Rev. Food Sci. Nutr.*, 2020, **60**, 740–759.
80. C. W. Simons and C. Hall, *Food Sci. Nutr.*, 2018, **6**, 77–84.
81. S. G. Nkhata, E. Ayua, E. H. Kamau and J. B. Shingiro, *Food Sci. Nutr.*, 2018, **6**, 2446–2458.
82. N. Takahashi and R. Sasaki, *Agric. Biol. Chem.*, 1979, **43**, 1883–1889.
83. S. Ben-Harb, A. Saint-Eve, M. Panouillé, I. Souchon, P. Bonnarme, E. Dugat-Bony and F. Irlinger, *Int. J. Food Microbiol.*, 2019, **293**, 124–136.
84. A. T. Torres, J. F. Frias, M. G. Granito and C. V. I. Alverde, *J. Agric. Food Chem.*, 2006, **54**, 6685–6991.

CHAPTER 9

Flavour of Cereal and Pseudocereal Proteins

JING ZHAO*, MADELINE GIBSON, AVA FROBLE AND CHANGQI LIU

San Diego State University, San Diego, California, USA
*E-mail: jzhao2@sdsu.edu

9.1 Introduction

Cereals are flowering plants of the grass family (Poaceae) cultivated for their edible grains. The word 'cereal' is derived from Ceres, the Roman goddess of agriculture. Cereals were among the first plants domesticated by humans, and their cultivation played a key role in transforming humans from living a hunter-gatherer lifestyle to forming an agrarian society, which then led to the rise of civilization.[1] Today, cereals are still the most important food source that sustains humankind. Globally, corn (*Zea mays*), rice (*Oryza sativa*), and wheat (*Triticum* spp.) are the largest cereal grains produced, accounting for over 90% of total cereal production.[2] In addition to the big three cereals, many other species are cultivated worldwide, such as barley (*Hordeum vulgare*), barnyard millet (*Echinochloa* spp.), finger millet (*Eleusine coracana*), fonio (*Digitaria* spp.), foxtail millet (*Setaria italica*), Job's tears (*Coix lacryma-jobi*), oat (*Avena sativa*), pearl millet (*Pennisetum glaucum*), proso millet (*Panicum miliaceum*), rye (*Secale cereale*), sorghum (*Sorghum bicolor*), teff (*Eragrostis tef*), triticale (× *Triticosecale*), and wild rice (*Zizania* spp.). Pseudocereals are broadleaf plants whose seeds are eaten like cereal grains. Examples of pseudocereals include amaranth

Food Chemistry, Function and Analysis No. 41
Flavour and Consumer Perception of Food Proteins
Edited by Jing Zhao and Changqi Liu

Published by the Royal Society of Chemistry, www.rsc.org

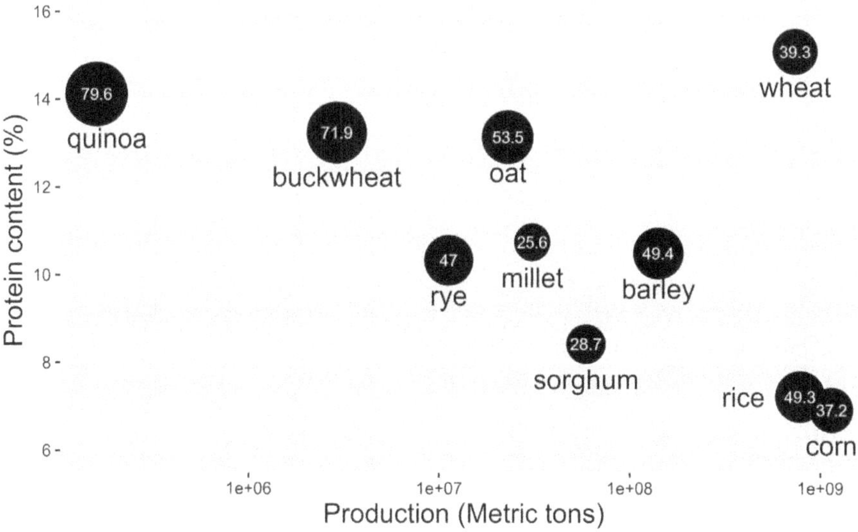

Figure 9.1 Global production of select cereal and pseudocereal crops and their protein content and quality.[3,4] The size of and the number within each bubble represent the digestible indispensable amino acid score of each grain.[5,6]

(*Amaranthus* spp.), buckwheat (*Fagopyrum esculentum*), chia (*Salvia hispanica*), quinoa (*Chenopodium quinoa*), and Tartary buckwheat (*Fagopyrum tataricum*). The global production of select cereal and pseudocereal grains in 2018 and their protein content and quality are summarized in Figure 9.1.

9.2 True Cereals

Cereals are staple foods and an important source of protein. As shown in Figure 9.1, the protein content of cereals ranges from 6 to 15%.[3] The protein content of corn and rice is about 6–9% and 7%, respectively.[7] Sorghum, millet, and rye have about 10–11% protein content. Wheat and barley contain about 13% protein, while oats contain 11–15%.[8] Besides being consumed as a whole food, some cereals are utilized for starch and syrup production (*e.g.*, corn and rice). Protein ingredients derived from cereals are often prepared from the by-products of starch or syrup production. Cereal proteins have a low quality compared to animal proteins due to the limited amounts of certain essential amino acids, especially lysine. In applications, cereal proteins are often combined with legumes/legume proteins to improve their nutritional value.

9.2.1 Flavour of Corn Proteins

The annual global corn production in 2021–2022 was 1216 million metric tonnes, making it the largest cereal crop in the world.[9] Corn contains about 6–20% protein, depending on the cultivar. Proteins from the whole corn kernel consist of 8% albumin, 9% globulin, 39% prolamin (zein), and 40% glutelin, while the endosperm contains 4% each of albumin and globulin, 47% of zein, and 39% of glutelins.[10] About 32% of the amino acids in corn proteins are essential amino acids, including about 13% of leucine. At about 6–10%, the leucine content in corn protein is higher than most proteins (such as those of oat, lupin, wheat, hemp, microalgae, soy, brown rice, pea, potato, whey, casein, egg, and muscle tissue).[11]

Zein, the ethanol-soluble fraction of corn proteins, is the major storage protein of corn. Due to its poor water solubility, zein was not commonly used as a food ingredient. Due to its hydrophobicity and resistance to microbial attack, zein is of commercial interest to be used in fibre, adhesive, coating, ceramic, ink, cosmetics, textiles, chewing gum, and biodegradable plastics.[10] It was reported that when zein was heated in the presence of lipids and carbohydrates, an increased number of volatiles were generated, including two lipid–protein–carbohydrate interaction products: 2-methyl-3(or 6)-pentylpyrazine and 2,5-dimethyl-3-pentylpyrazine.[12] When used in perfume, zein helped achieve a long-lasting, slow release of fragrance due to the trapping effect of zein films and possibly the interaction between zein and the fragrance.[13] Similarly, zein has been studied in the controlled delivery of perfumes on hair.[14] For food uses, zein has been used in antimicrobial edible films for cheese,[15] capsules of fish oil in mayonnaise,[16] and nanoemulsions.[17] The impact of zein on flavours due to its film formation and protein–flavour interaction in food products needs further investigation.

In one study that used hydrolysed corn gluten protein as bait for insects, the volatile compounds in the hydrolysates were found to be determined by pH. The acidic hydrolysates mainly contained phenylacetaldehyde and several aromatic oxygenated compounds. The basic hydrolysates contained primarily nitrogenous compounds, such as alkyl-substituted pyrazines and 3-methylbutylamine-derived imines.[18]

Corn protein isolates extracted from corn oil cake using the alkaline extraction isoelectric point precipitation method were formulated into a protein-based beverage drink. Based on the sensory evaluation by 15 trained panellists, the taste and flavour of the beverage were significantly affected by the packaging and storage conditions.[19]

In corn germ protein flour (CGPF), a by-product of corn milling, phenolic compounds have been associated with sour, bitter, and astringent off-flavours.[20] The taste thresholds for *o*-Coumaric, *p*-Coumaric, and ferulic acids in CGPF were 20, 48, and 90 ppm, respectively. The aroma characteristics of defatted, acid-treated, heat-treated, and ethanol-washed CGPFs were studied by sensory panellists and gas chromatography (GC). The study found that treated CGPF samples had fewer grainy, brown, corn-like, and sweet notes than untreated CGPF samples. However, there were more green, musty, and chemical notes in the treated CGPF samples.[21]

9.2.2 Flavour of Wheat Proteins

Wheat is the second largest cereal crop in the world. The worldwide wheat production in 2021–2022 was 778.6 million tonnes. The wheat kernel comprises bran, germ, and endosperm. White flour, which is derived from only the endosperm, contains about 74.6% carbohydrate, 12% protein, and 1.7% lipid.[3] In the bran layer, the protein content was found to be 16.80%.[22]

9.2.2.1 Wheat Flour

Wheat flour is widely used to make products such as bread, noodles, baked goods, and pancakes. The flavour formed from wheat flour during heating and extrusion cooking was determined by the cooking temperatures and moisture contents. According to gas chromatography-mass spectrometry analysis using the short-path thermal desorption technique, the volatile components of the heated and extruded wheat flour mainly comprised lipid-oxidation products, Maillard-type flavours, and some sulphur-containing compounds.[23]

A sensory study on extruded wheat flours showed that the extrudates from the lowest extrusion temperature (120 °C) and higher moisture levels (18% and 22%) were mainly characterized by porridge-like, cooked milk, and cooked vegetable terms. At the intermediate conditions (150 °C and 18% moisture), popcorn, toasted, and puffed wheat notes dominated. The popcorn aroma was lower in the extrudates prepared under the most severe conditions (180 °C and 14% moisture), where puffed wheat, burnt, and toasted aromas were scored at relatively high intensities.[24]

Riha *et al.* examined the effects of cysteine addition on the volatile formation in extruded wheat flour. The study found that the addition of cysteine up to 0.5% greatly increased volatile production. The volatiles included 18 sulphur-containing compounds, most of which are thiazoles derived from the thermal degradation of cysteine, four pyrazines, and one furan.[25] Bredie *et al.* performed a descriptive sensory evaluation on extruded wheat flour with and without cysteine and reducing sugars. The study found that the discriminative attributes in extruded flour were biscuity, cornflakes, sweet, and cooked milk, while the addition of cysteine and glucose resulted in stronger popcorn, nutty/roasted, and puffed wheat aromas. Replacing glucose with xylose enhanced the savoury and sulphur notes.[26]

9.2.2.2 Wheat Gluten

Wheat gluten is a main by-product of the production of wheat starch. It contains over 75% protein and has good nutritional value. However, wheat gluten has poor water solubility due to its high content of hydrophobic amino acids and the large hydrophobic regions within the molecules, which limits its application in the food industry. Ahmedna *et al.* examined the incorporation of solubilized wheat protein isolate in ice cream, chocolate chip cookies, banana nut muffins, and hamburger patties and found that wheat gluten isolate can be included in these products by up to 5, 10, 15, or 20%, respectively, without affecting consumer acceptance.

Wheat gluten is typically considered to be odourless. The volatiles in wheat gluten have not been characterized. On the other hand, numerous studies examined the flavour formation from wheat gluten hydrolysates through the Maillard reactions. Romero *et al.* studied the modification of wheat gluten using enzymatic hydrolysis (Umamizyme or Flavourzyme) alone or in combination with enzymatic deamidation (Glutaminase Daiwa).[27] When subjected to the Maillard reaction, the hydrolysates generated important flavour compounds such as pyrazines and furans. The major pyrazines identified were methylpyrazine, 2,5-dimethylpyrazine, 2-ethylpyrazine, and 2-methyl-6-propyl pyrazine, whereas the major furans included 2-furfural, 2-acetylfuran, 2-furfuryl alcohol, and 5-methyl-2-furfural. It was found that enzymatic treatments produced the same volatiles, while deamidation increased the quantities. In another study, Maillard peptides were prepared using a wheat gluten peptide-xylose-cysteine reaction system at pH 7 and heated at 120 °C for 120 minutes.[28] The key flavour substances generated were found to be 3-thiophenecarboxaldehyde, 2,5-thiophenedicarboxaldehyde, 4-amino-1-methyl-1H-benzotriazole, 2-acetylfuran, 3-methylthiophene, ethyl thioacetate, 2-acetylthiophene, 2-methylfuran, 1-hexanol, 1-pentanol, 2,5-dimethylthiophene, 5-methyl-2-thiophenecarboxaldehyde, 1,4,6-trimethyl-2(1*H*)-pyridone, 1-ethylidene-1H-indene, 3-heptanone, 2-furfurylthiol, 2-methyl-3-pentanethiol, heptanoic acid, hexanoic acid, thiirane, and furfural. When thermally reacted with glucose and fructose, pyrazines, furans, and sulphur-containing components were detected in wheat gluten hydrolysates and deamidated wheat gluten hydrolysates. The deamidation treatment increased the volatile components due to the release of free ammonia. Ammonia released from wheat protein *via* deamidation participated in the generation of diverse volatile components, including pyrazines, in the Maillard reaction.[29] Chen and Ho used a fungal protease to obtain wheat gluten hydrolysates, which were also deamidated or acid-treated before being subjected to the Maillard reactions with glucose. They also found that the deamidation reaction significantly increased the formation of aroma compounds.[30] Several studies reported meat-like flavours from treated wheat proteins.[31] Wu found that the flavour compounds obtained by the Maillard reaction of gluten hydrolysates and glucose analysed by solid-phase microextraction-gas chromatography-mass spectrometry (SPME-GC-MS) included pyrazines, furans, sulphur compounds, aldehydes, ketones, alcohols, and phenols.[32]

Besides volatile compounds, the hydrolysis of wheat gluten also generated taste-active components. Schlichtherle-Cerny and Amadò found that the enzymatic hydrolysate of an acid-deamidated wheat gluten was the least bitter and had an intense glutamate-like taste compared to other hydrolysates investigated.[33] Four pyroglutamyl peptides, pGlu-Pro-Ser, pGlu-Pro, pGlu-Pro-Glu, and pGlu-Pro-Gln, were also identified in this study. Liu *et al.* investigated the effects of deamidation with certain enzymes or acid treatment on the chemical composition, bitterness, and umami properties of wheat gluten hydrolysates. This study found that wheat gluten hydrolysates treated with glutaminase for 180 minutes had a significantly increased umami taste score and a decreased bitter taste score.[34] Liao *et al.* investigated the effect of acetic acid and hydrochloric acid-deamidation pre-treatment on the sensory characteristics of pancreatin wheat gluten hydrolysates. The hydrolysates prepared from acetic acid-deamidated wheat gluten were found to have a more intense glutamate-like and sauce-scented taste and better nutritional characteristics.[35]

Wang and Arntfield studied the interactions of homologous aldehydes (hexanal, heptanal, and octanal) and ketones (2-hexanone, 2-heptanone, and 2-octanone) with commercial wheat gluten using GC-MS.[36] Wheat gluten had a higher binding affinity for aldehydes than ketones.

9.2.3 Flavour of Rice Proteins

Rice is the third largest cereal crop in the world, with a global annual production of 513.7 million tonnes in 2021/22.[9] It is a staple food for more than 3.5 billion people around the world, particularly in Asia, Latin America, and parts of Africa. There are numerous varieties of rice.

Rice protein is vegan, gluten-free, and non-allergenic,[37,38] making it suitable for applications in infant formula, gluten-free products, and high-protein foods. However, the production and application of rice protein are limited by its objectionable flavour and poor solubility at neutral pH.[39] Rice protein powder is described by consumers as having an objectionable odour and a bitter taste, in spite of added flavours. Also, rice proteins are mainly composed of globulin (about 12%) and glutelin (referred to as oryzenin, about 80%), with minor amounts of albumin (5%) and prolamin (about 3%).[40] As the major component of rice protein is glutelin, which is only soluble in acidic or alkaline solutions, the solubility of rice proteins is normally low. As a result, some rice protein products are described as gritty, chalky, and unacceptable by consumers.

9.2.3.1 Brown Rice Proteins

Due to the large number of rice varieties, the flavour of rice and the rice protein derived from each variety differ significantly. The flavour of rice has been widely studied, while only a few studies have been done to examine the flavour profiles of rice proteins.

Zhao and Boatright identified the primary odorants in five commercial brown rice protein products using static headspace gas chromatography-olfactometry-mass spectrometry (GC-O-MS) analysis.[41] The volatile compounds present in an amount above their reported odour threshold values included acetaldehyde, sulphur-containing compounds (methanethiol, dimethyl disulphide, and dimethyltrisulphide), and lipid-oxidation-derived aldehydes and furans (hexanal, pentanal, benzaldehyde, and 2-pentylfuran). The study also demonstrated that the primary volatile compounds contributing to the odour of rice proteins varied substantially from one product to another. These differences could be caused by the different starting materials (*e.g.*, sprouted brown rice, rice bran, high-protein rice flour derived from rice starch, and syrup production) and processing methods. Singh *et al.* examined the volatiles in soy, pea, and brown rice proteins and reported that aldehydes (64%) and furans (18%) are the major volatile compounds in both brown rice protein and pea protein, with minor contents of alcohols, ketones, and other compounds.[42] Brown rice protein contained the highest intensity and number of volatile compounds among the three plant proteins.

9.2.3.2 Rice Bran Proteins

Rice bran is a by-product of rice milling. Rice bran contains a high amount of protein (11–17% versus 6–7% in white rice), 20% oil, dietary fibres (*e.g.*, beta-glucan, pectin, and gum), and many health-benefiting compounds, such as oryzanols, tocopherols/tocotrienols, and phytosterols.[43]

Enzymatic hydrolysis and acid hydrolysis were performed to produce desirable flavour substrates in rice bran protein concentrates due to the generation of peptides and amino acids.[44,45] Arsa and Theerakulkait reported that the most abundant amino acids in liquid Alcalase-hydrolysed rice bran protein concentrate were glutamic acid, arginine, aspartic acid, and leucine.[45] The intensity of desirable aromas, such as cereal-like, nut-like, milk-powder-like, sweet, and cocoa-like, was higher in spray-dried powder than in liquid and freeze-dried hydrolysed rice bran protein. Volatile compounds were prepared by solvent extraction and high-vacuum distillation and analysed by GC-MS. In the liquid hydrolysate, ethanol, hexanal, nonanal, and 2-ethylhexanol existed in high amounts, which contributed to the undesirable ethanol-like, pungent, grassy, and green notes. Similar to rice protein and other plant proteins, hexanal was reported to be the predominant aldehyde in rice bran protein hydrolysate, with a higher amount in the liquid sample.

The generated peptides and amino acids were further reacted with sugars through the Maillard reaction to produce desirable aromas in rice bran protein hydrolysates. Arsa *et al.* showed that the addition of fructose before spray drying increased the sweet, cocoa-like, and milk powder-like aroma in Alcalase-hydrolysed rice bran protein through sensory aroma characterization.[46] Arsa and Theerakulkait characterized the volatile compounds in fructose-added rice bran protein hydrolysates and reported higher levels of volatile compounds, such as aldehydes (2-methylbutanal, 3-methylbutanal, phenylacetaldehyde, and vanillin), pyrroles and pyrazines (2-acetylpyrrole and 2,5-dimethylpyrazine), and ketones (maltol), when fructose was added before spray drying.[45] Arsa *et al.* analysed the aroma profiles of rice bran protein hydrolysates under various preparation conditions (hydrolysis time, pH adjustment, and type of sugar addition before spray drying) using the SPME-MS-O.[47] Three major Strecker aldehydes, 2-methypropanal, 2-methylbutanal, and 3-methylbutanal, derived from Strecker degradation of valine, isoleucine, and leucine, respectively, were quantified using SIDS-GC-cryo-focusing technique. Arsa and Puechkamutr characterized the pyrazines in Alcalase-hydrolysed rice bran protein and identified 2-ethyl-3,5-dimethylpyrazine (coffee, nutty, and caramel aroma) as the odour compound with the highest odour active value.[48] It was stated that because pyrazine formation was primarily influenced by heat treatment and sugar addition, pyrazines might be derived from both the Strecker degradation and the Maillard reaction.

9.2.4 Flavour of Barley and Barley Proteins

Barley (*Hordeum vulgare*), a cereal plant of the grass family Poaceae, and its edible grain are the fourth largest grain crop in the world. The global production of barley amounted to about 145 million tonnes in the 2021–2022 crop year.[9] Barley is primarily grown as animal fodder and as a source of malt for beer and spirits. It is also commonly used in breads, soups, stews, and health products. The flavour of barley significantly affects the products made from it.

9.2.4.1 Barley

The flavour of barley is critical in determining the flavour and flavour stability of beers. One early study by Wang *et al.* examined the volatiles from roasted barley and identified volatile compounds such as dimethyl disulphide, acetic acid, propionic acid, *n*-butyric acid, isovaleric acid, *n*-valeric acid, isocaproic acid, *n*-caproic acid, ethanol, *n*-propanol, *n*-butanol, isobutanol, *n*-pentanol, *n*-hexanol, furfuryl alcohol, γ-octalactone, γ-nonalactone, and γ-decalactone.[49]

9.2.4.2 Barley Flour

Cramer *et al.* analysed the volatile compounds of barley flour from 12 barley cultivars using solid-phase microextraction and GC.[50] The major volatiles detected were aldehydes, ketones, alcohols, and a furan, with 1-octen-3-ol, 3-methylbutanal, 2-methylbutanal, hexanal, 2-hexenal, 2-heptenal, 2-nonenal, and decanal identified as the key odorants. It was also found that hulled barley had higher total volatiles than hull-less barley, indicating the importance of the husk in the barley grain aroma. The proanthocyanidin-free varieties showed higher total volatile and aldehyde contents than wild-type varieties.

9.2.4.3 *Barley Protein*

The protein content of barley flours ranged from 12.8% in the Steptoe cultivar to 16.5% in the SH97110 cultivar. The hulled barley cultivars contained significantly less protein than the hull-less varieties.[50] As barley is primarily used for malt and beer production, studies of barley protein ingredients, especially their flavour, are scarce. One study examined the use of barley protein as an egg replacer and protein enhancer in doughnuts.[51] At a 20% substitution level for wheat flour, barley protein-based doughnuts had lower hedonic scores for appearance, flavour, and overall liking compared to the control doughnuts, according to a consumer study of 61 participants. However, it was unknown how the volatiles changed in the experiment formulations or what compounds were responsible for the off-odours. Alu'datt *et al.* examined up to 15% use of barley protein isolate and barley flours in pita bread using both trained and consumer panels.[52] They found that increasing the levels of barley ingredients lowered the cooked white wheat aroma and increased the raw white wheat, toasted, and whole grain aroma, which corresponded to a decreased aromatic flavour rating. Barley protein isolate also caused an increase in the bitterness of the derived breads. Improvements in the flavour of barley ingredients are needed for their further application in foods.

9.2.4.4 *Interaction of Barley Protein with Other Flavours*

Houde *et al.* examined the binding of barley proteins to vanillin.[53] The binding between vanillin and barley proteins was found to be non-cooperative. The interaction was found to be significantly reduced under heat and high-pressure treatments.

9.2.5 Flavour of Sorghum and Sorghum Proteins

Global sorghum production in 2021–2022 was 62 million tonnes, ranking it fifth among cereal crops.[9] About 50% of sorghum production is used for human consumption.[54] Sorghum and millet are major contributors to the overall calorie intake in semiarid parts of Africa and India.[4] When used in foods, sorghum is used to make gluten-free products such as staple foods (noodles, breads, and porridge), snacks (cookies and biscuits), ready-to-eat cereals, and novel products like beverages and teas.

9.2.5.1 Sorghum Grain

The volatiles in the sorghum grain vary by cultivars. By comparing the fragrant and non-fragrant cultivars, it was found that the aroma of fragrant sorghum was mainly due to 2-acetyl-1-pyrroline, benzothiazole, 2,3,5-trimethylpyridine, (1*E*)-1-ethylidene-1H-indene, cedrene, 2,4-bis(2-methyl-2-propanyl)phenol, 2-hexyl-1-octanol, and 2-butyl-1-octanol.[55] In a study done by Cao *et al.*, sorghum samples [two from southwest China (S1, S2), two from northeast China (S3, S4), and two from Canada (S5, S6)] were analysed for their free and bound aroma compounds.[56] This study found 34 free aroma compounds, including 11 alcohols, 5 esters, 3 acids, 5 aldehydes and ketones, 8 benzoic compounds, 1 terpene, and 1 tetramethylpyrazine. 3-methyl-1-butanol was found to be the highest among the different alcohols in the six samples, with a range from 704.4 to 1394.9 μg kg^{-1}. Sample S4 (northeast of China) showed the highest proportion (53.9%) of alcohol among the six samples. For esters, ethyl hexanoate was detected in all samples except for sorghum S2. Ethyl acetate was found in sorghum samples S1 (45.5 μg kg^{-1}), S3, and S4 (>600 μg kg^{-1}). Ethyl palmitate was only found in sorghum samples S5 and S6. Acetic and hexanoic acids were present in almost all the sorghum samples, with pentanoic acid only found in S2. Of the aldehydes and ketones, nonanal was found only in the Canadian sorghum samples (S5, S6). Fan *et al.* characterized the volatile compounds of six Australian sorghum cultivars using both GC-MS and GC-ion mobility spectrometry.[57] In raw sorghum, esters, especially hexadecanoic acid ethyl ester, showed a high amount. Ketones and aldehydes are some other major volatiles in raw sorghum. Cooking significantly increased the amounts of aldehydes.

Off-flavours, such as musty, mouldy, and sour odours, have been reported in sorghum, especially after storage. Vázquez-Araújo *et al.* identified geosmin (0.1 mg kg^{-1}), 1,2-dimethoxybenzene (100 mg kg^{-1}), 3-octanone (100 mg kg^{-1}), and 1,2,4-trimethoxybenzene (100 mg kg^{-1}) as the compounds responsible for the musty off-odour in sorghum.[58] Infestation by certain insects (*e.g.*, red flour beetle) was also found to generate an off-odour in infested sorghum grain.[59]

9.2.5.2 Sorghum Proteins

The composition of sorghum varies among varieties. On average, sorghum has 72% carbohydrates, 11% proteins, and 3.5% lipids.[60] About 70% of sorghum storage protein is kafirin (prolamin) and is grouped as α-, β-, γ-, and δ-kafirin based on the molecular weight.[61] Sorghum proteins have low digestibility due to the tough inner protein wall, high disulphide content, and strong interaction with the rich phenolic compounds in sorghum.[62] This low protein digestibility and a deficiency in the essential amino acid lysine have significantly limited the utilization of sorghum protein as a food ingredient. As a result, few studies have examined the volatiles in isolated sorghum proteins.

9.2.6 Flavour of Millet Proteins

Millets are a group of small-seeded grasses that grow mainly in dry zones of Asia and Africa. The major types of millet include pearl millet, finger millet, fonio millet, proso millet, foxtail millet, teff, and other smaller varietals. According to the Food and Agriculture Organization (FAO) of the United Nations (UN), global millet production was estimated at 28.2 million tonnes in 2019, which increased to 30.5 million tonnes in 2020. India is the largest global producer, with a 33.3% global market share in 2020. About 80% of millet production is used for human consumption.[54] FAO and the UN have endorsed 2023 as an International Year of Millets to 'provide a unique opportunity to increase global production, ensure efficient processing and consumption, promote a better utilization of crop rotations, and encourage better connectivity throughout food systems to promote millets as a key component of the food basket'.[63]

Foxtail millet is one of the major sources of grain in China. Liu *et al.* analysed 13 foxtail millet cultivars from different parts of China.[64] Volatile compounds from commercial foxtail millet cultivars were extracted using simultaneous distillation and identified with GC-MS. Fifty-two volatile compounds were identified in all samples: 19 aldehydes, 5 alcohols, 10 ketones, 9 hydrocarbons, 6 benzene derivatives, and 3 others. Twenty-three constituents were found to be common to all the samples.

9.2.7 Flavour of Oat Proteins

The common oat (*Avena sativa*) is the largest produced oat species, with an annual production of 23 million tonnes.[4] Several minor species are also cultivated, such as red oat (*A. byzantina*), sand oat (*A. strigosa*), and Abyssinian oat (*A. abyssinica*).[2] While mostly used as a fodder crop, human demand for oats is rising due to an increased interest in their nutritional values.[65] Oat is a rich source of proteins, dietary fibres (particularly β-glucan), minerals, and various phytochemicals.[66,67] The protein content of oats is higher (12–24%) than that of other cereal grains.[67] Oat proteins are mainly composed of globulins (50–80%), which have a better nutritional quality than prolamins, the major storage proteins in many other cereals.[65,68] The predominant component in the globulin fraction of oat is a 12S legumin-like protein consisting of six disulphide-linked α- and β-subunits.[65,68] Although suffering from poor solubility and surface properties at pH 4–7, oat proteins have good water- and fat-binding properties and can serve as a binder or extender. Various chemical modifications can also be used to enhance the functional properties of oat proteins.[65]

The flavour of oats has been well characterized, with over 100 volatile compounds identified. Dry oat groats (dehulled oats) have a weak, raw grain-like, grassy odour that is mainly attributed to hexanal and less significantly to C_{10} terpenes, alkyl benzenes, alcohols, and ketones. Rehydration of the oat groats activates lipoxygenase and aldehyde oxidoreductase and generates additional lipid-oxidation products such as C_5 and C_6 alcohols and aldehydes.[69] Following heat processing, the oat flavour changes considerably. Two major flavour-generation pathways, namely the Maillard reaction and lipid oxidation, and the interaction between the two, are responsible for such changes. The typical Maillard reaction-derived products, such as pyrazines, pyrroles, furans, and sulphur-containing compounds, give heated oats their nutty, toasted flavour, while oxidative degradation of lipids generates aldehydes, ketones, and alcohols that cause rancidity. Besides the rancid odour, lipid oxidation can also contribute to a bitter off-taste due to the formation of hydroxy-fatty acid monoglycerides.[69] In addition, avenanthramides and saponins have been identified as bitter and astringent-tasting components in oats.[70] The bitter taste can be carried into oat proteins. When oat protein concentrates were fortified in acidic beverages and yoghurt, a bitter taste was detected.[71,72] The oat protein concentrates also gave rise to an oat flavour when added to yoghurt, indicating the presence of other characteristic flavour volatiles in oats.[72] Ye *et al.* investigated the effect of defatting on the aromatic profile of oat proteins.[73] They identified 17 volatiles, including acetoin, 1-hexanol, isoamyl alcohol, 4-methoxy-1-butanol, 1-methyl-2-(1-methylethyl)-benzene, pyridine, 1-methyl-4-(1-methylethyl)-1,4-cyclohexadiene, nonanal, 1-dodecanol, 3-pentadecanol, 6-dodecanone, 4-octanol, 2-octanone, 2-decanone, 5-methyl-2-hexanone, 2-pentylfuran, and hexanal, but their occurrence differed among oat varieties and treatments. The dominant volatiles in full-fat samples included 1-decanol, 1-methyl-4-(1-methylethyl)-1,4-cyclohexadiene, nonanal, 4-octanol, 2-octanone, and 2-decanone, which were likely derived from lipid oxidation and decomposition of unsaturated fatty acids. On the other hand, acetoin, 1-hexanol, and isoamyl alcohol were the main flavour components in defatted samples, among which 1-hexanol was the most common oat-flavour component, while acetoin and isoamyl alcohol were probably associated with activation of latent metabolism enzymatic activity.

9.3 Flavour of Pseudocereal Proteins

Buckwheat, quinoa, amaranth, and chia are gluten-free dicotyledonous grains known as pseudocereals that have a high protein content.[74] Amaranth, buckwheat, and quinoa have a protein content of approximately 12.0–18.9%, with the main proteins being globulins and albumins[74,75] and chia has a protein content of approximately 18.2–24.1% of edible matter.[76] Pseudocereals have become increasingly popular due to the rising global demand for plant-based proteins and more sustainable agriculture, and their highly soluble proteins are desirable for use in many functional foods.[74] Pseudocereal protein-enriched fractions can be prepared using dry or wet fractionation techniques.[74] While protein-rich fractions from pseudocereals have potential as ingredients in functional foods, little scientific literature exists about their flavours specifically. This section will discuss some of the already-existing literature on buckwheat, quinoa, amaranth, and chia protein flavours and identify opportunities for future research.

9.3.1 Flavour of Amaranth Proteins

Amaranths are annual or short-lived perennial plants in the *Amaranthus* genus (Amaranthaceae family). The three main amaranth species cultivated for their seeds are *Amaranthus cruentus*, *A. hypocondricus*, and *A. caudatus*.[77] Amaranth is considered a valuable food crop because of its fast growth, adaptability, and nutritional values.[78] The protein content of different species of amaranth seeds varies considerably, from 12.5% to 22%.[77] Amaranth seed proteins are composed of approximately 51% albumins, 31% glutelins, 16% globulins, and 2% prolamins.[79] Amaranth protein has a balanced amino acid composition and is particularly rich in lysine.[77]

Amaranth protein is of great interest as a plant-based protein source because it is gluten-free and has a more neutral or pleasant flavour than beany-flavoured soy proteins,[80] but research on the flavour profile of amaranth protein is limited. There are some studies, though, that have evaluated the application of amaranth proteins as a functional ingredient in food products and their consumer acceptability thereafter. A study reported that pasta enriched with amaranth protein concentrate and hydrolysate scored lower overall on a sensory test, but the colour, odour, texture, and taste were not significantly different from the control pasta.[81] Another study found that amaranth protein isolate can be a good stand-in for gluten, contributing to the desirable texture and crust colour of gluten-free muffins.[82] When incorporated into sourdough, amaranth protein hydrolysates improve the taste, flavour, texture, colour, and overall acceptance of the bread.[83] The small peptides and free amino acids in the amaranth protein hydrolysates not only directly contributed to the enhanced bread flavour but also increased the production of lactic acid and alcohol and the conversion of glutamine to glutamate by promoting the growth of lactic acid bacteria.

The interactions between protein ingredients and other food components (*e.g.*, lipids) may influence the volatile composition of a food. Gürbüz *et al.* reported that emulsions stabilized by amaranth proteins showed lower oxidative stability and a distinct volatile profile in comparison to those stabilized by Tween 20.[84] The amaranth protein-stabilized emulsions had a higher abundance of 2,4-hexadienal, 5-pentyl-2(5*H*)-furanone, and nonanal after seven days, while the Tween 20-stabilized emulsions were richer in 2-heptenal, 2,4-heptadienal, and 2-pentenal. The presence of protein radicals likely affected lipid oxidation and the formation of secondary lipid-oxidation products. Such oxidative instability should be considered when using amaranth protein as an emulsifying agent.

9.3.2 Flavour of Buckwheat Proteins

Buckwheat is a pseudocereal of the Polygonaceae family. Two buckwheat species are cultivated, namely common buckwheat (*Fagopyrum esculentum*) and Tartary buckwheat (*Fagopyrum tataricum*). Tartary buckwheat is also known as bitter buckwheat due to its bitter taste. In addition, various wild species, such as perennial buckwheat (*Fagopyrum cymosum*), are consumed.[2,85]

Buckwheat proteins are gluten-free and contain a considerable amount of essential amino acids, including lysine, the limiting amino acid in cereals. They have excellent functional properties, such as solubility, emulsifying properties, and foaming properties. In addition, numerous bioactivities of buckwheat proteins and peptides have been reported, such as cholesterol-lowering, antihypertensive, antimicrobial, antioxidant, and anti-tumour properties.[86] The aroma profiles of buckwheat and its products have been characterized, but much less work has been done for buckwheat proteins. Important aroma compounds of common buckwheat include salicylaldehyde, (*E*,*E*)-2,4-decadienal, (*E*)-2-nonenal, 2,5-dimethyl-4-hydroxy-3(2*H*)-furanone, 2-methoxy-4-vinylphenol, and 2-phenylacetaldehyde. The aroma profile of Tartary buckwheat is different, with hexanal, nonanal, decanal, (*E*)-2-nonenal, (*E*,*E*)-2,4-nonadienal, (*E*,*E*)-2,4-decadienal, and phenylethyl alcohol as major odorants.[87] Whether these components are retained in buckwheat proteins is largely unknown and warrants further investigation.

The bitter taste of Tartary buckwheat is mainly attributed to flavonoids such as rutin, quercitrin, and quercetin.[88] These bitter components can be removed during protein extraction by isoelectric precipitation.[89] Another study reported dry fractionation of Tartary buckwheat by surface abrasion, which generated protein-rich fractions with as much as 36% protein content.[90] The high protein fractions contain approximately 55 mg g^{-1} of polyphenols and may, thus, have a bitter and astringent taste. Over 47 volatiles were detected in the protein fractions, including alkanes, aldehydes, alcohols, ketones, terpenes, acids, and furanoids, with alkanes being the dominant compounds that give rise to fruity and grassy aromas. Hydrothermal processing increased the volatile compounds in the protein fractions, including nonanal, 1-octen-3-ol, and naphthalene, which are responsible for the characteristic aroma of Tatary buckwheat.[90]

The bran-aleurone fraction of buckwheat seeds is rich in protein but also has a high phenolic content. Syringic acid, *p*-hydroxybenzoic acid, and vanillic acid occurred in bound form in this fraction.[91] In addition, flavan-3,4-diols and pelargonidin- and cyanidin-based soluble-condensed tannins were present.[91] The presence of tannins and phenolic acids may contribute to a bitter taste and an astringent mouthfeel in buckwheat protein concentrates prepared from the bran-aleurone fraction.[91]

The flavour of buckwheat proteins may be altered by processes such as enzymatic hydrolysis. Kawakami *et al.* isolated buckwheat globulin *via* isoelectric precipitation and prepared protein hydrolysates using different proteases, including pronase, pepsin, papain, trypsin, subtilisin, thennolysin, and α-chymotrypsin.[92] Sensory evaluations of the buckwheat globulin hydrolysates demonstrated that all the protein hydrolysates had a sulphurous odour, except for the pepsin hydrolysates, and exhibited a bitter taste.[92] The pepsin and trypsin hydrolysates seemed to exhibit the most intense bitterness, and α-chymotrypsin hydrolysates demonstrated the weakest bitterness.[92] While the protein hydrolysates of buckwheat globulin appeared to be mostly unsuitable as food additives due to the sulphurous odour and bitter taste, the hydrolysates prepared with pepsin appeared to have a more favourable parched-buckwheat grain-like flavour and could be of interest to manufacturers of functional foods.[92]

9.3.3 Flavour of Chia Proteins

Chia (*Salvia hispanica* L.) is a pseudocereal of the Lamiceae family. The protein content of chia seeds is approximately 19–23%. About half of the proteins in chia are globulins, with glutelins, albumins, and prolamins making up the rest in about equal fractions.[93] Not only is chia a good source of protein, but it also contains rich amounts of α-linolenic acid, dietary fibre, and antioxidants, such as phenolic compounds, and has shown health benefits such as regulating intestinal transit, improving lipid profiles, and reducing the risk of cardiovascular disease, type 2 diabetes, obesity, and some cancers.[93] Chia's many health benefits and its proven functional capabilities—as a thickening agent and edible coating ingredient—make it a pseudocereal with great food formulation potential.[93]

The flavour of chia has not been well characterized. One study analysed volatile compounds in dry fermented sausage with ground chia seeds incorporated.[94] They found that the incorporation of chia promoted the production of volatiles derived from lipid and amino acid degradation, such as 1-pentanol, butanal, pentanal, hexanal, 3,5-octadien-2-one, 3-methylbutanal, 3-ehylbenzeneacetaldehyde, benzaldehyde, and benzeneacetaldehyde. Chia also increased the accumulation of volatiles such as ethyl propanoate, trimethylene oxide, 3,3-dimethyl-1,2-epoxybutane, 2-ethylfuran, 2-pentylfuran, acetoin, methanethiol, and various terpenoids. However, it is unclear which compounds were solely derived from chia. Overall, the sausages with chia seeds exhibited positive sensory features. The incorporation of chia seed also improved the texture, flavour, juiciness, and processing properties of a restructured ham-like product.[95] Besides meat products, chia has shown great potential in baked goods. For example, Sharma *et al.* incorporated chia seeds into a muffin mix and found that muffins with 20% of wheat flour replaced by chia seeds, as compared to the control muffins made with only wheat flour, scored higher on sensory tests.[96] On the other hand, a study with the goal of increasing protein levels in pineapple jam found that incorporating chia seeds at a concentration as low as 6.25% significantly reduced flavour scores and overall acceptability scores as compared to a control jam with no chia seeds.[97] Results like those of the jam study perhaps point to future research opportunities: when attempting to increase protein in products like jams and jellies, perhaps chia protein isolates would prove more acceptable to consumers than whole chia seeds.

Chia protein isolate has good oil absorption capacity and is suggested to improve mouthfeel and flavour retention in food formulation.[98] However, another study reported a sandy, gritty mouthfeel from chia protein.[99] In comparison to other plant proteins, chia protein has a low level of astringency. It has a strong aroma intensity with a distinctive green/grassy note. Following rehydration and heat treatment, such flavours increased, while a painty flavour was also detected. The compounds responsible for such flavours are yet to be characterized.

9.3.4 Flavour of Quinoa Proteins

Chenopodium quinoa is a pseudocereal in the Amaranthaceae family.[100] Quinoa seeds are composed of 12–23% protein with an excellent essential amino acid composition.[101] Quinoa protein is particularly rich in lysine, histidine, and methionine and registered the highest digestible indispensable amino acid score among cereals and pseudocereal grains (Figure 9.1). In comparison to wheat protein, quinoa protein demonstrated a higher water and oil absorption capacity, which may contribute to improved mouthfeel and flavour retention in certain food products.[102] However, when pasta was supplemented with quinoa protein isolate, a 4–8% supplementation reduced the scores of all sensory attributes, and a 12% supplementation made the pasta unacceptable.[103]

Different quinoa varieties differ in aroma, taste, flavour, and texture.[104] Most quinoa seeds, however, contain triterpene saponins that lead to bitterness and an undesirable flavour. Saponin levels can be reduced by processes such as washing, pearling, fermentation, germination, and various heat treatments.[105] Food processing may also alter the flavour of quinoa proteins. Li *et al.* found that fermentation of quinoa seeds with *Lactobacillus casei* promoted the formation of desirable aroma components such as alcohols, aldehydes, ketones, esters, and acids.[106] Among the fermentation-induced volatiles, 2-phenylmethanol and 2-phenylethanol with a floral smell were derived from phenylalanine decomposition, and isoamyl alcohol with malty and chocolate notes was produced by the catabolism of leucine. Quinoa protein hydrolysates have been used as a precursor for generating thermal process flavours through the Maillard reaction with xylose. At pHs 5 and 6, the reaction favoured the formation of furans, furanones, and cyclopentanones with a strong caramel flavour. As the pH increased from 7 to 9, an increase in Strecker aldehydes and pyrazines was observed, which contributed to chocolate, roasted, and nutty notes.[107] This suggests that the flavour of quinoa protein could be somewhat customized for different applications. More research is needed to fully investigate the flavour properties of quinoa proteins and tap into their potential for use in product formulations.

The interaction of quinoa protein with other food ingredients in a food system and its impact on the flavour of the product should also be considered. Like amaranth protein, quinoa protein isolate exhibited poor oxidative stability in an oil-in-water emulsion.[84] On days 1–7, the lipid oxidation-derived volatiles 2-pentylfuran, hexanal, 2-octenal, 3,5-octadien-2-one, and 2,4-hexadienal were more abundant in the quinoa protein-stabilized emulsion than in Tween 20- or amaranth protein-stabilized emulsions. This was likely due to the early onset of oxidative reactions in quinoa proteins. The formation of protein-free radicals resulted in more advanced lipid oxidation and the generation of volatile compounds.

9.4 Summary

Due to the poor functional properties of cereal and pseudocereal proteins, proteins from these sources have not been widely used as an ingredient in the food industry. As a result, information on the flavour profile of proteins isolated from cereals and pseudocereals is limited. In some cases, even the flavour of the grain (*e.g.*, chia) has not been characterized. On the other hand, hydrolysates from cereal and pseudocereal proteins have been investigated as precursors of taste-active peptides and aromas. In addition, a few studies have investigated the change in flavour profile and sensorial attributes of complex food systems after incorporating cereal proteins. To better understand off-flavour formation and improve their applications in foods, the flavour properties of cereal and pseudocereal proteins and their interactions with other food components need further investigation.

Abbreviations

CGPF	Corn germ protein flour
FAO	Food and Agriculture Organization
GC	Gas chromatography
MS	Mass spectrometry
O	Olfactometry
SPME	Solid-phase microextraction

UN United Nations

References

1. S. O. Serna-Saldivar, in *Cereal Grains: Properties, Processing, and Nutritional Attributes*, CRC Press, Boca Raton, Florida, USA, 2010.
2. C. W. Wrigley, H. Corke, K. Seetharaman and J. Faubion, in *Encyclopedia of Food Grains*, Academic Press, Waltham, Massachusetts, USA, 2016.
3. USDA, *FoodData Central*, 2020.
4. FAO, *FAOSTAT: Food and agriculture data*, 2020.
5. P. J. Woolf, L. L. Fu and A. Basu, *PLoS One*, 2011, **6**, e18836.
6. S. K. Cervantes-Pahm, Y. Liu and H. H. Stein, *Br. J. Nutr.*, 2014, **111**, 1663–1672.
7. Y. Miao, D. J. Mulla, J. A. Hernandez, M. Wiebers and P. C. Robert, *Soil Sci. Soc. Am. J.*, 2007, **71**, 1490–1499.
8. P. Rasane, A. Jha, L. Sabikhi, A. Kumar and V. Unnikrishnan, *J. Food Sci. Technol.*, 2015, **52**, 662–675.
9. USDA, *Grain: World Markets and Trade*, United States Department of Agriculture Foreign Agricultural Service, 2022.
10. R. Shukla and M. Cheryan, *Ind. Crops Prod.*, 2001, **13**, 171–192.
11. S. H. Gorissen, J. J. Crombag, J. M. Senden, W. Waterval, J. Bierau, L. B. Verdijk and L. J. van Loon, *Amino Acids*, 2018, **50**, 1685–1695.
12. T. C. Huang, L. J. Bruechert, T. G. Hartman, R. T. Rosen and C. T. Ho, *J. Agric. Food Chem.*, 1987, **35**, 985–990.
13. A. Cabin-Flaman, A. Delaune, P. Poutrain, Y. G. Gangwe Nana, B. Jourdain, D. Gibouin, J. Paris, S. Trestour, J. Seigneuret and A. Léopoldès de Vendômois, *Int. J. Cosmet. Sci.*, 2018, **40**, 575–582.
14. A. Tinoco, F. Gonçalves, A. F. Costa, D. S. Freitas, A. Cavaco-Paulo and A. Ribeiro, *Ind. Crops Prod.*, 2021, **159**, 113067.
15. S. Ghasemi, N. H. S. Javadi, M. Moradi and K. Khosravi-Darani, *Afr. J. Biotechnol.*, 2015, **14**, 2014–2021.
16. G. A. Miguel, C. Jacobsen, C. Prieto, P. J. Kempen, J. M. Lagaron, I. S. Chronakis and P. J. García-Moreno, *J. Food Eng.*, 2019, **263**, 348–358.
17. Y.-H. Wang, Z.-L. Wan, X.-Q. Yang, J.-M. Wang, J. Guo and Y. Lin, *Food Hydrocolloids.*, 2016, **54**, 40–48.
18. R. A. Flath, K. E. Matsumoto, R. G. Binder, R. T. Cunningham and T. R. Mon, *J. Agric. Food Chem.*, 1989, **37**, 814–819.
19. J. Premkumar, M. S. Kumar, R. RaghulPrashanth, G. Eshwar and A. Kumar, *J. Food Process. Preserv.*, 2022, **46**, e16257.
20. C. Huang and J. Zayas, *J. Food Sci.*, 1991, **56**, 1308–1310.

21. C. Huang and J. Zayas, *J. Food Sci.*, 1996, **61**, 258–261.
22. W. H. Idris, E. E. Babiker and A. H. El Tinay, *Food/Nahrung.*, 2003, **47**, 425–429.
23. H.-I. Hwang, T. G. Hartman, M. V. Karwe, H. V. Izzo and C.-T. Ho, in *Lipids in Food Flavors*, ed. C.-T. Ho and T. G. Hartman, ACS Publications, 1994, pp. 144–157.
24. W. Bredie, D. Mottram, G. Hassell and R. Guy, *J. Cereal Sci.*, 1998, **28**, 97–106.
25. W. E. Riha, C.-F. Hwang, M. V. Karwe, T. G. Hartman and C.-T. Ho, *J. Agric. Food Chem.*, 1996, **44**, 1847–1850.
26. W. Bredie, G. Hassell, R. Guy and D. Mottram, *J. Cereal Sci.*, 1997, **25**, 57–63.
27. M. V. Romero, S. Raghavan and C.-T. Ho, in *Nutrition, Functional and Sensory Properties of Foods*, ed. C. Mussinan, F. Shahidi and E. T. Contis, Royal Society of Chemistry, Cambridge, UK, 2013, pp. 57–65.
28. S. S. Xu, PhD thesis, *Preparation and flavor characteristics of Maillard peptides derived from wheat gluten*, Jiangnan University, 2018.
29. S. E. Lee, H. Chung and Y.-S. Kim, *Food Chem.*, 2012, **131**, 1248–1254.
30. Q. Chen and C.-T. Ho, in *Biotechnology for Improved Foods and Flavors*, ed. G. R. Takeoka, R. Teranishi, P. J. Williams and A. Kobayashi, ACS Publications, 1996, pp. 88–96.
31. L. Sun and Y. Zhuang, *Food Bioprocess. Technol.*, 2012, **5**, 1287–1294.
32. S. Wu, PhD thesis, *Study on the preparation of meat-like flavors by Maillard reaction of gluten hydrolysates*, Jiangnan University, 2013.
33. H. Schlichtherle-Cerny and R. Amadò, *J. Agric. Food Chem.*, 2002, **50**, 1515–1522.
34. B.-Y. Liu, K.-X. Zhu, X.-N. Guo, W. Peng and H.-M. Zhou, *J. Sci. Food Agric.*, 2017, **97**, 3181–3188.
35. L. Liao, C.-y. Qiu, T.-x. Liu, M.-m. Zhao, J.-y. Ren and H.-f. Zhao, *J. Cereal Sci.*, 2010, **52**, 395–403.
36. K. Wang and S. D. Arntfield, *Food Chem.*, 2014, **157**, 364–372.
37. M. Reche, C. Pascual, A. Fiandor, I. Polanco, M. Rivero-Urgell, R. Chifre, S. Johnston and M. Martín-Esteban, *Pediatr. Allergy Immunol.*, 2010, **21**, 577–585.
38. Q. Zhao, C. Selomulya, H. Xiong, X. D. Chen, X. Li, S. Wang, C. Bai, H. Peng, Q. Zhou and W. Sun, *Int. J. Food Prop.*, 2014, **17**, 1791–1804.
39. I. Paraman, N.Hettiarachchy and C. Schaefer, *Cereal Chem.*, 2008, **85**, 76–81.
40. R. B. Yadav, B. S. Khatkar and B. S. Yadav, *Int. J. Food Prop.*, 2013, **16**, 1776–1788.
41. J. Zhao and W. L. Boatright, *Food Chem.*, 2017, **221**, 345–350.

42. A. Singh, Y. Shi, P. Magreault, D. D. Kitts, M. Jarzębski, P. Siejak and A. Pratap-Singh, *Mol.*, 2021, **26**, 4104.
43. M. Nagendra Prasad, K. Sanjay, M. Shravya Khatokar, M. Vismaya and S. Nanjunda Swamy, *J. Nutr. Food Sci.*, 2011, **1**, 108.
44. K. Kaewka, C. Therakulkait and K. R. Cadwallader, *Int. J. Nutr. Food Sci. Technol.*, 2011, **46**, 1654–1660.
45. S. Arsa and C. Theerakulkait, *J. Sci. Food. Agric.*, 2018, **98**, 4479–4487.
46. S. Arsa and C. Theerakulkait, *J. Food. Sci. Technol.*, 2015, **52**, 5285–5291.
47. S. Arsa, C. Theerakulkait and K. R. Cadwallader, *J. Agric. Food Chem.*, 2019, **67**, 8205–8211.
48. S. Arsa and Y. Puechkamutr, *J. Food. Sci. Technol.*, 2022, **59**, 890–897.
49. P.-S. Wang, H. Kato and M. Fujimaki, *Agric. Biol. Chem.*, 1970, **34**, 561–567.
50. A.-C. J. Cramer, D. S. Mattinson, J. K. Fellman and B.-K. Baik, *J. Agric. Food. Chem.*, 2005, **53**, 7526–7531.
51. E. Eckert, W. Wismer, K. Waduthanthri, O. Babii, J. Yang and L. Chen, *J. Am. Oil Chem. Soc.*, 2018, **95**, 1027–1040.
52. M. H. Alu'datt, T. Rababah, G. J. Al-Rabadi, K. Ereifej, S. Gammoh, N. Masadeh and P. J. Torley, *J. Food Qual.*, 2014, **37**, 329–338.
53. M. Houde, N. Khodaei and S. Karboune, *LWT-Food Sci. Technol.*, 2018, **91**, 133–142.
54. J. M. Awika, in *Advances in cerealscience: implications to food processing and health promotion*, ed. J. M. Awika, V. Piironen and S. Bean, ACS Publications, 2011, pp. 1–13.
55. R. Zanan, K. Khandagale, V. Hinge, M. Elangovan, R. J. Henry and A. Nadaf, *Mol. Breed.*, 2016, **36**, 1–12.
56. S. Cao, L. Wang, Q. Wu, D. Ni, Y. Xu and L. Lin, *J. Inst. Brew.*, 2020, **126**, 306–315.
57. X. Fan, X. Jiao, J. Liu, M. Jia, C. Blanchard and Z. Zhou, *Food Res. Int.*, 2021, **140**, 109975.
58. L. Vázquez-Araújo, E. Chambers IV and D. B. Funk, *J. Cereal Sci.*, 2011, **54**, 460–466.
59. L. Seitz and D. Sauer, *Cereal Chem.*, 1996, **73**, 744–750.
60. Y. Xiong, P. Zhang, R. D. Warner and Z. Fang, *Compr. Rev. Food Sci. Food Saf.*, 2019, **18**, 2025–2046.
61. P. Belton, I. Delgadillo, N. Halford and P. Shewry, *J. Cereal Sci.*, 2006, **44**, 272–286.
62. K. Duodu, J. Taylor, P. Belton and B. Hamaker, *J. Cereal Sci.*, 2003, **38**, 117–131.
63. UN, *International Year of Millets 2023 - Building momentum for the year*, 2022.

64. J. Liu, S. Li, A. Zhang, W. Zhao, Y. Liu and Y. Zhang, *Cereal Chem.*, 2017, **94**, 170–176.
65. F. Webster and P. J. Wood, in *Oats: Chemistry and Technology*, AACC International, St. Paul, Minnesota, USA, 2011.
66. M. S. Butt, M. Tahir-Nadeem, M. K. I. Khan, R. Shabir and M. S. Butt, *Eur. J. Nutr.*, 2008, **47**, 68–79.
67. R. Singh, S. De and A. Belkheir, *Crit. Rev. Food Sci. Nutr.*, 2013, **53**, 126–144.
68. C. Klose and E. K. Arendt, *Crit. Rev. Food Sci. Nutr.*, 2012, **52**, 629–639.
69. R. J. McGorrin, *J. Agric. Food Chem.*, 2019, **67**, 13778–13789.
70. K. Günther-Jordanland, C. Dawid, M. Dietz and T. Hofmann, *J. Agric. Food Chem.*, 2016, **64**, 9639–9652.
71. J. Cluskey, Y. Wu, G. Inglett and J. Wall, *J. Food Sci.*, 1976, **41**, 799–804.
72. M. Brückner-Gühmann, A. Benthin and S. Drusch, *Food Hydrocolloids*, 2019, **86**, 146–153.
73. J. Yue, Z. Gu, Z. Zhu, J. Yi, J.-B. Ohm, B. Chen and J. Rao, *Food Hydrocolloids*, 2021, **112**, 106368.
74. L. Alonso-Miravalles and J. A. O'Mahony, *Foods*, 2018, **7**, 73.
75. C. Mota, M. Santos, R. Mauro, N. Samman, A. S. Matos, D. Torres and I. Castanheira, *Food Chem.*, 2016, **193**, 55–61.
76. M. Grancieri, H. S. D. Martino and E. Gonzalez de Mejia, *Compr. Rev. Food Sci. Food Saf.*, 2019, **18**, 480–499.
77. P. R. Venskutonis and P. Kraujalis, *Compr. Rev. Food Sci. Food Saf.*, 2013, **12**, 381–412.
78. A. Rastogi and S. Shukla, *Crit. Rev. Food Sci. Nutr.*, 2013, **53**, 109–125.
79. M. Segura-Nieto, N. Vazquez-Sanchez, H. Rubio-Velazquez, L. E. Olguin-Martinez, C. E. Rodriguez-Nester and L. Herrera-Estrella, *J. Agric. Food Chem.*, 1992, **40**, 1553–1558.
80. H. Corke, in *Using Cereal Science and Technology for the Benefit of Consumers*, ed. S. P. Cauvain, S. S. Salmon and L. S. Young, Woodhead Publishing, 2005, p. 68.
81. E. E. Valdez-Meza, A. Raymundo, O. G. Figueroa-Salcido, G. I. Ramírez-Torres, P. Fradinho, S. Oliveira, I. de Sousa, M. Suárez-Jiménez, F. I. Cárdenas-Torres and A. R. Islas-Rubio, *Foods*, 2019, **8**, 282.
82. K. Shevkani and N. Singh, *Int. J. Food Sci. Technol.*, 2014, **49**, 2237–2244.
83. N. Karimi, F. Zeynali, M. Rezazad Bari, M. Nikoo, F. Mohtarami and M. Kadivar, *Food Sci. Nutr.*, 2021, **9**, 6683–6691.
84. G. Gürbüz, V. Kauntola,J. M. R. Diaz, K. Jouppila and M. Heinonen, *Eur. Food Res. Technol.*, 2018, **244**, 469–479.
85. K. Ikeda, *Adv. Food Nutr. Res.*, 2002, **44**, 395–434.

86. J. Jin, I. C. Ohanenye and C. C. Udenigwe, *Crit. Rev. Food Sci. Nutr.*, 2020, 1–13.
87. M. Starowicz, G. Koutsidis and H. Zieliński, *Crit. Rev. Food Sci. Nutr.*, 2018, **58**, 1767–1779.
88. N. Fabjan, J. Rode, I. J. Kosir, Z. Wang, Z. Zhang and I. Kreft, *J. Agric. Food Chem.*, 2003, **51**, 6452–6455.
89. A. Kawakami, H. Kayahara, K. Tadasa and A. Ujihara, *Nippon Shokuhin Kogyo Gakkaishi*, 1994, **41**, 481–484.
90. X.-W. Chen, D.-Y. Luo, Y.-J. Chen, J.-M. Wang, J. Guo and X.-Q. Yang, , *Food Chem.*, 2019, **285**, 414–422.
91. A. B. Durkee, *J. Agric. Food Chem.*, 1977, **25**, 286–287.
92. A. Kawakami, T. Inbe, H. Kayahara and A. Horii, *Current Advances in Buckwheat Research*, 1995, **2**, 927–934.
93. M. R. Sandoval-Oliveros and O. Paredes-López, *J. Agric. Food Chem.*, 2013, **61**, 193–201.
94. P. Borrajo, M. Karwowska and J. M. Lorenzo, *Mol.*, 2022, **27**, 652.
95. Y. Ding, H.-W. Lin, Y.-L. Lin, D.-J. Yang, Y.-S. Yu, J.-W. Chen, S.-Y. Wang and Y.-C. Chen, *J. Food Drug. Anal.*, 2018, **26**, 124–134.
96. V. Sharma, R. Mogra and A. Mordia, *Nutr. Food Sci.*, 2020, **50**, 1151–1161.
97. J. M. Nduko, R. W. Maina, R. K. Muchina and S. K. Kibitok, *Food Sci. Nutr.*, 2018, **6**, 2344–2349.
98. D. N. López, R. Ingrassia, P. Busti, J. Wagner, V. Boeris and D. Spelzini, *LWT-Food Sci. Technol.*, 2018, **97**, 523–529.
99. S. Nishku, MSc thesis, *Development of a lexicon for plant protein powders*, North Carolina State University, 2020.
100. B. L. Graf, P. Rojas-Silva, L. E. Rojo, J. Delatorre-Herrera, M. E. Baldeón and I. Raskin, *Compr. Rev. Food Sci. Food Saf.*, 2015, **14**, 431–445.
101. S. Dakhili, L. Abdolalizadeh, S. M. Hosseini, S. Shojaee-Aliabadi and L. Mirmoghtadaie, *Food Chem.*, 2019, **299**, 125161.
102. S. Elsohaimy, T. Refaay and M. Zaytoun, *Ann. Agric. Sci.*, 2015, **60**, 297–305.
103. A. Gupta, S. Sharma and V. K. R. Surasani, *LWT-Food Sci. Technol.*, 2021, **135**, 110045.
104. G. Wu, C. F. Ross, C. F. Morris and K. M. Murphy, *J. Food Sci.*, 2017, **82**, 993–1005.
105. D. Suárez-Estrella, L. Torri, M. A. Pagani and A. Marti, *J. Sci. Food Agric.*, 2018, **98**, 4033–4041.
106. S. Li, C. Chen, Y. Ji, J. Lin, X. Chen and B. Qi, *J. Cereal Sci.*, 2018, **84**, 83–89.
107. S. N. Lotfy, R. Saad, K. F. El-Massrey and H. H. Fadel, *LWT-Food Sci. Technol.*, 2021, **145**, 111328.

CHAPTER 10

Flavour of Novel Food Proteins

CASSANDRA MAYA[a], SHRUTI SHERTUKDE[b] AND CHANGQI LIU*[a]

[a]School of Exercise and Nutritional Sciences, San Diego State University, San Diego, California, USA; [b]Friedman School of Nutrition Science and Policy, Tufts University, Boston, Massachusetts, USA
*E-mail: changqi.liu@sdsu.edu

10.1 Introduction

The unprecedented food security challenge of the 21st century—feeding over 9 billion people by the year 2050—requires a 100% increase in food production and a 110% increase in protein production from 2005 to 2050.[1] This occurs alongside water scarcity, farmland shortages, and climate change, which have negatively impacted food crops' yields and nutrient profiles.[2,3] The COVID-19 pandemic aggravated this further by disrupting agricultural production, the food supply chain, and trade.[4] Notably, the pandemic has caused a decrease in the supply of protein-rich foods, such as meat, dairy, fish, and soybean.[5] Agricultural expansion and intensification may further increase the risk of the emergence and spread of infectious diseases that are detrimental to global economies and public health.[6] Improving food sustainability is, thus, imperative to meet the ever-growing food demand while preserving the natural ecosystem.

Shifting towards more sustainable food and protein sources is a recommended strategy for sustaining the food system within environmental limits.[7,8] Globally, 83% of the agricultural land is used for producing meat, aquaculture, eggs, and dairy. These animal products contribute to 56–58%

Food Chemistry, Function and Analysis No. 41
Flavour and Consumer Perception of Food Proteins
Edited by Jing Zhao and Changqi Liu

Published by the Royal Society of Chemistry, www.rsc.org

of greenhouse gas emissions while providing only 37% of dietary protein and 18% of calories.[9] In addition, overconsumption of animal-source foods, particularly red and processed meat, may lead to adverse health outcomes.[7,10] Numerous novel proteins, such as algal proteins, cultured meat, fungal proteins, insect proteins, leaf proteins, oilseed proteins (not including soy protein), pulse proteins, and tuber proteins, are emerging as sustainable alternatives to traditional animal source proteins.[11] Although some of these proteins have been studied for decades (*e.g.*, leaf proteins and oilseed proteins), we have included them in this chapter because they have not been widely utilized in the food industry. Pulse proteins are not included, as they have been discussed in Chapter 8.

Despite the potential health and environmental benefits of alternative proteins, poor consumer acceptance remains a major barrier to their widespread applications.[12] Among the various drivers for consumer acceptance of novel proteins, taste is of vital importance, particularly with pulse proteins, algal proteins, insect proteins, and plant-based meat alternatives.[12,13] Characterizing the flavour profile of these proteins is therefore critical for developing food products that are well accepted by consumers. Unfortunately, very limited literature on this subject is available. In this chapter, we summarize the existing knowledge regarding the flavour of novel food proteins and identify the needs for future research.

10.2 Flavour of Algal Proteins and Peptides

10.2.1 Flavour of Algae

Algae are a polyphyletic group of photosynthetic organisms that include both prokaryotic and eukaryotic members. They are classified into macroalgae and microalgae based on their size. Macroalgae are large plant-like organisms commonly referred to as seaweed. Most macroalgae used as food belong to Rhodophyta (red algae), Ochrophyta (brown algae), or Chlorophyta (green algae). Microalgae are microscopic, unicellular organisms. *Arthrospira platensis*, *Chlamydomonas reinhardtii*, *Auxenochlorella protothecoides*, *Chlorella vulgaris*, *Dunaliella bardawil*, and *Euglena gracilis* are microalgae species 'Generally Recognized as Safe' by the United States Food and Drug Administration.[14,15] Both macroalgae and microalgae are promising sources of various food ingredients, such as proteins, carbohydrates, polyunsaturated fatty acids, pigments, and bioactive compounds.[16]

The global production of macroalgae was 32.4 million tonnes in 2018.[17] Microalgae are produced on a smaller scale, with an underestimated biomass yield of 87 thousand tonnes.[17] The protein yield per unit area for macroalgae and microalgae is 2.5–7.5 tonnes Ha^{-1} $year^{-1}$ and 4–15 tonnes Ha^{-1} $year^{-1}$, respectively, higher than that of terrestrial crops (0.6–2 tonnes Ha^{-1} $year^{-1}$).[18] The essential amino acid index of macroalgae and microalgae (66–96% and 64–98%, respectively) is superior or comparable to plant proteins in general.[19,20]

The flavour of algae has been well characterized. The main aroma compounds of algae and their odour thresholds and olfactory descriptors have been summarized by Francezon *et al.*[21] Macroalgae are rich in aspartic acid and glutamic acid, which contribute to a strong umami taste.[18] Some macroalgae species are also rich in mannitol, which gives the seaweed a sweet taste.[22] Besides these tastants, over 200 volatiles, including hydrocarbons, ketones, aldehydes, alcohols, halogen- or sulphur-containing compounds, acids, esters, furans, and phenolics, have been identified in macroalgae species.[23] Particularly, the presence of amines, pyridines, furans, halogens (*e.g.*, dibromoethane, dibromochloromethane, bromodichloromethane, chloroiodomethane, iodoethane, and iodopentane), and isoprene (2-methyl-1,3-butadiene) contributes to the crustacean, fish, marine, and seafood notes of macroalgae.[23]

Similarly, over 200 volatile compounds have been identified in various microalgae species, including aldehydes, alkanes, alcohols, ketones, sulphuric compounds, benzenes, furans, acids, terpenes, and esters.[24–26] Hosoglu reported 1-octen-3-ol (mushroom), (*Z*)-1,5-octadien-3-one (geranium), 2-isopropyl-3methoxypyrazine (plastic/dirty), and (*E*,*Z*)-2,6-nonadienal (cucumber) as common aroma-active compounds in three marine microalgae and two freshwater microalgae.[26] Van Durme *et al.* found that the seafood-like odour of some species was associated with sulphur-containing compounds (dimethyl disulphide, dimethyl trisulphide, and methional), diketones, α-ionone, and β-ionone while fresh green, fruity flavours were linked with aldehydes such as 2,4-alkadienals and 2,4,6-alkatrienals.[24] The volatile composition varies between different microalgae species and growth phases.[25] Cyanobacteria are known as a major source of volatile compounds (*e.g.*, geosmin and 2-methylisoborneol). These compounds are commonly associated with an offensive taste and odour in drinking water and seafood.[27] Manipulating the growing conditions of microalgae may help reduce these undesirable odours. For example, the addition of nitrogen reduced the production of 2-methylisoborneol, while copper suppressed the production of geosmin.[28]

10.2.2 Flavour of Algal Proteins

Incorporating a high concentration of algal biomass into food often negatively affects the colour, flavour, and overall acceptance of the products.[29,30] Using isolated algal protein may help reduce unfavourable flavour compounds and improve consumer acceptance.[29] However, it remains unclear whether some of the volatile compounds in algae would be carried into isolated algal proteins. Deodorization techniques such as acid-base neutralization, steaming, physical adsorption, microencapsulation, and microbial fermentation have been developed to remove or mask the undesirable odours of algal biomass.[31] Ethanol extraction has also been reported to largely remove lipid oxidation products, pyridines, and pyrazines from *Tetraselmis chuii* biomass and leave predominantly hydrocarbons with high odour thresholds.[32] Such techniques may be applied prior to or after protein extraction. Another strategy is to use algal protein in foods with a strong flavour-masking capacity[33] or in products that welcome seafood-like flavours.[34]

Grossmann *et al.* evaluated the sensorial attributes of a less-refined protein extract from *Chlorella protothecoides* and found an overall low sensory acceptance of the protein in aqueous dispersions.[35] At neutral pH, a 2% *C. protothecoides* protein extract dispersion was dominated by umami, salty, and bitter tastes. Adding sucrose and organic acids greatly enhanced the sensory properties of the dispersions and increased the acceptable protein extract concentration to 4%. Zamuz *et al.* incorporated Spirulina and Chlorella proteins with 70% and 60% purity, respectively, into meat products such as pork liver pâté, turkey patties, cooked turkey breast, pork sausages, and marinated turkey breast.[36] Products with the addition of algal proteins registered the poorest acceptability in comparison to those incorporated with vegetable or dairy proteins. The low sensory quality of the algal proteins was attributed to their high content of bitter-tasting amino acids.

10.2.3 Flavour of Macroalgae Protein Hydrolysates

Macroalgae, or seaweeds, are well known for their rich glutamic acid content and umami taste. In fact, the umami taste was first identified in kombu seaweed.[22] The umami taste of seaweeds can be further enhanced through protein hydrolysis due to the release of umami amino acids. However, the selection of proteases is very important. Fitzgerald *et al.* reported that *Palmaria palmata* protein hydrolysates prepared by papain gave a strong, bitter aftertaste when incorporated in bread at a concentration of 4%.[37] Laohakunjit *et al.* reported that the hydrolysis of the red alga, *Gracilaria fisheri*, by bromelain released more bitter and sweet amino acids than umami amino acids.[38] The ratio of umami/bitter/sweet amino acids was 1.00/8.00/7.58. Flavourzyme, on the other hand, generates considerably less bitter amino acids and more umami amino acids. Brown algae, *Undaria pinnatifida* and *Saccharina japonica*, hydrolysed by Flavourzyme, had an umami/bitter/sweet amino acid ratio of 1.00/0.32/1.58 and 1.00/0.59/1.56, respectively.[39,40]

Many volatile compounds in macroalgae are also found in their protein hydrolysates. The primary odorants identified include hexanal, hexanoic acid, nonanoic acid, and dihydroactinidiolide in *G. fisheri* protein hydrolysates; hexanal, cedrol, nonanal, 2-heptenal, acetoin, and heptanal in *U. pinnatifida* protein hydrolysates; and hexanal, (*E*)-2-octenal, nonanal, pentanal, heptanal, and 4-ethylcyclohexanol in *S. japonica* protein hydrolysates.[38–40] Most of these volatiles are derived from the oxidative degradation of polyunsaturated fatty acids[41] and are also found in many other seafoods.[38,39]

Amin *et al.* identified 54 volatiles in dried sea grape (*Caulerpa racemosa*) and 58 volatiles in sea grape protein hydrolysates.[42] Some volatiles detected in the algae, such as chloroform, hexadecane, 2,6,10,14-tetramethylhexadecane, 1-heptyl-2-methylcyclopropane, 2-methyloctacosane, 2,4-di-*tert*-butylphenol, and 2-butyloctan-1-ol, were no longer detected after protein hydrolysis, while 7-methyl-3-methylideneocta-1,6-diene, 3,7-dimethylocta-1,3,6-triene, 2,6-dimethylocta-2,4,6-triene, 1-methyl-4-propan-2-ylidenecyclohexene, 4,7,7-trimethylbicyclo[4.1.0]hept-2-ene, 1,3-di-*tert*-butylbenzene, 3,3-dimethylhexan-2-one, and octane-2,3-dione were only detected in the protein hydrolysates and were likely derived from the thermal degradation of fatty acids and the Maillard reaction. Common volatiles in seaweed protein hydrolysates reported by several studies are summarized in Table 10.1. In general, seaweed protein hydrolysates are characterized by rich umami tastes and seafood-like flavours and have great potential as flavouring ingredients.

10.3 Flavour of Bacterial Proteins

Besides microalgae, proteins can be produced by other microorganisms, such as bacteria. Single-cell proteins produced from bacteria typically contain 50–80% protein on a dry weight basis, with an essential amino acid content comparable to or higher than that of the Food and Agriculture Organization (FAO) recommendations.[43] In addition, bacterial proteins can be produced sustainably by utilizing food and agriculture waste materials and the greenhouse gas methane.[43,44] The synthesis of aroma compounds by microbial cultivation can be leveraged to enrich the flavour of the bacterial biomass.[45] Genetic engineering can also be utilized to produce high-demand flavour metabolites.[46]

Sakarika *et al.* analysed volatiles of single-cell proteins produced by *Cupriavidus necator*, *Acinetobacter* sp., *Paracoccus* sp., an acetate-enriched culture (dominated by *Acinetobacter* and *Comamonas*), and a formate-enriched culture (dominated by *Paracoccus* and *Pseudomonas*). They found that sweet, fruity, and fishy were the dominant odour descriptors of bacterial proteins.[47] The volatile profiles of the proteins were greatly influenced by species and the culture substrate. For example, the formate-grown *C. necator* produced 4.1-fold more benzyl alcohol (fruity) and 4.5-fold more dimethyl sulphoxide (cabbage-like) than the acetate-grown sample. Overall, the acetate-grown sample produced more sweet aromas, while the formate-grown sample produced more cabbage odours. In enriched cultures, trimethylamine (fish-like) was 95 times higher in the formate enrichment than in the acetate-enriched sample. Furthermore, a 1.1- to 4.2-fold higher content of butanone, trichloroethane, and acetoin was observed in the acetate-enriched culture, contributing to a 3.5-fold higher sweet fraction of the aroma. The broad range of species in the enriched cultures produced more volatiles with higher peak areas compared to isolated cultures.

Table 10.1 Major volatiles identified in seaweed protein hydrolysates.[38–40,42]

Volatiles	Class	Algae	Descriptor
Octanoic acid	Acid	*Gracilaria fisheri*, *Caulerpa racemosa*	Fatty, rancid
(2*E*,4*E*)-Deca-2,4-dienal	Aldehyde	*Undaria pinnatifida*, *Laminaria japonica*, *Caulerpa racemosa*	Chicken, fish, beef
(*E*)-Oct-2-enal	Aldehyde	*Undaria pinnatifida*, *Laminaria japonica*, *Caulerpa racemosa*	Green, fatty
Heptanal	Aldehyde	*Undaria pinnatifida*, *Laminaria japonica*, *Caulerpa racemosa*	Fatty, citrus, rancid
Hexanal	Aldehyde	*Gracilaria fisheri*, *Undaria pinnatifida*, *Laminaria japonica*, *Caulerpa racemosa*	Grassy, green
Nonanal	Aldehyde	*Undaria pinnatifida*, *Laminaria japonica*, *Caulerpa racemosa*	Citrus, fatty
Benzaldehyde	Aldehyde, aromatic	*Gracilaria fisheri*, *Caulerpa racemosa*	Sweet, nutty
Icosane	Alkane	*Undaria pinnatifida*, *Caulerpa racemosa*	Fume, waxy
Pentadecane	Alkane	*Undaria pinnatifida*, *Laminaria japonica*, *Caulerpa racemosa*	Fume, alkane
2,4-Di-*tert*-butylphenol	Phenol	*Gracilaria fisheri*, *Caulerpa racemosa*	Fermented sausage

10.4 Flavour of Fungal Proteins and Peptides

The uses of fungi as a common nutritive source and fermentable medium have been well documented worldwide.[48,49] With the emergence of industrialized microbiology over the last century, along with a greater understanding of food mycology, there has been a progressive rise in studies on the use of mushrooms, yeasts, and filamentous fungi within foods.[50] Due to their high protein content, good protein quality, and low cost of production, many fungal species have great proteinaceous potential.[50–52] For example, a wide variety of fungi have been used to produce single-cell proteins with products from *Saccharomyces*, *Fusarium*, and *Torulopsis* commercially available.[43] In addition, the filamentous mycelium structure of certain fungi can simulate the fibrous nature of meat and are thus good candidates as meat substitutes (*e.g.*, mycoprotein Quorn produced from *Fusarium venenatum*).[50,52] Despite the great potential of fungi as a novel protein source, studies that characterize the flavour of fungal proteins are scarce.

10.4.1 Flavour of Proteins from Filamentous Fungi

Mycelia of both mushrooms and moulds have been cultivated for food production.[50,53] The flavour of mushroom mycelium ranges from flavourless to equivalent or even preferable to that of the fruiting bodies. Species, growing media, and cooking have been reported to affect the flavour.[54] Besides serving as a proteinaceous food, fungal mycelium can also be used as a source of food flavours.[53,55] For instance, through *de novo* synthesis or biotransformation, various Basidiomycetes species have been used to generate a myriad of aroma compounds, such as benzoate-type, phenylacetate-type, cinnamate-type, and butanone-type flavours.[56] A well-known example of protein produced by filamentous mould is Quorn, a product that can mimic the taste and consistency of meat.[57] It has been used as an alternative for chicken nuggets, meat balls, and minced meat since the 1960s. A consumer study showed that Quorn received a higher overall liking than several plant-based meat alternatives.[58] New start-ups are now growing mycelia with the goal of mimicking higher-quality cuts of meat (*e.g.*, steak).[59]

10.4.2 Flavour of Mushroom Proteins and Peptides

The fruiting bodies of mushrooms have been well appreciated by consumers for their rich umami taste and unique aromas. The aroma of some species is so intense that they are primarily used as flavouring agents (*e.g.*, truffles).[50] The umami taste of mushrooms comes from glutamate, aspartate, 5′-ribonucleotides, and a variety of umami peptides.[60] The volatile compounds of mushrooms are grouped into alcohols, ketones, esters, aldehydes, hydrocarbons, acids, and heterocyclic and aromatic compounds, with C_8 volatiles being the major contributors to the mushroom odour.[60] Cultivation conditions, maturity, processing, and storage are known to influence the flavour of mushrooms.[60] Such factors need to be taken into consideration when isolating fungal proteins, as flavour compounds may be associated with the isolated proteins. Moreover, the high sulphur-containing amino acid content in some mushroom species is conducive to the production of a meaty flavour.[61] Although several studies have investigated the use of mushrooms as meat or egg substitutes,[61,62] none have used isolated mushroom proteins.

A few studies used enzymes to produce free amino acids and peptides from mushroom proteins or used mushroom protein hydrolysates for flavour production. Mushrooms are a rich source of glutamic acid and aspartic acid.[60] However, these amino acids only exhibit umami flavour when free in solution; glutamate within a protein structure, for example, has no umami taste.[63] Separation of umami amino acids from fungal proteins may allow for greater potency of the umami flavour. Hydrolysis of mushroom proteins may also generate bioactive and taste-active peptides (Table 10.2).[60,75] Poojary *et al.* described the extraction of umami amino acids from several cultivated mushrooms using enzymes.[76] The authors utilized endo-1,3(4)-β-glucanase, an enzyme containing cellulolytic activity, to allow for the degradation of the chitinous cell wall associated with mushrooms. A commercial composite of endo- and exopeptidases, known as Flavourzyme, was used for protein hydrolysis and the release of umami amino acids. Overall, the enzyme extraction methods increased free amino acids significantly compared to conventional means. White and brown mushrooms were found to contain the greatest abundance of umami-free amino acids (26.8 ± 1.2 and 25.6 ± 0.9 mg g^{-1} dry mass, respectively). Another study produced mushroom protein hydrolysates from shiitake, oyster, bunashimeji, and enoki mushrooms using stem bromelain and found that incorporation of bunashimeji and oyster mushroom protein hydrolysates in chicken soup increased the aroma, taste, mouthfeel, and overall preference scores of the soup.[77] Gao *et al.* analysed the umami taste substances of morel mushroom (*Morchella sextelata*) hydrolysates derived from different enzymatic systems. They found that morel hydrolysates produced from the Neutrase–Flavourzyme combination contained the highest concentrations of small peptides, free amino acids, and 5′-nucleotides, which gave the best overall flavour properties.[78]

Many peptides produced from mushroom protein hydrolysis have exhibited the umami or kokumi taste. Xu *et al.* identified two umami peptides, ASNMSDL and LQPLNAH, in *Volvariella volvacea* mushrooms produced from high-pressure cooking.[67] Kong *et al.* used different enzymes to hydrolyse shiitake mushrooms and identified five umami peptides: GE, GCG, CM, VF, and EPE.[68] Feng *et al.* identified two naturally occurring kokumi peptides, GLPD and GHGD, in *Agaricus bisporus* mushrooms.[73]

Table 10.2 Taste active peptides identified in novel protein sources.

Peptides	Tastes	Treatments	Sources
AAPY	Umami	Enzyme hydrolysis	*Bombyx mori*[64]
ADA	Umami	Thermal treatment	*Saccharomyces cerevisiae*[65]
AEA	Umami	Thermal treatment	*Saccharomyces cerevisiae*[65]
AL	Bitter, kokumi	Naturally occurring	*Saccharomyces cerevisiae*[66]
ASNMSDL	Umami	High-pressure cooking	*Volvariella volvacea*[67]
AY	Bitter	Naturally occurring	*Saccharomyces cerevisiae*[66]
CM	Umami	Enzyme hydrolysis	*Lentinus edode*[68]
DD	Salty, umami, sour	Naturally occurring	*Saccharomyces cerevisiae*[69]
DDD	Salty, umami	Naturally occurring	*Saccharomyces cerevisiae*[69]
EA	Umami, sour	Thermal treatment	*Saccharomyces cerevisiae*[70]
ECG	Sour, kokumi	Naturally occurring	*Saccharomyces cerevisiae*[66]
ED	Salty, umami, sour	Thermal treatment	*Saccharomyces cerevisiae*[65,69]
EL	Umami, bitter, kokumi	Naturally occurring, thermal treatment	*Saccharomyces cerevisiae*[66,70]
EMH	Umami	Thermal treatment	*Saccharomyces cerevisiae*[65]
EN	Umami	Thermal treatment	*Saccharomyces cerevisiae*[70]
EPE	Umami	Enzyme hydrolysis	*Lentinus edode*[68]
EPLCNQ	Umami	Naturally occurring	*Stropharia rugosoannulata*[71]
ES	Umami	Thermal treatment	*Saccharomyces cerevisiae*[70]
ESCAPQL	Umami	Naturally occurring	*Stropharia rugosoannulata*[71]
EV	Kokumi	Naturally occurring	*Saccharomyces cerevisiae*[66]
EY	Bitter, kokumi	Naturally occurring	*Saccharomyces cerevisiae*[66]
FI	Salty, bitter	Naturally occurring	*Saccharomyces cerevisiae*[69]
FQ	Bitter, sour	Thermal treatment	*Saccharomyces cerevisiae*[72]
FT	Bitter, sour	Thermal treatment	*Saccharomyces cerevisiae*[72]
GCG	Umami	Enzyme hydrolysis	*Lentinus edode*[68]
GE	Umami	Enzyme hydrolysis	*Lentinus edode*[68]
GFP	Umami	Enzyme hydrolysis	*Bombyx mori*[64]
GGPG	Umami	Thermal treatment	*Saccharomyces cerevisiae*[65]

(*continued*)

Table 10.2 (*continued*)

Peptides	Tastes	Treatments	Sources
GGY	Umami, bitter	Thermal treatment	*Saccharomyces cerevisiae*[70]
GHGD	Kokumi	Naturally occurring	*Agaricus bisporus*[73]
GLPD	Kokumi	Naturally occurring	*Agaricus bisporus*[73]
HV	Umami, bitter	Thermal treatment	*Saccharomyces cerevisiae*[70]
KG	Umami, bitter, sour	Thermal treatment	*Saccharomyces cerevisiae*[70]
LA	Bitter, kokumi	Naturally occurring	*Saccharomyces cerevisiae*[66]
LE	Bitter, kokumi	Naturally occurring	*Saccharomyces cerevisiae*[66]
LK	Bitter, kokumi	Naturally occurring	*Saccharomyces cerevisiae*[66]
LQ	Bitter, kokumi	Naturally occurring	*Saccharomyces cerevisiae*[66]
LQPLNAH	Umami	High-pressure cooking	*Volvariella volvacea*[67]
LT	Bitter, kokumi	Naturally occurring	*Saccharomyces cerevisiae*[66]
LV	Umami, bitter	Thermal treatment	*Saccharomyces cerevisiae*[70]
LVG	Umami	Thermal treatment	*Saccharomyces cerevisiae*[70]
NY	Umami, sour	Thermal treatment	*Saccharomyces cerevisiae*[70]
PAA	Umami	Thermal treatment	*Saccharomyces cerevisiae*[70]
pEC	Kokumi	Thermal treatment + enzyme hydrolysis	*Agaricus bisporus*[74]
pED	Umami	Thermal treatment + enzyme hydrolysis	*Agaricus bisporus*[74]
pEE	Umami	Thermal treatment + enzyme hydrolysis	*Agaricus bisporus*[74]
pEP	Umami	Thermal treatment + enzyme hydrolysis	*Agaricus bisporus*[74]
PET	Umami	Thermal treatment	*Saccharomyces cerevisiae*[70]
pEV	Kokumi	Thermal treatment + enzyme hydrolysis	*Agaricus bisporus*[74]
PGD	Umami	Thermal treatment	*Saccharomyces cerevisiae*[65]
PHEMQ	Umami	Naturally occurring	*Stropharia rugosoannulata*[71]
PL	Bitter	Thermal treatment	*Saccharomyces cerevisiae*[72]
PSG	Umami	Thermal treatment	*Saccharomyces cerevisiae*[65]
QL	Umami, kokumi	Thermal treatment	*Saccharomyces cerevisiae*[70]

(*continued*)

Table 10.2 (*continued*)

Peptides	Tastes	Treatments	Sources
QPS	Umami	Thermal treatment	*Saccharomyces cerevisiae*[65]
RL	Bitter, sour	Thermal treatment	*Saccharomyces cerevisiae*[72]
SEPSHF	Umami	Naturally occurring	*Stropharia rugosoannulata*[71]
SGCVNEL	Umami	Naturally occurring	*Stropharia rugosoannulata*[71]
SPE	Salty, sour	Naturally occurring	*Saccharomyces cerevisiae*[69]
TAY	Umami	Enzyme hydrolysis	*Bombyx mori*[64]
VAV	Umami	Thermal treatment	*Saccharomyces cerevisiae*[70]
VEH	Umami	Thermal treatment	*Saccharomyces cerevisiae*[65]
VF	Umami	Enzyme hydrolysis	*Lentinus edode*[68]
VPY	Umami	Enzyme hydrolysis	*Bombyx mori*[64]
VV	Umami, bitter	Thermal treatment	*Saccharomyces cerevisiae*[70]
YE	Bitter	Naturally occurring	*Saccharomyces cerevisiae*[66]

Additional taste-modulating pyroglutamyl dipeptides were identified in thermally treated and enzymatically hydrolysed *A. bisporus* mushrooms, including pyroglutamyl cysteine (pEC), pEV, pED, pEE, and pEP.[74] Chen *et al.* identified five naturally occurring umami peptides, EPLCNQ, SGCVNEL, PHEMQ, SEPSHF, and ESCAPQL, in *Stropharia rugosoannulata* mushrooms.[71] Through *in-silico* molecular docking analysis, the authors suggested that the peptides can enter the binding pocket in the Venus flytrap domain of the T_1R_3 cavity of the umami taste receptor and form stable hydrogen bonds to stimulate a strong umami taste. Glu45, Ser104, and His145 may play important roles in the binding of the peptides.

Mushroom protein hydrolysates can also undergo the Maillard reaction and produce additional flavour compounds. Lofty *et al.* reported the production of beef-like flavour from Flavourzyme and Alcalase-hydrolysed *Agaricus bisporus* mushroom proteins by heating the hydrolysates with xylose, cysteine, taurine, and thiamine.[79] The resulting volatile profile of the beef-like flavour was dominated by thiol-containing compounds, particularly 2-methyl-3-furanthiol. Additional sulphur-containing volatiles generated included disulphide compounds, thiophenes, thiazoles, and mercaptoketones. These compounds have also been reported as important volatiles in cooked meat and were likely formed through the reaction of hydrogen sulphide (released from the sulphur-containing precursors) with sugar-degradation products. Interestingly, another study reported that thermal treatment of hydrolysed *A. bisporus* mushroom protein and cysteine produced odorants that can enhance saltiness perception.[80] The key odorants identified included 2-furfurylthiol, 1-(2-furyl)ethanethiol, 3-sulfanylpentan-2-one, sotolon, indole, 2-methyl-3-(methyldithio)furan, and *p*-cresol. Chen *et al.* hydrolysed Finnish mushrooms (*Craterellus tubaeformis*) using cellulase and neutral protease and heated the hydrolysates with D-xylose and L-cysteine.[81] The Maillard reaction products with caramel-like (3-phenylfuran and 2-octylfuran), mushroom-like [1-octen-3-ol, (*E*)-2-octen-1-ol, and geranyl acetone], and meat-like (2-thiophenecarboxaldehyde, 2,5-thiophenedicarboxaldehyde, and 3-methylbutanal) flavours were generated. The common Maillard reaction-generated volatiles prepared from novel protein hydrolysates are summarized in Table 10.3.

10.4.3 Flavour of Yeast Proteins and Peptides

Yeasts have been used to produce single-cell proteins,[43] as well as various flavour compounds.[45,46] Yeast products such as dried yeasts (*e.g.*, nutritional yeast, Torula), yeast extracts, and yeast spreads (*e.g.*, Marmite and Vegemite) are rich in proteins (28–69%), B vitamins, and flavour compounds and are widely used as nutritional and flavouring ingredients.[92] The predominant tastes associated with yeast extracts include salty, umami, bitter, kokumi, and sour. The strong umami taste of yeast extracts is attributed to their rich content of amino acids, peptides, and nucleotides.[70] Over 180 volatiles have been identified in yeast extracts.[65,86,90,93–100] The volatile profiles vary considerably among different types of yeast extracts and production processes. The major volatiles repeatedly reported include pyrazines, furans, sulphur compounds, aromatic compounds, acids, aldehydes, ketones, pyrroles, and alcohols (Table 10.4). They contribute to the meaty, nutty, roasted, floral, fruity, malty, cheesy, sweet, potato-like, and popcorn-like notes of yeast extracts. Besides the desirable odorants, 4-methylphenol (medicine, phenolic), 3-methylpyridine (medicine), 3-methylbutanoic acid (cheesy, rancid), propanoic acid (rancid, pungent), and 1*H*-indole (animal) have been identified as the key volatiles responsible for the yeasty off-flavours of yeast extracts.[97,100]

Heat treatment has a profound influence on the flavour of yeast extracts. Münch *et al.* reported that heat treatment resulted in an increase in furan-2-ylmethanethiol and 2-methoxyphenol, giving the yeast extract roasty and burnt notes.[93] Alim *et al.* found that both the number and concentrations of volatiles increased as temperatures rose. The nutty, roasty, and meaty aroma became more intense due to increased concentrations of heterocyclic compounds (*e.g.*, furans, pyrazines, thiophenes, and thiazoles) and branched-chain sulphur compounds.[65,98] Maillard reaction precursors such as sugars, cysteine, thiamine, and glutathione can be added to further enhance the meaty aroma of yeast extracts (Table 10.3).[86,90]

Extraction methods also influence the volatile profile. Mahadevan and Farmer compared simultaneous distillation extraction (SDE) and

Table 10.3 Major Maillard reaction products of novel protein hydrolysates.

Volatiles	Class	Descriptor	Ref.
(*E*)-Oct-2-enal	Aldehyde	Grassy, fatty	81–85
2-Methylbutanal	Aldehyde	Cocoa, malty	79,80,82,86,87
3-Methylbutanal	Aldehyde	Fruity, green, malty	79–82,88
3-Methylsulfanylpropanal	Aldehyde, sulphur	Potato	79,80,82,84,85
Decanal	Aldehyde	Green, fatty	81,84,85,88,89
Heptanal	Aldehyde	Citrus, green, fatty	81,84,85,88–90
Hexanal	Aldehyde	Green	80–82,84–86,88
Nonanal	Aldehyde	Citrus, green, fatty	81,84–86,88–90
Octanal	Aldehyde	Citrus, green, fatty	81,82,86,88
1-(Furan-2-yl)ethanone	Furan, ketone	Balsamic, sweet	84,86,88,90
2-Methyl-3-(methyldisulfanyl)furan	Furan, sulphur	Meaty, beefy, roasty	79,83,90
2-Methyl-3-[(2-methylfuran-3-yl)disulfanyl]furan	Furan, sulphur	Roasted meat	79,81,83,90
2-Methylfuran-3-thiol	Furan, sulphur	Meaty	79,83,85,86,90
2-Pentylfuran	Furan	Buttery, earthy, green	81,84–86,88,89
Furan-2-carbaldehyde	Furan, aldehyde	Sweet, roasty, nutty	79,83–86,88–90
Furan-2-ylmethanethiol	Furan, sulphur	Coffee, roasty, meaty	79,80,83–86,88–90
Furan-2-ylmethanol	Furan, alcohol	Sweet	84–88
2,3,5-Trimethylpyrazine	Pyrazine	Roasty, potato	79,82,86,90
2,5-Dimethylpyrazine	Pyrazine	Roasted beef, grassy, popcorn	83,84,86,88,90
2-Ethyl-3,5-dimethylpyrazine	Pyrazine	Roasty, rice, popcorn-like	80,82,83,86
2-Ethyl-6-methylpyrazine	Pyrazine	Nutty, grassy, roasted potato	81,83,86,90
2-Methylpyrazine	Pyrazine	Roasty, potato, popcorn	79,81,86,88,90
3-Ethyl-2,5-dimethylpyrazine	Pyrazine	Popcorn, potato, roasty	79,81,82,84,86,90
2-Methylthiophene	Thiophene	Roasted beef, sulphurous, onion	79,83,86,88,90
3-Methylthiophene-2-carbaldehyde	Thiophene, aldehyde	Grassy, fatty, burnt	81,83–85,88,89

(*continued*)

Table 10.3 (*continued*)

Volatiles	Class	Descriptor	Ref.
5-Methylthiophene-2-carbaldehyde	Thiophene, aldehyde	Burnt, sulphurous, meaty	81,83–85,88–90
Thiophene-2,5-dicarbaldehyde	Thiophene, aldehyde	Meaty	81,83–85,89
Thiophene-2-carbaldehyde	Thiophene, aldehyde	Burnt	81,84,85,88
1*H*-Indole	Aromatic	Animal, burnt, floral	80,82,84,86,89,91
2,4-Di-*tert*-butylphenol	Aromatic, phenol	Phenolic	84,85,90,91
2-Methoxyphenol	Aromatic, phenol	Smoky, aromatic, burnt	80,82,84,86,90,91
2-Phenylacetaldehyde	Aromatic, aldehyde	Fruity, nutty, sweet	80–82,84,85,90,91
Benzaldehyde	Aromatic, aldehyde	Almond, burnt sugar	81,84–86,89–91
3-Methylbutanoic acid	Acid	Sweaty, rancid	80,82,84,85,90
Acetic acid	Acid	Pungent, vinegar	80,84,85,91
Decanoic acid	Acid	Rancid, fat	84,85,89,91
Octanoic acid	Acid	Sweaty, cheese	81,82,84,85,89,91
Hexanoic acid	Acid	Sweaty, rancid, cheesy	80,82,84,85,88,91
Hexan-1-ol	Alcohol	Green, fruity	81,84,85,88,89
Oct-1-en-3-ol	Alcohol	Mushroom	81,83–85,88
Octan-1-ol	Alcohol	Fatty, citrus	81,84,85,88,89
Heptan-2-one	Ketone	Green, sweet	84,85,88,89
(Methyltrisulfanyl)methane	Sulphur	Sulphurous, meaty	79,80,84,90
1-(1,3-Thiazol-2-yl)ethanone	Thiazole, ketone	Roasty, nutty, sulphurous	79–81,84,85,88–90
1-(1*H*-Pyrrol-2-yl)ethanone	Pyrrole, ketone	Bread, nutty, sweet	84,86,88,91

dynamic headspace (DHS) analysis and found that SDE was more efficient in extracting less volatile compounds, such as (*E*)-β-damascenone and 2-methoxyphenol, while DHS detected more (methyltrisulfanyl)methane.[95] Lin *et al.* found that pentane-2,3-dione, ethenyl acetate, octan-2-one, 1-(3,4-dihydro-2*H*-pyrrol-5-yl)ethanone, (methyltrisulfanyl)methane, (*E*)-oct-2-enal, (*E*)-non-2-enal, (2*E*,4*E*)-deca-2,4-dienal, and 5-methylthiophene-2-carbaldehyde were only extracted by SDE, while acetic acid, propanoic acid, hexanoic acid, and (*E*)-β-damascenone were only found following DHS extraction.[96] Alim *et al.* reported that more volatile compounds were extracted by solid-phase microextraction (SPME) than by solvent-assisted flavour evaporation (SAFE). More pyrazines, furans, and thiophenes were

Table 10.4 Major volatiles identified in yeast extracts.

Volatiles	Class	Descriptor	Ref.
2,3,5,6-Tetramethylpyrazine	Pyrazine	Beef, nutty	90,96,99
2,3,5-Trimethylpyrazine	Pyrazine	Potato	65,86,90,93,95–100
2,3-Dimethylpyrazine	Pyrazine	Nutty	96,97,99,100
2,5-Dimethylpyrazine	Pyrazine	Roasted beef, nutty	65,86,90,96,97,100
2,6-Dimethylpyrazine	Pyrazine	Roasted beef, nutty	65,86,96–100
2-Ethyl-3,5-dimethylpyrazine	Pyrazine	Potato	65,86,93,96
2-Ethyl-5-methylpyrazine	Pyrazine	Fruity, sweet	65,86,98,100
2-Ethyl-6-methylpyrazine	Pyrazine	Nutty, roasted potato	86,90,96,100
2-Ethylpyrazine	Pyrazine	Peanut butter	65,86,96,97,100
2-Methylpyrazine	Pyrazine	Popcorn	65,86,90,96–100
3-Ethyl-2,5-dimethylpyrazine	Pyrazine	Potato	65,86,90,95,98–100
Pyrazine	Pyrazine	Nutty	90,96,100
5-Pentyloxolan-2-one	Furanone, ketone	Coconut	65,98,99
1-(Furan-2-yl)ethanone	Furan, ketone	Balsamic	65,86,90,96,98,100
2-Pentylfuran	Furan	Green, rancid	86,95,96
5-Methylfuran-2-carbaldehyde	Furan, aldehyde	Burnt sugar	65,86,95,96,98
Furan-2-carbaldehyde	Furan, aldehyde	Bread, sweet	65,86,90,95–98,100
Furan-2-ylmethanol	Furan, alcohol	Burnt	65,86,96,98
2-Methylfuran-3-thiol	Furan, sulphur	Meaty	86,90,94,95,97,99
Furan-2-ylmethanethiol	Furan, sulphur	Coffee, roasty	86,90,93–96,98
(Methyldisulfanyl)methane	Sulphur	Sulphurous	65,96,100
(Methyltrisulfanyl)methane	Sulphur	Sulphurous	90,95–98
3-Methylsulfanylpropan-1-ol	Sulphur, alcohol	Meaty	96,99,100
3-Methylsulfanylpropanal	Sulphur, aldehyde	Potato	65,93–98,100
5-Methylthiophene-2-carbaldehyde	Sulphur, thiophene, aldehyde	Sulphurous	65,90,95,96,98

(continued)

Table 10.4 (*continued*)

Volatiles	Class	Descriptor	Ref.
2-Methylthiophene	Sulphur, thiophene	Sulphurous	65,86,90,95,98
2-Methylthiolan-3-one	Sulphur, thiophene, ketone	Potato, onion	90,95,98
1-(4,5-Dihydro-1,3-thiazol-2-yl)ethanone	Sulphur, thiazole, ketone	Popcorn, burnt	93–96
Benzothiazole	Sulphur, thiazole, aromatic	Sulphurous	90,95,100
2-Phenylethanol	Aromatic, alcohol	Floral, honey	65,93,94,96–100
Phenylmethanol	Aromatic, alcohol	Floral, sweet	98–100
2-Methoxyphenol	Aromatic, phenol	Smoky	65,86,90,93,95,100
2-Methylphenol	Aromatic, phenol	Smoky	90,95,100
2-Phenylacetaldehyde	Aromatic, aldehyde	Honey, floral	65,90,93–97,99,100
Benzaldehyde	Aromatic, aldehyde	Almond, burnt sugar	65,86,90,96–100
1-Phenylethanone	Aromatic, ketone	Floral	90,97,98,100
1*H*-Indole	Pyrrole, aromatic	Burnt	65,86,97–99
1-(1*H*-Pyrrol-2-yl)ethanone	Pyrrole, ketone	Nutty, bread	65,86,96–100
1-(3,4-Dihydro-2*H*-pyrrol-5-yl)ethanone	Pyrrole, ketone	Roasty, popcorn	93–96
(*E*)-β-Damascenone	Ketone	Honey, floral	93,95,96
Butane-2,3-dione	Ketone	Buttery	93,94,96–98
Pentane-2,3-dione	Ketone	Sweet	93,95,96
2-Methylbutanoic acid	Acid	Cheesy	93,94,97,99
2-Methylpropanoic acid	Acid	Cheesy, buttery	90,93,96,97,99,100
3-Methylbutanoic acid	Acid	Sweaty, cheesy	65,90,93,94,96,97,99,100
Acetic acid	Acid	Sour	93,96–98,100
Butanoic acid	Acid	Cheesy, sweaty	65,93,96,99,100
Propanoic acid	Acid	Rancid, pungent	93,96,97,100
2-Methylbutanal	Aldehyde	Cocoa	65,86,94,96,98

(*continued*)

Table 10.4 (*continued*)

Volatiles	Class	Descriptor	Ref.
3-Methylbutanal	Aldehyde	Malty, chocolate	65,93,94,96–98
Heptanal	Aldehyde	Fatty	65,90,96,99
Nonanal	Aldehyde	Green, citrus	65,86,90,96–100
Octanal	Aldehyde	Green	65,86,96–98,100
3-Methylbutan-1-ol	Alcohol	Malty, yeasty	93,94,96

detected in the SPME extract, while more sulphur-containing odours appeared in the SAFE extract.[98]

As shown in Table 10.2, numerous umami-, bitter-, kokumi-, salty-, and sour-tasting peptides have been identified in yeast extracts,[65,66,69,70,72] among which umami peptides are the most predominant. Alim *et al.* further categorized the umami peptides into two groups based on their structures.[70] Type I peptides are mainly composed of aliphatic and aromatic amino acids. They induce complex tastes by interacting with the umami taste receptor T_1R_1 and other taste receptors (*e.g.*, the bitter receptor). Type II peptides mainly consist of acidic, amide, basic, and hydroxy amino acids. They stimulate both T_1R_1 and T_1R_3 receptors and exhibit strong umami tastes. Bitter oligopeptides and bitter amino acids are responsible for the bitter taste of yeast extracts. These bitter tastants are generated by enzymatic hydrolysis and thermal treatment.[72,98] Keeping the thermal reaction temperature under 120 °C can maximize the umami taste while controlling the bitter taste.[72] Several peptides in yeast extracts exhibit a sour taste, which is likely attributed to the release of protons from the carboxylic groups. The sourness is stronger in water than in broth, suggesting that the buffering capacity of the broth can help reduce or eliminate the sour taste.[70,72] Salty peptides have also been identified in yeast extract, which can function as salt-reducing agents.[69] Besides the peptides that activate the basic tastes, several glutamyl and leucyl peptides in yeast extracts have been found to elicit a long-lasting, taste-enhancing kokumi sensation.[66]

These peptides can also generate beef-like aromas through the Maillard reaction (Table 10.3). Alim *et al.* reacted yeast extract peptides with xylose, cysteine, and thiamine and generated beef-like aromas.[86] Among the volatiles, 2,5-dimethylpyrazine, 2,6-dimethylpyrazine, ethylpyrazine, 3-ethyl-2,5-dimethylpyrazine, 2-methylthiophene, 2-methyl-3-furanthiol, and 2-furfurylthiol were only detected when cysteine was added, while 5-ethenyl-4-methylthiazole was generated when thiamine was added. Raza *et al.* heated glutathione-enriched yeast extract with ribose and cysteine and generated beef-like aromas such as 2,5-dimethylfuran, 2,5-dimethyl-pyrazine, thiazole, 2-methyl-3-furanthiol, dimethyl trisulphide, 3,5-diethyl-2-methyl-pyrazine, 3,3′-dithiobis[2-methyl-furan], and 2-methyl-3-(methylthio) furan.[90]

10.5 Flavour of Insect Proteins and Peptides

Edible insects have great potential as a novel protein source due to their optimal nutritional value, high feed conversion efficiency, low environmental footprint, and ability to be reared on agricultural waste materials.[101–103] Although insect consumption is commonly practised in many cultures,[104,105] consumer acceptance in Western countries remains low.[106,107] Flavour is an important factor that affects consumer acceptance of this novel food.[108] Depending on the species and developmental stages, some insects are known to possess obnoxious odours.[109,110] Characterizing the flavour profiles of different edible insects is thus important for product formulation. So far, the flavours of only a few edible insects have been reported.[111] The flavour characterization of insect proteins is even more scarce.

Lee *et al.* reported volatile profiles of proteins isolated from *Protaetia brevitarsis* larvae.[112] A total of 50 volatile compounds were identified following the defatting and heating of the proteins, including 3 alcohols, 11 aldehydes, 4 esters, 7 ketones, 23 hydrocarbons, a sulphur compound, and a phenol. The defatting process greatly reduced alcohols, aldehydes, esters, ketones, and hydrocarbons, many of which were lipid oxidation products with off-flavours (*e.g.*, oct-1-en-3-ol with a fishy odour). In contrast, the levels of several monoterpenes, such as *o*-cymene (smoke), D-limonene (citrus), γ-terpinene (citrus), *p*-cymene (citrus), and α-terpinene (berry), which impart pleasant flavours, increased following defatting. Heating increased the aldehyde content, likely due to accelerated lipid oxidation. Heating also increased the contents of γ-terpinene, *p*-cymene, and α-terpinene, possibly due to the thermal conversion of D-limonene to these compounds.

Zhou *et al.* formulated protein bars with hempseed and *Tenebrio molitor* larvae proteins.[113] Protein bars with a hempseed protein to mealworm protein ratio of 7:3 exhibited the best sensory acceptance. Increasing the content of mealworm protein decreased alcohols and increased aldehydes and aliphatic hydrocarbons. Storage greatly increased the heterocyclic compounds due to the Maillard reaction while reducing aldehyde and ketone off-odours.

Grossmann *et al.* studied the flavours of cricket and mealworm proteins, their Flavourzyme and Protease A hydrolysates, and the Maillard reaction products.[82] Protein hydrolysis led to an increase in both bitter and umami tastes. Both protein hydrolysis and the Maillard reaction with xylose changed the volatile profiles of the insect proteins. A total of 38 odour-active compounds were identified, including 1 alcohol, 5 acids, 11 aldehydes, 5 ketones, and 16 heterocyclic compounds (Table 10.3). The volatiles 2-/3-methylbutanoic acid, phenylacetaldehyde, 2,3-butanedione, 2-acetyl-1-pyrroline, furaneol, trimethylpyrazine, and *p*-cresol were found in both cricket and mealworm proteins under all treatments. The compounds 2-phenyl-2-butenal, 2,3-pentanedione, 2-acetyl-2-thiazoline, 2,3-diethyl-5-methylpyrazine, 5-vinyl-2-methoxyphenol, indole, and g-undecalactone were only detected in the cricket samples, while 2-methylpropanoic acid, (*E*)-2-nonenal, *trans*-4,5-epoxy-(*E*)-2-decenal, 4-hydroxy-5-methyl-3(2*H*)-furanone, 2-methoxyphenol, 2-methoxy-4-vinylphenol, and g-octalactone were only identified in mealworm samples. Many of these volatiles have also been reported in meat products such as cooked beef, roasted pork, and roasted livers of beef, duck, and pork.

Yu *et al.* hydrolysed silkworm (*Bombyx mori*) pupa protein by Flavourzyme and Alcalase and produced umami-tasting oligopeptides VPY, TAY, AAPY, and GEP (Table 10.2). The peptides VPY, TAY, and AAPY also exhibited an umami-enhancing effect on a monosodium glutamate-sodium chloride solution. Interestingly, these peptides are composed of sweet and bitter amino acids instead of umami amino acids.[64]

10.6 Flavour of Leaf Proteins

Leaf proteins are concentrated proteins extracted from the leaves of green plants and have great potential as an alternative protein source. For example, ribulose-1,5-bisphosphate carboxylase-oxygenase (RuBisCO) in leaves is estimated to be the most abundant protein in the world.[114] Leaf proteins can be accessed wherever it is possible to grow green plants, are derived from any leaf that is safe for humans to consume, and have the potential to reduce waste from commonly unused plants and plant parts. They were suggested as a promising source of protein as early as 1942, but interest and research were stagnant, and acceptability and widespread use as human food remained low.[115–117]

Leaf proteins have long been used for animal feed, but their high protein content, balanced amino acid composition, and other nutritional benefits have driven the consideration of using leaf proteins as a future food source for humans. Various studies have shown that leaf proteins contain sufficient quantities of all essential amino acids and meet the FAO/WHO (World Health Organization) requirements.[118–121] Besides their nutritional values, leaf proteins have good functional properties. RuBisCO protein isolates from sugar beet leaves exhibited superior foaming and gelling properties, and comparable emulsifying properties in comparison to whey protein isolate and soy protein isolate.[122] The gelling property of duckweed protein concentrate was superior to soy protein and comparable to egg white protein.[123] When enriched in wheat-based foods, RuBisCO actively participated in protein network formation and did not impair dough viscoelasticity.[124]

Leaf proteins are classified as water-insoluble thylakoid proteins (green proteins) and water-soluble cytoplastic proteins (white proteins).[117] The green proteins have a dark green colour and a grassy taste, while the white proteins are white cream in colour and are almost tasteless and odourless.[125] Saponins, polyphenols, and their oxidized products may give rise to off-flavours such as bitterness and astringency.[117] Removal of such undesired compounds is a strenuous step during leaf protein extraction.[116] Firdaous *et al.* reported that alfalfa white protein concentrate prepared by ultrafiltration followed by diafiltration and drying is still rich in polyphenols and, thus, has poor organoleptic quality. Further purification of alfalfa white proteins could be achieved by a column adsorption process.[126]

10.7 Flavour of Oilseed Proteins

10.7.1 Flavour of Canola/Rapeseed Proteins and Peptides

Rapeseed (*Brassica napus*) is the second most produced oil crop, following soybean. The annual production of rapeseed meal is approximately 38 million tonnes,[127] which translates into 14 million metric tonnes of crude proteins.[128] 11S cruciferin and 2S napin are the two major proteins in rapeseed, which respectively account for 32–60% and 25–45% of total rapeseed proteins.[129] These two protein fractions differ considerably in physicochemical and functional properties and perform best when utilized separately.[129] The nutritional value of rapeseed protein is favourable, with a high level of indispensable and particularly sulphur-containing amino acids.[130] Although predominantly used for animal feed, rapeseed proteins also have great potential to serve as a high-quality protein source for humans.

The utilization of rapeseed protein can be limited by its intense bitter off-taste. The bitterness was likely attributed to residual phytometabolites such as glucosinolates and phenolics.[131] Recently, kaempferol 3-*O*-(2‴-*O*-sinapoyl-β-sophoroside) (see Scheme 10.1) has been identified as a key contributor to the bitterness of rapeseed proteins with a low taste threshold of 3.4 μM.[132] The concentration of this compound varies in different fractions of rapeseed proteins. It is much more abundant in cruciferin-rich protein isolate (475.6 μmol kg^{-1}) compared to napin-rich protein isolate (32.9 μmol kg^{-1}), which explains the lower bitterness of the latter product.[132]

Membrane processing technologies such as ultrafiltration and diafiltration have been used to remove the bitter-tasting components in rapeseed proteins.[133–135] Xu and Diosady studied the interactions between rapeseed proteins and phenolic compounds in aqueous media and reported that about 70% of rapeseed proteins were extracted at pH 12.0, along with over 80% of phenolic acids and more than 90% of condensed tannins.[135] Although free phenolic acids and condensed tannins can readily permeate the membrane and be removed by diafiltration, approximately half of the extracted phenolic compounds were bound to the rapeseed proteins. The phenolic acids were predominately bound *via* ionic bonding, while hydrophobic forces were mainly involved in the tannin–protein interactions. Reducing pH and adding salt helped release the bound phenolic compounds.[135]

Like other proteins, rapeseed proteins are capable of binding volatile compounds and changing the flavour profile. The intrinsic characteristics of the flavour and protein molecules and extrinsic conditions can

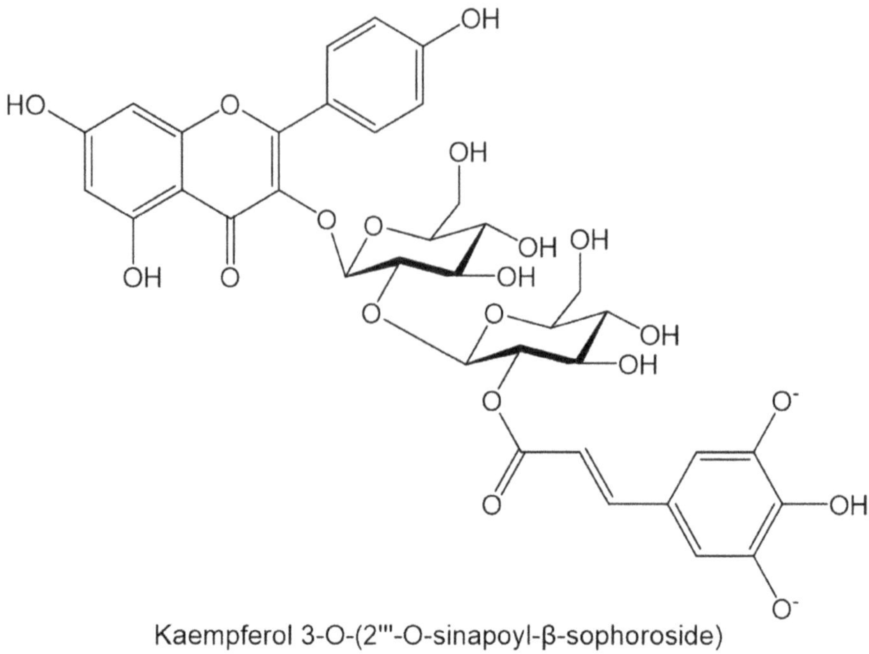

Scheme 10.1 Kaempferol 3-*O*-(2‴-*O*-sinapoyl-β-sophoroside).

greatly influence the interactions between the proteins and the flavour compounds. For example, hydrophobic interactions play a major role in the interactions among rapeseed proteins, aldehydes, and ketone flavours. Therefore, a greater carbon number would result in stronger flavour retention.[136] However, the competition between the flavour compounds may change this general rule. For example, in a ternary system where a mixture of hexanal, heptanal, and octanal is introduced to rapeseed proteins, the binding of hexanal increased while the binding of heptanal and octanal decreased compared to the binding of a single compound.[137] The larger heptanal and octanal molecules may have limited access to certain binding sites due to steric hindrance.

The functional group of a flavour molecule also determines how it interacts with proteins. Rapeseed proteins have a higher affinity for aldehydes than ketones. Aldehydes not only have a higher hydrophobicity but can also form irreversible covalent bonds with functional groups of proteins. In comparison to wheat gluten and pea proteins, rapeseed proteins have the highest binding capacity to aldehydes yet the lowest affinity to ketones. This may be attributed to the different protein quaternary structures and the abundance of sulphur-containing amino acids. The higher cysteine and methionine contents of rapeseed proteins may enable more efficient aldehyde binding, while the hydrophobic cavity of

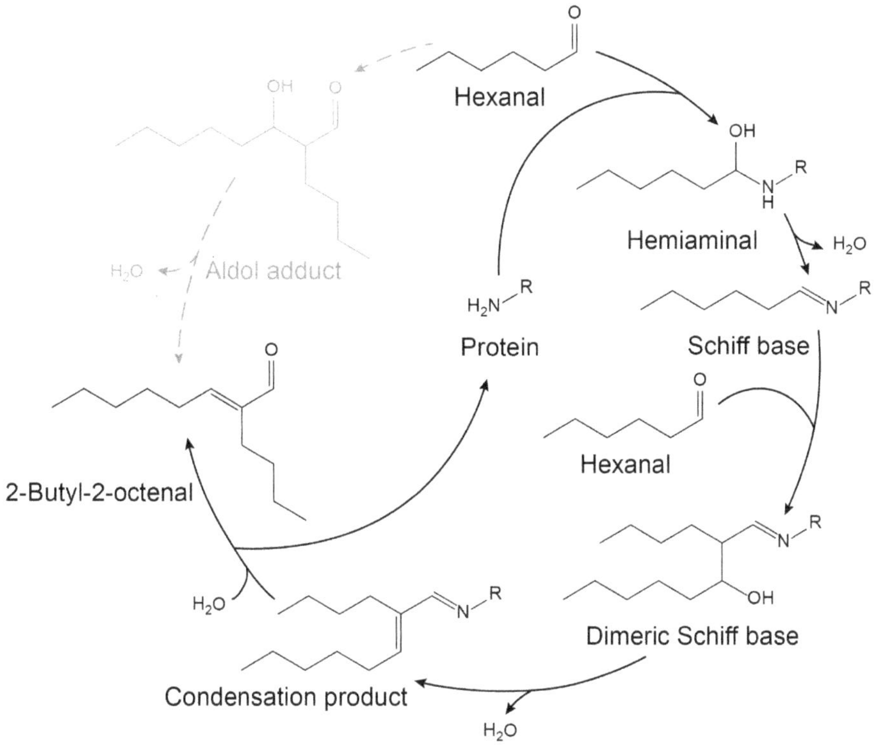

Figure 10.1 Rapeseed protein-catalysed aldolization of hexanal and formation of dimeric aldehyde. Dashed lines: aldol addition and condensation of two hexanals. Solid lines: rapeseed protein-catalysed pathway.

cruciferin may be less accessible to ketones compared to wheat gluten and pea proteins.[136]

Protein structure is important for flavour binding. Alkaline-extracted rapeseed proteins exhibited lower binding capacity to aldehyde compared to salt-extracted proteins, possibly due to severe changes in protein conformation and solubility, which limited the access of the flavour compounds to the binding sites.[136] When interacting with proteins, aldehydes may induce the unfolding of the protein structure and the exposure of buried binding sites. This may, in turn, enhance the binding of other flavour compounds. For example, the binding of 2-hexanone to rapeseed proteins increased in the presence of hexanal.[137] The interactions between aldehydes and proteins may also give rise to new volatiles. When hexanal and heptanal were mixed with rapeseed proteins, 2-butyl-2-octenal and 2-pentyl-2-nonenal were formed, respectively. The aldehydes likely reacted with the amino groups of the protein and formed Schiff bases, followed by dimerization, condensation, and hydrolysis to generate the new dimeric aldehydes (see Figure 10.1).[136,138]

Temperature is another key factor. Upon heating at 95 °C, the binding of hexanal to rapeseed proteins increased gradually over the first 10 min and then remained constant. The enhanced retention of hexanal was likely due to heat-induced exposure of binding sites on the rapeseed proteins. A similar increase in the binding of 2-octanone was observed during the first 2 min of heating, but the binding capacity gradually decreased afterwards. Aldehydes predominantly form covalent cross-links with proteins, while ketones reversibly bind to proteins through hydrophobic interactions. The decreased binding capacity of 2-octanone during extended heating was possibly due to the competition between protein–flavour interaction and protein aggregation.[137]

Hydrolysates of rapeseed proteins have also been used to generate meat-like flavours through the Maillard reaction. The meaty aroma was largely attributed to the formation of sulphur-containing compounds such as thiazoles and dimethyl disulphide (Table 10.3).[87] Rapeseed proteins are rich in sulphur-containing amino acids and can produce more sulphur-containing volatiles than soy proteins.

10.7.2 Flavour of Cottonseed Proteins

Cottonseed is a sustainable source of plant proteins. The annual production of cottonseed was 45 million metric tonnes in 2019, which translates into approximately 10 million metric tonnes of proteins.[139,140] The cottonseed proteins mainly consist of salt-soluble globulins (33–67%), water-soluble albumins (20.8–32.2%), and alkali-soluble glutelins (9.2–28%).[140] They are known to possess superior protein quality and functional properties.[140] Despite the great potential, a major obstacle to using cottonseed protein is the presence of gossypol (see Scheme 10.2), a toxic polyphenol with a pronounced bitter taste.[141] This compound is secreted by the resin glands present in the petals, leaves, root bark, bolls, and seeds of cotton plants and is primarily found in cottonseed kernels. Numerous physical, chemical, and biological methods have been developed to reduce gossypol content to a minimum amount and achieve a bland flavour.[140]

Berardi and Cherry isolated cottonseed proteins from glandless cottonseed flour by alkaline extraction followed by acid precipitation.[142] Heating the water suspension of the cottonseed protein isolates at a pH of 4.5–9.0 and 90 °C produced texturized proteins with the mouthfeel and chewiness of cooked meat. Incorporation of 10–20% textured cottonseed protein isolates into meat loaves resulted in acceptable texture, colour, and flavour, while blandness was perceived in products containing 30% texturized cottonseed proteins. Gas chromatography analysis showed that hexanal, hexanol, xylene, styrene, dimethyl ethyl benzene, benzaldehyde, 2-furyl methyl ketone, and phenethyl alcohol were high in cottonseed storage protein isolate and supplemented meat loaves.[143] Among these volatiles, hexanal and hexanol exhibited positive correlations with organoleptic characteristics. These flavour compounds

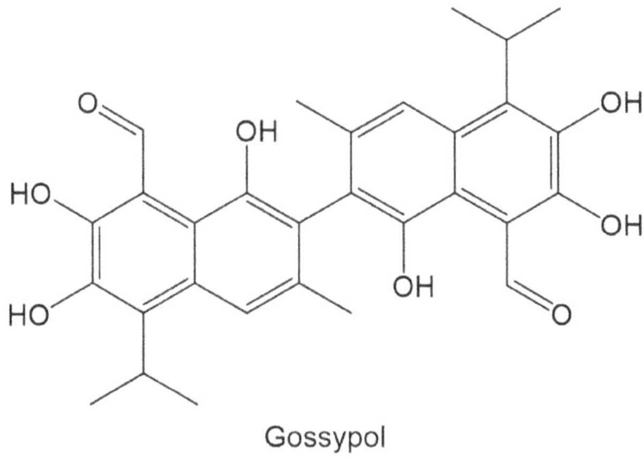

Scheme 10.2 Gossypol.

have also been identified in beef products.[143] Therefore, cottonseed proteins can serve as a good filler in meat products due to their chewy texture and compatible flavour profile. Ziprin *et al.* also reported that substituting 10% of the meat in ground beef patties with rehydrated cottonseed protein isolates did not change the sensory quality of the patties and hindered lipid oxidation.[144] Similarly, replacing 15% of meat with cottonseed proteins did not affect moistness, off-flavour, texture, and the overall satisfaction of frankfurters.[145] In frozen desserts, 60% of non-fat milk solids can be replaced by cottonseed storage protein isolate without changing the flavour,[146,147] odour,[147] and overall acceptability.[147] Cottonseed whey proteins, a by-product of the protein isolation process, can also be utilized as a food ingredient following an ultrafiltration treatment.[148] An orange-flavoured 'ade' drink and an instant fruit drink fortified with 3% cottonseed whey proteins exhibited acceptable organoleptic properties. The whipping of the cottonseed whey proteins resulted in stable and pleasant-tasting foams. Cottonseed whey without membrane treatment, however, registered a bitter taste.

10.7.3 Flavour of Flaxseed Protein Hydrolysates

Flaxseed, or linseed (*Linum usitatissimum*), is a potential source of functional ingredients, including high-quality proteins.[149] The amino acid composition of flaxseed protein is comparable to that of soy protein.[150] Various biological properties of flaxseed proteins and peptides have also been reported.[151] The flavour profile of flaxseed protein has not been characterized. A few studies have used flaxseed protein hydrolysates to generate flavours through the Maillard reaction (Table 10.3). Flavour-enhancing properties were observed in flaxseed protein hydrolysates with a molecular weight above 1000 Da, while smaller peptides (128–1000 Da) mainly contributed to umami and bitter tastes and the development of a meat-like flavour.[152] The introduction of cysteine resulted in a more pronounced meaty flavour, but adding too much cysteine may lead to a sulphurous odour. The key components responsible for the meaty aroma included sulphur-containing volatiles, such as thiophenes, thiazoles, thiols, sulphur-substituted furans, and aliphatic sulphur compounds, and nitrogen-containing volatiles, such as pyrroles, pyrazines, and pyridines.[153] Adding chicken fat promoted the Maillard reaction of flaxseed protein hydrolysates, generating more aliphatic aldehydes and alcohols but less sulphur compounds such as thiols and thiophenes. The Maillard reaction products had a meaty and characteristic chicken flavour.[88] Ni *et al.* found that the Maillard reaction consumed 41 flaxseed peptides, particularly DLSFIP and ELPGSP, which accounted for 42.22% and 20.41% of total consumption, respectively. The formation of sulphur-containing volatiles depended on cysteine, while nitrogen-containing heterocycles were mainly derived from peptides. The following volatiles were identified as the major flavour compounds in the Maillard reaction products of flaxseed proteins: 2-methylthiophene, 2-methylfuran-3-thiol, furfural, 2-furfurylthiol, 3-thiophenethiol, thieno[3,2-*b*] thiophene, 2,5-thiophenedicarboxaldehyde, 2-methylthieno[2,3-*b*] thiophene, 1-(2-methyl-3-furylthio)-ethanethiol, 2-methylthieno[3,2-*b*] thiophene, and bis(2-methyl-3-furyl)-disulphide.[83]

10.7.4 Flavour of Hempseed Proteins

Hemp (*Cannabis sativa*) is a versatile crop that has been cultivated for thousands of years to meet various utilitarian needs. The two most notable uses of the plant are to produce fibre and psychoactive content. To a lesser extent, hempseeds are used as oilseeds. It should be noted that hempseeds are not true seeds but achenes of the plant.[154] These small fruits contain approximately 25–30% of lipid, 25–30% of protein, 30–40% of fibre, and a myriad of bioactive phytochemicals. Hempseed oil predominantly comprises unsaturated fatty acids (> 90%) with a desirable ω-6/ω-3 balance (2.5–3.5/1).[155] Hempseed proteins have two major fractions: an 11S globulin (edestin) that accounts for 60–80% of the total protein content and an albumin.[156] The amino acid profile of hempseed proteins is comparable to that of other high-quality proteins, such as egg white, casein, and soy proteins.[155,156] Due to their excellent nutritional values, the demand for hempseed protein is on the rise.

Recently, the flavour profile of hempseed protein has been characterized.[157] Among the 36 volatile compounds identified, four were terpenes, namely β-caryophyllene, β-myrcene, α-farnesene, and D-limonene (see Scheme 10.3). Terpenes are molecules that give *Cannabis* plants their classic odour and are produced in the glandular trichomes on the leaves and floral parts. These secretory epidermal glands are absent on the achenes, and their presence in the hempseed protein was likely due to contamination.[154] Among

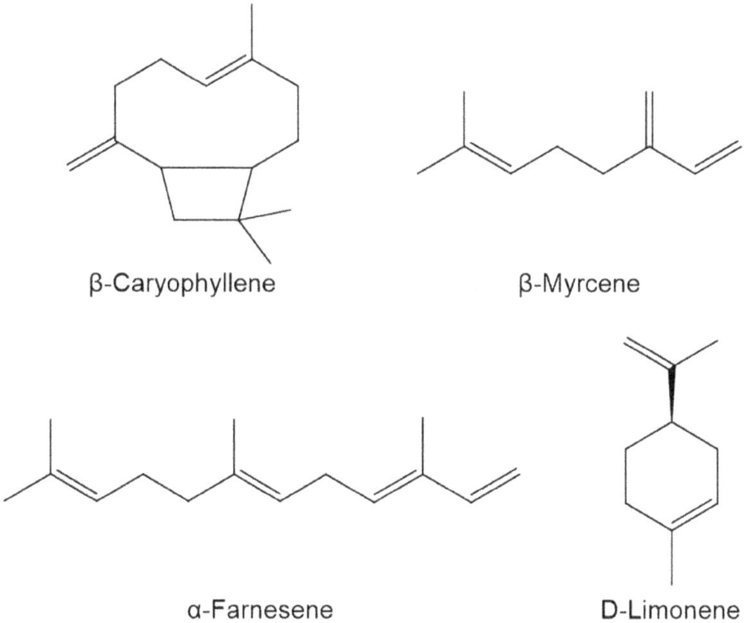

Scheme 10.3 Terpenes in hempseed protein.

the terpenes, D-limonene has a citrus odour; β-myrcene has an earthy, fruity, and clove-like odour; α-farnesene has a vegetable and flowery odour; and β-caryophyllene has a woody, spicy, and tenacious odour.

In addition to terpenes, a variety of aldehydes, alcohols, acids, and ketones derived from lipid oxidation were detected in hempseed protein isolate, among which 1-hexanol, hexanal, and (*E*)-2-octenal are the most abundant volatiles. Due to their high polyunsaturated fatty acid content, hempseeds are prone to lipid oxidation. Although most lipids are removed during protein isolation, these secondary oxidation products can bind to hempseed proteins through hydrophobic interactions, ionic bonding, and covalent linkages.[157] Dehulling has a profound influence on the flavour of hempseed protein. In general, dehulling increased the terpene content, particularly D-limonene and β-myrcene, and decreased lipid oxidation products such as hexanal and nonanal. This was likely due to the altered chemical composition (*e.g.*, less lipid and carbohydrate) and flavour-binding capacity of the hempseed protein isolate after the dehulling process.[157]

10.7.5 Flavour of Mustard Seed Proteins

Oilseeds from Brassicaceae often possess a strong flavour. The flavour components, such as isothiocyanates from glucosinolate breakdown and phenolics, may bind to the proteins and be carried into the isolates.

Thus, additional procedures, such as membrane processing, are needed to remove undesirable components.[158] Marnoch and Diosady reported the preparation of protein isolate from Oriental mustard seed [*Brassica juncea* (L.) Czern.] using alkaline extraction followed by ultrafiltration, diafiltration, and acid precipitation.[159] The mustard seed protein isolate contained 96% crude protein, no detectable glucosinolates or phytates, and 168 mg/100 g phenolics. It had little flavour and a somewhat astringent taste, possibly due to the residual phenolic compounds.

10.7.6 Flavour of Sesame Proteins

Sesame (*Sesamum indicum*) seed is an important oilseed that contains 50% oil and 25% protein. Dehulling sesame is important because the hull contains 2–3% oxalic acid, which chelates calcium and elicits a bitter taste. Dehulled and defatted sesame meal contains 60% protein and has a bland taste. Sesame proteins are heat stable and contain large quantities of methionine but a limited amount of lysine.[160] Although sesame proteins have poor solubility, they exhibit superior surface properties.[161] Rivas *et al.* prepared sesame protein isolates using both alkaline extraction and salt extraction.[162] The resulting isolates contained approximately 95% protein with a bland smell and taste; these isolates were also free of the nutty sesame flavour. Laohakunjit *et al.* produced sesame protein hydrolysates using bromelain and Flavourzyme.[91] Bromelain hydrolysis resulted in higher levels of free and sweet amino acids, while Flavourzyme hydrolysis produced more umami and bitter amino acids. The major volatiles in Flavourzyme-hydrolysed sesame proteins were 2-phenylethanal, sorbic acid, (*E*)-geranyl acetone, and benzaldehyde, while those in bromelain-produced hydrolysates were 2-phenylethanal, 9-hexadecenoic acid, and benzaldehyde. Both hydrolysates registered a chocolate note. The bromelain-based hydrolysates were mixed with valine, leucine, glucose, and tannic acid and heated to further generate a processed chocolate flavour through the Maillard reaction (Table 10.3). The following volatiles were found to be responsible for the chocolate-like aroma: 3-methyl-1-butanol, furfural, 2-ethyl-1-hexanol, ionone, 1-dodecanol, eugenol, and vinylguaiacol. This study demonstrated that sesame protein hydrolysate is a suitable precursor for developing chocolate flavour.

10.7.7 Flavour of Sunflower Seed Protein Hydrolysates

Several studies have used sunflower seed protein hydrolysates to generate flavours through the Maillard reaction. The Maillard reaction products of sunflower protein hydrolysates registered greater mouthfullness and taste continuity than those prepared from soy protein hydrolysates.[163] This was largely attributed to the size of the resulting peptides. Sunflower proteins hydrolysed by Alcalase and Flavourzyme generated more peptides with a molecular weight of 1000 to 5000 Da than soy proteins. The Maillard peptides in this molecular weight range had strong kokumi effects,[164] while those above or below this range may exert negative impacts on sensory attributes.[165]

10.8 Flavour of Tuber Proteins and Peptides

The global production of tubers, including potatoes, cassava, sweet potatoes, and yams, totalled 826 million metric tonnes in 2020.[139] This successful large-scale production of tubers highlights their frequent use as a key non-grain food commodity, especially in parts of the developing world.[166] The high consumption rates of tubers can be attributed to their high productivity, affordability, ease of access, and nutritive values. While tubers are known to be carbohydrate-rich and vitamin- and mineral-dense, the average protein concentration is estimated to be quite low (<2%).[167] Despite this low concentration, many tuber proteins are versatile in their amino acid composition and functionality and, thus, have the potential to serve as functional ingredients in the food industry.[168–170]

About 159 volatiles have been identified in raw potatoes, which are mainly derived from oxidative and enzymatic reactions.[171] Pęzksa *et al.* analysed flavour components of different potato protein preparations and identified 41 volatiles, including aldehydes, alcohols, esters, and alkanes.[172] Among these volatile compounds, 2-pentylfuran was the predominant odorant with an unpleasant, green, and beany smell. The production of 2-pentylfuran drastically decreased when calcium lactate was used to remove pectins from the potato juice. While the use of antioxidants and calcium salts, in general, reduced the major off-flavour hexanal (green, grassy), its concentration was the lowest when ascorbic acid (as an antioxidant agent) and calcium hydrogen phosphate were used in combination. All treatments resulted in an increased odour activity value of 2-methylbutanal, a desirable volatile with a musty, chocolate, nutty, and malted aroma.

Tanger *et al.* used potato protein microparticles as a fat replacer in a model milk dessert.[173] Incorporating the potato protein microparticles increased undesirable bitter and astringent notes and decreased the creaminess and mouthfullness of the dessert. In addition, the potato proteins introduced rancid, grassy, green, and beany off-flavours. The bitter, astringent off-taste and intense rancidity in potato protein-containing desserts were attributed to increased concentrations of free fatty acids, such as butyric, pentanoic, hexanoic, decanoic, dodecanoic, and tetradecanoic acids. Patatin, the major protein in potatoes, is a lipase and likely responsible for the generation of these free fatty acids.[174]

Potato and sweet potato proteins have also been used to prepare protein hydrolysates and Maillard reaction products. The hydrolysis of potato proteins into smaller peptides allows for a great improvement in their functional properties and bioactivities, yet this hydrolysis method can be limiting because of the development of undesirable flavours. Potato proteins are rich in hydrophobic amino acids and tend to produce bitter peptides when hydrolysed.[175] The evident bitterness associated with potato protein hydrolysates makes it difficult to incorporate these compounds into food products. The bitterness of potato peptides was related to their molecular weights; peptides above 6000 Da did not possess the bitterness the smaller peptides had. The addition of gelatin to the hydrolysis reaction decreased the occurrence of peptides smaller than 6000 Da, which resulted in a decreased bitter off-taste.

Potato and sweet potato protein hydrolysates have also been used to generate aroma compounds through the Maillard reaction (Table 10.3). Habinshuti *et al.* reacted potato and sweet potato protein hydrolysates with different sugars and cysteines. The Maillard reaction increased the umami, sweet, and sour taste of the hydrolysates while decreasing the bitterness; this was likely due to the reduction in bitter amino acids and peptides.[84,85,89] Over 170 volatile compounds were identified, including acids, alcohols, aldehydes, alkanes, alkenes, esters, furans, imines, ketones, pyrazoles, and thiophenes. Thiophenes, such as 3-methylthiophene-2-carbaldehyde, were found to be associated with the umami perception, most likely due to their meaty and fish-like notes.[84,89]

10.9 Flavour of Cultured Meat

Cultured meat, also known as *in vitro*, lab-grown, synthetic, artificial, or clean meat, is a promising alternative to conventional meat products due to its slaughter-free harvest, environmental sustainability, and clean production.[176,177] Cultured meat is produced from animal stem cells by inducing them to proliferate, differentiate into muscle cells, and grow on scaffolds or microcarriers in a bioreactor.[176,177] Little information is available regarding the flavour of cultured meat. Benjaminson *et al.* cultured the dorsal abdominal skeletal muscle of goldfish to assess the feasibility of the technology for the production of space food.[178] The cultured fish muscle mass was marinated, breaded, deep-fried, and evaluated for odour and appearance by a panel of four. The cooked product was described as glistening, firm, and odourless. In 2013, the world's first cultured beef burger was cooked and tasted by a panel of three in a press conference.[179] To simulate the texture and colour of a conventional meat patty, breadcrumbs, binders, beetroot juice, saffron, and caramel were added to the cultured burger, and it was sizzled with butter and oil. The burger registered a familiar mouthfeel but carried no meaty flavour. Recently, Ben-Arye *et al.* cultured bovine skeletal muscle tissue using textured soy protein as a scaffold.[180] Three volunteers tasted the cooked product and noted a pleasant meaty flavour and sensory attributes.

Scalable meat culturing is still a great technical challenge. It took approximately 3 months and $330 000 to produce the 140-g cultured burger,[179,181] which explains the limited information on the sensorial characteristics of cultured meat. Mimicking the flavour of traditional meat products would be another complication. Many flavours and flavour precursors in muscle are derived from feed and not formed in the muscle.[182] These compounds would be missing in cultured meat unless they were added to the medium and accumulated by the cells. Moreover, it is unclear whether cultured cells can undergo a similar post-mortem metabolism as muscle tissue, a process that generates tastants such as lactic acid and 5′-inosine monophosphate (IMP). IMP is also a source of Maillard reaction precursors such as ribose and ribose 5-phosphate.[182] The Maillard reaction and associated Strecker degradation are important routes to the formation of volatiles in cooked meat, such as furfural and furanone derivatives, dicarbonyls, aldehydes, ketones, thiols, hydrogen sulphide, and ammonia.[183] Lipid degradation upon cooking is another reaction that produces a myriad of volatiles, including aldehydes, lactones, ketones, alcohols, furans, hydrocarbons, acids, and esters.[183,184] In addition, heterocyclic compounds such as pyridines, pyrroles, pyrazines, thiazoles, thiazolines, thiophenes, trithiolanes, and oxazoles are formed through the interaction between lipid oxidation products and the Maillard reaction products.[183] These volatiles collectively contribute to the meat's flavour. Due to the lack of fat, the flavour of cultured meat may not resemble that of conventional meat products. Adding fat to the final product or co-culturing the muscle cells with adipocytes may help to improve the flavour.[184] Myoglobin is another component missing in cultured meat, which deprives the meat of its red colour. Adding myoglobin to bovine myosatellite cells has shown improved cell proliferation and tissue colour.[185] The iron-complexed haemoglobin in myoglobin also plays an important role in catalysing the development of meat flavours upon cooking.[186] Therefore, an iron- or haem-fortified medium may also be needed for flavour enhancement.[176,182]

10.10 Flavour Formation Catalysed by Leghaemoglobin

Meat production and consumption have been associated with environmental problems and public health concerns.[10] The search for sustainable alternatives has led to the development of many meat substitutes. However, these products often differ considerably in flavour and texture from meat and cannot accomplish the same satisfaction as meat consumption. Therefore, they have only limited market success among vegetarian or vegan consumers.[187,188] To attract a broader consumer base and accelerate the shift away from animal products, meat analogues are formulated to

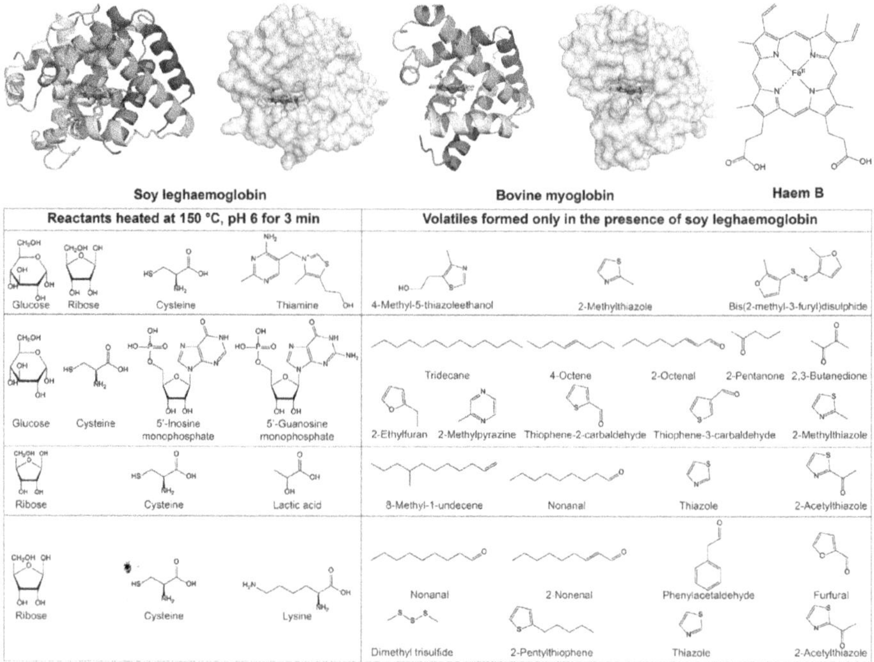

Figure 10.2 Soy leghaemoglobin, bovine myoglobin, haem B group, and soy leghaemoglobin catalysed flavour formation.

mimic not only the nutritional profiles but also the sensorial characteristics of the meat. One such example is the Impossible Burger, a plant protein-based product that incorporates soy leghaemoglobin to imitate the colour and flavour of meat. Leghaemoglobins are haemoproteins found in the nodules of leguminous plants. They play a crucial role in buffering the oxygen concentration in the nodules to ensure proper nitrogen fixation.[189] Leghaemoglobins have a similar structure to myoglobin, the major pigment in meat (see Figure 10.2), and impart a signature meat colour when added to the burger. This protein is mass produced by fermentation from genetically modified *Pichia pastoris* Bg11 yeasts that carry the recombinant gene for soy leghaemoglobin. The haem biosynthesis pathway of the yeast is upregulated to generate sufficient haem-bound leghaemoglobin.[190]

Besides its ability to imitate meat's colour, leghaemoglobin is also an important catalyst for the generation of a meat-like aroma. The incorporation of soy leghaemoglobin into a vegan burger recipe resulted in a beefier, bloodier, more savoury, and overall preferred product.[186] These flavours are likely formed *via* the same routes that occur in cooked meat, namely the Maillard reaction, lipid degradation, and a combination of both.[183] Several experiments were detailed in the patent for Impossible Burger.[186] In a model system where cysteine was heated with glucose or ribose, different flavour compounds were formed in the absence and presence of leghaemoglobin. Adding thiamine, nucleotides (IMP and 5′-guanosine monophosphate), lactic acid, and lysine to the reaction mixture generated additional volatiles with beefy, meaty, and savoury notes. These compounds were only formed in the presence of leghaemoglobin (see Figure 10.2). Introducing lipids to the leghaemoglobin-Maillard reaction system produced fatty, nutty, and roasted aromas.[186] The haem in leghaemoglobin is a well-known catalyst of lipid oxidation and likely promoted the oxidative breakdown of lipids and the interaction between lipid degradation and the Maillard reaction.[183] Interestingly, adding different types of haemoproteins to a reaction mixture resulted in distinct volatile profiles, which could be further explored for flavour development.

10.11 Concluding Remarks

To meet the ever-growing protein demand without disrupting the already stressed ecosystem, considerable attention has been paid to the development of novel, sustainable protein sources. However, little research has been performed to characterize the flavour profiles of these new protein ingredients. To promote the wider acceptance and utilization of novel proteins, more investigations are warranted to identify their major flavour components, understand how they interact with other food ingredients, and characterize their changes during processing.

Abbreviations

DHS	Dynamic headspace
FAO	Food and Agriculture Organization
IMP	5′-inosine monophosphate
RuBisCO	Ribulose-1,5-bisphosphate carboxylase-oxygenase
SAFE	Solvent-assisted flavour evaporation
SDE	Simultaneous distillation extraction
SPME	Solid-phase microextraction
WHO	World Health Organization

References

1. D. Tilman, C. Balzer, J. Hill and B. L. Befort, *Proc. Natl. Acad. Sci.*, 2011, **108**, 20260–20264.
2. D. K. Ray, P. C. West, M. Clark, J. S. Gerber, A. V. Prishchepov and S. Chatterjee, *PLoS One*, 2019, **14**, e0217148.
3. M. R. Smith and S. S. Myer, *Nat. Clim. Change*, 2018, **8**, 834–839.

4. D. Laborde, W. Martin, J. Swinnen and R. Vos, *Sci.*, 2020, **369**, 500–502.
5. FAO, *Food Outlook - Biannual Report on Global Food Markets: June 2020*, Rome, Italy, 2020.
6. J. R. Rohr, C. B. Barrett, D. J. Civitello, M. E. Craft, B. Delius, G. A. DeLeo, P. J. Hudson, N. Jouanard, K. H. Nguyen and R. S. Ostfeld, *Nat. Sustain.*, 2019, **2**, 445–456.
7. W. Willett, J. Rockström, B. Loken, M. Springmann, T. Lang, S. Vermeulen, T. Garnett, D. Tilman, F. DeClerck and A. Wood, *The Lancet*, 2019, **393**, 447–492.
8. T. Searchinger, R. Waite, C. Hanson, J. Ranganathan, P. Dumas, E. Matthews and C. Klirs, *Creating a sustainable food future: a menu of solutions to feed nearly 10 billion people by 2050*, World Resources Institute, 2019.
9. J. Poore and T. Nemecek, *Sci.*, 2018, **360**, 987–992.
10. H. C. J. Godfray, P. Aveyard, T. Garnett, J. W. Hall, T. J. Key, J. Lorimer, R. T. Pierrehumbert, P. Scarborough, M. Springmann and S. A. Jebb, *Sci.*, 2018, **361**, eaam5324.
11. S. M. Loveday, *Annu. Rev. Food Sci. Technol.*, 2019, **10**, 311–339.
12. M. Onwezen, E. Bouwman, M. Reinders and H. Dagevos, *Appetite*, 2021, **159**, 105058.
13. S. Berger, C. Bärtsch, C. Schmidt, F. Christandl and A. M. Wyss, *Front. Nutr.*, 2018, **5**, 88.
14. L. Barsanti and P. Gualtieri, in *Algae: Anatomy, Biochemistry, and Biotechnology*, CRC Press, Boca Raton, USA, 2014.
15. Y. Torres-Tiji, F. J. Fields and S. P. Mayfield, *Biotechnol. Adv.*, 2020, **41**, 107536.
16. S. Ścieszka and E. Klewicka, *Crit. Rev. Food Sci. Nutr.*, 2019, **59**, 3538–3547.
17. FAO, *The State of World Fisheries and Aquaculture 2020. Sustainability in Action*, Food and Agriculture Organization of the United Nations, Rome, 2020.
18. S. Bleakley and M. Hayes, *Foods*, 2017, **6**, 33.
19. C. Dawczynski, R. Schubert and G. Jahreis, *Food Chem.*, 2007, **103**, 891–899.
20. M. L. Amorim, J. Soares, J. S. D. R. Coimbra, M. de O. Leite, L. F. T. Albino and M. A. Martins, *Crit. Rev. Food Sci. Nutr.*, 2020, **61**, 1976–2002.
21. N. Francezon, A. Tremblay, J.-L. Mouget, P. Pasetto and L. Beaulieu, *J. Agric. Food Chem.*, 2021, **69**, 11753–11772.
22. L.-E. Rioux, L. Beaulieu and S. L. Turgeon, *Food Hydrocolloids*, 2017, **68**, 255–265.

23. E. G. Vilar, M. G. O'Sullivan, J. P. Kerry and K. N. Kilcawley, *Algal Res.*, 2020, **45**, 101740.
24. J. Van Durme, K. Goiris, A. De Winne, L. De Cooman and K. Muylaert, *J. Agric. Food Chem.*, 2013, **61**, 10881–10890.
25. L. Zhou, J. Chen, J. Xu, Y. Li, C. Zhou and X. Yan, *J. Sci. Food Agric.*, 2017, **97**, 761–769.
26. M. I. Hosoglu, *Food Chem.*, 2018, **240**, 1210–1218.
27. S. B. Watson, P. Monis, P. Baker and S. Giglio, *Harmful Algae*, 2016, **54**, 112–127.
28. I. Milovanović, A. Mišan, J. Simeunović, D. Kovač, D. Jambrec and A. Mandić, *J. Chem.*, 2015, **2015**, 969542.
29. P. Geada, C. Moreira, M. Silva, R. Nunes, L. Madureira, C. M. Rocha, R. N. Pereira, A. A. Vicente and J. A. Teixeira, *Bioresour. Technol.*, 2021, **332**, 125125.
30. T. Lafarga, *Algal Res.*, 2019, **41**, 101566.
31. J. Wang, M. Zhang and Z. Fang, *Trends Food Sci. Technol.*, 2019, **88**, 251–259.
32. W. M. Qazi, S. Ballance, A. K. Uhlen, K. Kousoulaki, J.-E. Haugen and A. Rieder, *LWT-Food Sci. Technol.*, 2021, **143**, 111115.
33. L. M. R. Almeida, L. F. da Silva Cruz, B. A. S. Machado, I. L. Nunes, J. A. V. Costa, E. de Souza Ferreira, P. V. F. Lemos, J. I. Druzian and C. O.de Souza, *Algal Res.*, 2021, **58**, 102387.
34. B. Coleman, C. Van Poucke, B. Dewitte, A. Ruttens, T. Moerdijk-Poortvliet, C. Latsos, K. De Reu, L. Blommaert, B. Duquenne and K. Timmermans, *Future Foods*, 2022, **5**, 100139.
35. L. Grossmann, V. Wörner, J. Hinrichs and J. Weiss, *J. Sci. Food Agric.*, 2020, **100**, 1344–1349.
36. S. Zamuz, L. Purriños, F. Galvez, N. Zdolec, V. Muchenje, F. J. Barba and J. M. Lorenzo, *J. Food Process. Preserv.*, 2019, **43**, e13940.
37. C. Fitzgerald, E. Gallagher, L. Doran, M. Auty, J. Prieto and M. Hayes, *LWT-Food Sci. Technol.*, 2014, **56**, 398–405.
38. N. Laohakunjit, O. Selamassakul and O. Kerdchoechuen, *Food Chem.*, 2014, **158**, 162–170.
39. H. Qi, Z. Xu, Y.-b. Li, X.-l. Ji, X.-f. Dong and C.-x. Yu, *Int. J. Food Prop.*, 2017, **20**, 2867–2876.
40. X. Zhang, D. Jiang, D. Li, C. Yu, X. Dong and H. Qi, *J. Food Meas. Charact.*, 2019, **13**, 1185–1194.
41. M.-A. Le Pape, J. Grua-Priol, C. Prost and M. Demaimay, *J. Agric. Food Chem.*, 2004, **52**, 550–556.
42. M. Amin, M. Rizky Distiawan, M. Herlina, R. Cholifatun, F. Rohim, C. Anggraeni, M. Alamsjah, T. Wahju and S. Sri, *Iran. J. Fish. Sci.*, 2021, **20**, 1097–1113.

43. A. Ritala, S. T. Häkkinen, M. Toivari and M. G. Wiebe, *Front. Microbiol.*, 2017, **8**, 2009.
44. M. Spiller, M. Muys, G. Papini, M. Sakarika, M. Buyle and S. E. Vlaeminck, *Water Res.*, 2020, **171**, 115406.
45. D. B. Lopes, J. V. M. Júnior, L. V. de Castro Reis, K. M. M. Leão and G. A. Macedo, in *Microbial production of food ingredients and additives*, ed. A. M. Holban and A. M. Grumezescu, Elsevier, 2017, pp. 1–32.
46. M. G. Pessôa, B. N. Paulino, G. Molina and G. M. Pastore, in *Bioprocessing for Biomolecules Production*, ed. V. K. Gupta, B. N. Singh, N. Gathergood and G. Molina, Wiley, 2019pp. 93–123.
47. M. Sakarika, D. A. Tzompa-Sosa, M. Depoortere, H. Rottiers, R. Ganigué, K. Dewettinck and K. Rabaey, *FEMS Microbiol. Lett.*, 2020, **367**, fnaa138.
48. S.-T. Chang and P. G. Miles, in *Mushrooms: Cultivation, nutritional value, medicinal effect, and environmental impact*, CRC press, Boca Raton, USA, 2004.
49. P. J. Punt, N. van Biezen, A. Conesa, A. Albers, J. Mangnus and C. van den Hondel, *Trends Biotechnol.*, 2002, **20**, 200–206.
50. S. Ghorai, S. P. Banik, D. Verma, S. Chowdhury, S. Mukherjee and S. Khowala, *Food Res. Int.*, 2009, **42**, 577–587.
51. A. González, M. Cruz, C. Losoya, C. Nobre, A. Loredo, R. Rodríguez, J. Contreras and R. Belmares, *Food Funct.*, 2020, **11**, 7400–7414.
52. X. Dai, M. Sharma and J. Chen, in *Fungi in sustainable food production*, Springer, Switzerland, 2021.
53. G. Bakratsas, A. Polydera, P. Katapodis and H. Stamatis, *Future Foods*, 2022, **4**, 100086.
54. S. S. Block, *J. Biochem. Microbiol. Technol. Eng.*, 1960, **2**, 243–252.
55. H. Rathore, S. Prasad, M. Kapri, A. Tiwari and S. Sharma, *J. Funct. Foods*, 2019, **56**, 182–193.
56. A. Lomascolo, C. Stentelaire, M. Asther and L. Lesage-Meessen, *Trends Biotechnol.*, 1999, **17**, 282–289.
57. F. Hashempour-Baltork, K. Khosravi-Darani, H. Hosseini, P. Farshi and S. F. S. Reihani, *J. Cleaner Prod.*, 2020, **253**, 119958.
58. J. E. Elzerman, A. C. Hoek, M. A. van Boekel and P. A. Luning, *Food Qual. Prefer.*, 2011, **22**, 233–240.
59. N. R. Rubio, N. Xiang and D. L. Kaplan, *Nat. Commun.*, 2020, **11**, 1–11.
60. L.-b. Sun, Z.-y. Zhang, G. Xin, B.-x. Sun, X.-j. Bao, Y.-y. Wei, X.-m. Zhao and H.-r. Xu, *Trends Food Sci. Technol.*, 2020, **96**, 176–187.
61. A. Pérez-Montes, E. Rangel-Vargas, J. M. Lorenzo, L. Romero and E. M. Santos, *Curr. Opin. Food Sci.*, 2021, **37**, 118–124.
62. A. Muniz and X. Du, *ACS Food Sci. Technol.*, 2021, **1**, 1629–1641.
63. K. Kurihara, *Biomed. Res. Int.*, 2015, **2015**, 189402.

64. Z. Yu, H. Jiang, R. Guo, B. Yang, G. You, M. Zhao and X. Liu, *Food Res. Int.*, 2018, **108**, 144–150.
65. A. Alim, H. Song and T. Zou, *Eur. Food. Res. Technol.*, 2020, **246**, 2119–2133.
66. J. Liu, H. Song, Y. Liu, P. Li, J. Yao and J. Xiong, *J. Sci. Food Agric.*, 2015, **95**, 3183–3194.
67. X. Xu, R. Xu, Z. Song, Q. Jia, T. Feng, M. Huang and S. Song, *J. Chromatogr. A*, 2019, **1596**, 96–103.
68. Y. Kong, L.-L. Zhang, J. Zhao, Y.-Y. Zhang, B.-G. Sun and H.-T. Chen, *Food Res. Int.*, 2019, **121**, 463–470.
69. Y. Zheng, L. Tang, M. Yu, T. Li, H. Song, P. Li, K. Li and J. Xiong, *J. Food Sci. Technol.*, 2021, **58**, 1199–1208.
70. A. Alim, C. Yang, H. Song, Y. Liu, T. Zou, Y. Zhang and S. Zhang, *Food Res. Int.*, 2019, **120**, 534–543.
71. W. Chen, W. Li, D. Wu, Z. Zhang, H. Chen, J. Zhang, C. Wang, T. Wu and Y. Yang, *J. Food Compos. Anal.*, 2022, **110**, 104530.
72. A. Alim, H. Song, C. Yang, Y. Liu, T. Zou, Y. Zhang and S. Zhang, *J.Sci. Food Agric.*, 2019, **99**, 4651–4658.
73. T. Feng, Y. Wu, Z. Zhang, S. Song, H. Zhuang, Z. Xu, L. Yao and M. Sun, *Foods*, 2019, **8**, 43.
74. A. Moore, C. R. Luckett and J. P. Munafo, *J. Agric. Food Chem.*, 2021, **69**, 11947–11959.
75. J. Zhou, M. Chen, S. Wu, X. Liao, J. Wang, Q. Wu, M. Zhuang and Y. Ding, *Food Res. Int.*, 2021, **134**, 109230.
76. M. M. Poojary, V. Orlien, P. Passamonti and K. Olsen, *Food Chem.*, 2017, **234**, 236–244.
77. S.-S. Ang and M. R. Ismail-Fitry, *Food Technol. Biotechnol.*, 2019, **57**, 472–480.
78. J. Gao, D. Fang, B. M. Kimatu, X. Chen, X. Wu, J. Du, Q. Yang, H. Chen, H. Zheng and X. An, *Food Chem.*, 2021, **362**, 130192.
79. S. N. Lotfy, H. H. Fadel, A. H. El-Ghorab and M. S. Shaheen, *Food Chem.*, 2015, **187**, 7–13.
80. J. Lopez, T. Kerley, L. Jenkinson, C. R. Luckett and J. P. Munafo, *J. Agric. Food Chem.*, 2019, **67**, 11444–11453.
81. X. Chen, J. Yu, H. Cui, S. Xia, X. Zhang and B. Yang, *Mol.*, 2018, **23**, 247.
82. K. K. Grossmann, M. Merz, D. Appel, M. M. De Araujo and L. Fischer, *Food Chem.*, 2021, **364**, 130336.
83. Z.-J. Ni, C.-K. Wei, A.-R. Zheng, K. Thakur, J.-G. Zhang and Z.-J. Wei, *Food Chem. X*, 2022, **13**, 100224.
84. I. Habinshuti, M. Zhang, H. N. Sun and T. H. Mu, *Int. J. Food Sci. Technol.*, 2021, **56**, 6086–6099.

85. I. Habinshuti, M. Zhang, H. N. Sun and T. H. Mu, *Int. J. Food Sci. Technol.*, 2022, **57**, 3652–3664.
86. A. Alim, H. Song, Y. Liu, T. Zou, Y. Zhang, S. Zhang and A. Raza, *LWT-Food Sci. Technol.*, 2019, **112**, 108239.
87. X. Guo, S. Tian and D. M. Small, *Food Chem.*, 2011, **119**, 167–172.
88. C.-K. Wei, Z.-J. Ni, K. Thakur, A.-M. Liao, J.-H. Huang and Z.-J. Wei, *Food Chem.*, 2020, **306**, 125560.
89. I. Habinshuti, T.-H. Mu and M. Zhang, *Food Chem.*, 2021, **361**, 130090.
90. A. Raza, H. Song, J. Raza, P. Li, K. Li and J. Yao, *Food Funct.*, 2020, **11**, 8583–8601.
91. N. Laohakunjit, O. Kerdchoechuen, R. Kaprasob and F. B. Matta, *J. Food Process. Preserv.*, 2017, **41**, e13075.
92. D. Tomé, *ACS Food Sci. Technol.*, 2021, **1**, 487–494.
93. P. Münch, T. Hofmann and P. Schieberle, *J. Agric. Food Chem.*, 1997, **45**, 1338–1344.
94. P. Münch and P. Schieberle, *J. Agric. Food Chem.*, 1998, **46**, 4695–4701.
95. K. Mahadevan and L. Farmer, *J. Agric. Food Chem.*, 2006, **54**, 7242–7250.
96. M. Lin, X. Liu, Q. Xu, H. Song, P. Li and J. Yao, *J. Sci. Food Agric.*, 2014, **94**, 882–889.
97. Y. Zhang, H. Song, P. Li, J. Yao and J. Xiong, *LWT-Food Sci. Technol.*, 2017, **82**, 184–191.
98. A. Alim, H. Song, Y. Liu, T. Zou, Y. Zhang and S. Zhang, *J. Sci. Food Agric.*, 2018, **98**, 3774–3783.
99. A. Raza, N. Begum, H. Song, K. Li and P. Li, *J. Food Sci.*, 2019, **84**, 2031–2041.
100. Z. Wang, Q. Xiao, J. Zhuang, T. Feng, C.-T. Ho and S. Song, *J. Agric. Food Chem.*, 2020, **68**, 267–278.
101. C. Liu and J. Zhao, in *Encyclopedia of Food Chemistry*, ed. L. Melton, F. Shahidi and P. Varelis, Elsevier, 2019, vol. **3**, pp. 428–436.
102. A. van Huis, B. Rumpold, C. Maya and N. Roos, *Annu. Rev. Nutr.*, 2021, **41**, 551–576.
103. C. Liu, J. Masri, V. Perez, C. Maya and J. Zhao, *Foods*, 2020, **9**, 151.
104. K. J. Hurd, S. Shertukde, T. Toia, A. Trujillo, R. L. Pérez, D. L. Larom, J. J. Love and C. Liu, *Ann. Entomol. Soc. Am.*, 2019, **112**, 552–559.
105. D. Raheem, C. Carrascosa, O. B. Oluwole, M. Nieuwland, A. Saraiva, R. Millán and A. Raposo, *Crit. Rev. Food Sci. Nutr.*, 2019, **59**, 2169–2188.
106. E. Woolf, Y. Zhu, K. Emory, J. Zhao and C. Liu, *LWT-Food Sci. Technol.*, 2019, **102**, 100–105.
107. S. Mancini, R. Moruzzo, F. Riccioli and G. Paci, *Food Res. Int.*, 2019, **122**, 661–678.

108. E. Woolf, C. Maya, J. Yoon, S. Shertukde, T. Toia, J. Zhao, Y. Zhu, P. C. Peter, C. Liu and P. Peter, *J. Insects Food Feed*, 2021, **7**, 129–139.
109. C. Liu and J. Zhao, *J. Insects Food Feed*, 2019, **5**, 241–242.
110. J. R. Aldrich, *Annu. Rev. Entomol.*, 1988, **33**, 211–238.
111. C. Perez-Santaescolastica, A. De Winne, J. Devaere and I. Fraeye, *Trends Food Sci. Technol.*, 2022, **127**, 352–367.
112. J. H. Lee, J.-Y. Cha, T.-K. Kim, Y.-S. Choi and H. W. Jang, *LWT-Food Sci. Technol.*, 2021, **151**, 112095.
113. X. Zhou, Y. Zheng, Y. Zhong, D. Wang, J. Xu, R. Liu and Y. Deng, *J. Food Process. Preserv.*, 2022, **46**, e16276.
114. R. J. Ellis, *Trends Biochem. Sci.*, 1979, **4**, 241–244.
115. N. Pirie, *Nat.*, 1942, **149**, 251.
116. E. Di Stefano, D. Agyei, E. N. Njoku and C. C. Udenigwe, *J. Am. Oil Chem. Soc.*, 2018, **95**, 1063–1074.
117. S. Pérez-Vila, M. A. Fenelon, J. A. O'Mahony and L. G. Gómez-Mascaraque, *Food Hydrocolloids*, 2022, **133**, 107902.
118. G. Zengin, A. Aktumsek, G.-O. Guler, Y.-S. Cakmak, J. Girón-Calle, M. Alaiz and J. Vioque, *Food Chem.*, 2012, **135**, 1360–1364.
119. M. P. Hojilla-Evangelista, G. W. Selling, R. Hatfield and M. Digman, *J. Sci. Food Agric.*, 2017, **97**, 882–888.
120. K.-J. Appenroth, K. S. Sree, M. Bog, J. Ecker, C. Seeliger, V. Böhm, S. Lorkowski, K. Sommer, W. Vetter and K. Tolzin-Banasch, *Front. Chem.*, 2018, **6**, 483.
121. M. Ducrocq, M.-H. Morel, M. Anton, V. Micard, S. Guyot, V. Beaumal, V. Solé-Jamault and A. Boire, *Food Chem.*, 2022, **381**, 132254.
122. A. H. Martin, O. Castellani, G. A. de Jong, L. Bovetto and C. Schmitt, *J. Sci. Food Agric.*, 2019, **99**, 1568–1576.
123. M. Nieuwland, P. Geerdink, N. P. Engelen-Smit, A. Fabisiak, I. M. Van Der Meer, A. H. America, J. J. Mes, A. M. J. Kooststra, J. T. Henket and W. J. Mulder, *ACS Food Sci. Technol.*, 2021, **1**, 908–916.
124. M. Ducrocq, A. Boire, M. Anton, V. Micard and M.-H. Morel, *Food Hydrocolloids*, 2020, **109**, 106101.
125. R. Fiorentini and C. Galoppini, *Plant Food Hum. Nutr.*, 1983, **32**, 335–350.
126. L. Firdaous, B. Fertin, O. Khelissa, M. Dhainaut, N. Nedjar, G. Chataigné, L. Ouhoud, F. Lutin and P. Dhulster, *Sep. Purif. Technol.*, 2017, **178**, 29–39.
127. USDA, *Oilseeds: World Markets and Trade (May 2020)*, 2020.
128. J. M. Bell, *Can. J. Anim. Sci.*, 1993, **73**, 689–697.
129. J. P. Wanasundara, *Crit. Rev. Food Sci. Nutr.*, 2011, **51**, 635–677.

130. C. Bos, G. Airinei, F. Mariotti, R. Benamouzig, S. Bérot, J. Evrard, E. Fénart, D. Tomé and C. Gaudichon, *J. Nutr.*, 2007, **137**, 594–600.
131. M. Naczk, R. Amarowicz and F. Shahidi, in *Developments in Food Science*, Elsevier, 1998, vol. **40**, pp. 597–613.
132. C. Hald, C. Dawid, R. Tressel and T. Hofmann, *J. Agric. Food Chem.*, 2018, **67**, 372–378.
133. Y. M. Tzeng, L. L. Diosady and L. J. Rubin, *J. Food Sci.*, 1990, **55**, 1147–1151.
134. L. Xu and L. Diosady, *J. Am. Oil Chem. Soc.*, 1994, **71**, 935.
135. L. Xu and L. Diosady, *Food Res. Int.*, 2000, **33**, 725–731.
136. K. Wang and S. D. Arntfield, *Food Chem.*, 2014, **157**, 364–372.
137. K. Wang and S. D. Arntfield, *Food Hydrocolloids*, 2015, **43**, 410–417.
138. J. Pokorný, W. Janitz, I. Víden, J. Velíšek, H. Valentová and J. Davídek, *Food/Nahrung*, 1987, **31**, 63–70.
139. FAO, *FAOSTAT*, 2020.
140. M. Kumar, M. Tomar, S. Punia, S. Grasso, F. Arrutia, J. Choudhary, S. Singh, P. Verma, A. Mahapatra and S. Patil, *Trends Food Sci. Technol.*, 2021, **111**, 100–113.
141. W. Withers and F. Carruth, *J. Agric. Res.*, 1915, **5**, 261–288.
142. L. C. Berardi and J. P. Cherry, *J. Food Sci.*, 1980, **45**, 377–380.
143. S. P. Fore, M. G. Legendre, J. P. Cherry, L. C. Berardi and C. H. Vinnett, *J. Food Sci.*, 1980, **45**, 912–917.
144. Y. Ziprin, K. Rhee, Z. Carpenter, R. Hostetler, R. Terrell and K. Rhee, *J. Food Sci.*, 1981, **46**, 58–61.
145. R. Terrell, R. Swasdee, P. Wan and E. Lusas, *J. Food Sci.*, 1981, **46**, 845–849.
146. J. Lawhon, N. Golightly and E. Lusas, *J. Am. Oil Chem. Soc.*, 1980, **57**, 302–306.
147. R. Simmons, J. Green, C. Payne, P. Wan and E. Lusas, *J. Food Sci.*, 1980, **45**, 1505–1508.
148. J. Lawhon, S. Lin, L. Rooney, C. Cater and K. Mattil, *J. Food Sci.*, 1974, **39**, 183–187.
149. B. D. Oomah, *J. Sci. Food Agric.*, 2001, **81**, 889–894.
150. B. Oomah and G. Mazza, *Food Chem.*, 1993, **48**, 109–114.
151. S. Wu, X. Wang, W. Qi and Q. Guo, *Trends Food Sci. Technol.*, 2019, **92**, 184–193.
152. C.-K. Wei, K. Thakur, D.-H. Liu, J.-G. Zhang and Z.-J. Wei, *Food Chem.*, 2018, **263**, 186–193.
153. C.-K. Wei, Z.-J. Ni, K. Thakur, A.-M. Liao, J.-H. Huang and Z.-J. Wei, *Int. J. Food Prop.*, 2019, **22**, 84–99.
154. E. Small, *Bot. Rev.*, 2015, **81**, 189–294.

155. W. Leonard, P. Zhang, D. Ying and Z. Fang, *Compr. Rev. Food Sci. Food Saf.*, 2020, **19**, 282–308.
156. Q. Wang and Y. L. Xiong, *Compr. Rev. Food Sci. Food Saf.*, 2019, **18**, 936–952.
157. P. Shen, Z. Gao, M. Xu, J.-B. Ohm, J. Rao and B. Chen, *Food Hydrocolloids*, 2020, 105889.
158. L. Xu, F. Lui, H. Luo and L. Diosady, *Food Res. Int.*, 2003, **36**, 849–856.
159. R. Marnoch and L. L. Diosady, *J. Amer. Oil Chem. Soc.*, 2006, **83**, 65–69.
160. L. Johnson, T. Suleiman and E. Lusas, *J. Am. Oil Chem. Soc.*, 1979, **56**, 463–468.
161. A. Achouri, V. Nail and J. I. Boye, *Food Res. Int.*, 2012, **46**, 360–369.
162. N. Rivas R., J. E. Dench and J. C. Caygill, *J. Sci. Food Agric.*, 1981, **32**, 565–571.
163. K. Eric, L. V. Raymond, M. Huang, M. J. Cheserek, N. D. Hayat, R. Savio, M. Amédée and X. Zhang, *Food Res. Int.*, 2013, **54**, 1437–1447.
164. M. Ogasawara, T. Katsumata and M. Egi, *Food Chem.*, 2006, **99**, 600–604.
165. K. Eric, L. V. Raymond, S. Abbas, S. Song, Y. Zhang, K. Masamba and X. Zhang, *Food Res. Int.*, 2014, **57**, 242–258.
166. G. J. Scott, *Int. J. Food Sci. Technol.*, 2021, **56**, 1093–1114.
167. A. Chandrasekara and T. Josheph Kumar, *Int. J. Food Sci.*, 2016, **2016**, 3631647.
168. T.-H. Mu, S.-S. Tan and Y.-L. Xue, *Food Chem.*, 2009, **112**, 1002–1005.
169. V. Bártová, J. Bárta, A. Brabcová, Z. Zdráhal and V. Horáčková, *J. Food Compos. Anal.*, 2015, **40**, 78–85.
170. Y. Fu, W. N. Liu and O. P. Soladoye, *Int. J. Food Sci. Technol.*, 2020, **55**, 2314–2322.
171. J. F. Dresow and H. Böhm, *Landbauforschung-vTI Agriculture and Forestry Research*, 2009, **59**, 309–338.
172. A. Pęksa, J. Miedzianka, A. Szumny, J. Łyczko, A. Nemś and A. Kita, *Int. J. Food Sci. Technol.*, 2020, **55**, 2323–2334.
173. C. Tanger, F. Utz, A. Spaccasassi, J. Kreissl, J. Dombrowski, C. Dawid and U. Kulozik, *ACS Food Sci. Technol.*, 2021, **2**, 169–179.
174. R. E. Spelbrink, H. Lensing, M. R. Egmond and M. L. Giuseppin, *Appl. Biochem. Biotechnol.*, 2015, **176**, 231–243.
175. K.Ney, *J. Am. Oil Chem. Soc.*, 1979, **56**, 295–297.
176. Z. F. Bhat, J. D. Morton, S. L. Mason, A. E.-D. A. Bekhit and H. F. Bhat, *Compr. Rev. Food Sci. Food Saf.*, 2019, **18**, 1192–1208.
177. G. Zhang, X. Zhao, X. Li, G. Du, J. Zhou and J. Chen, *Trends Food Sci. Technol.*, 2020, **97**, 443–450.

178. M. A. Benjaminson, J. A. Gilchriest and M. Lorenz, *Acta. Astronaut.*, 2002, **51**, 879–889.
179. K. Kupferschmidt, *Sci.*, 2013, **341**, 602–603.
180. T. Ben-Arye, Y. Shandalov, S. Ben-Shaul, S. Landau, Y. Zagury, I. Ianovici, N. Lavon and S. Levenberg, *Nat. Food*, 2020, **1**, 210–220.
181. M. J. Post, *J. Sci. Food Agric.*, 2014, **94**, 1039–1041.
182. I. Fraeye, M. Kratka, H. Vandenburgh and L. Thorrez, *Front. Nutr.*, 2020, **7**, 35.
183. F. Shahidi, in *Flavor of Meat, Meat Products and Seafoods,* Blackie Academic and Professional, London, UK, 2nd edn, 1998.
184. K. D. Fish, N. R. Rubio, A. J. Stout, J. S. Yuen and D. L. Kaplan, *Trends Food Sci. Technol.*, 2020, **98**, 53–67.
185. R. Simsa, J. Yuen, A. Stout, N. Rubio, P. Fogelstrand and D. L. Kaplan, *Foods*, 2019, **8**, 521.
186. *U.S. Pat.*, 10,314,325, 2019.
187. O. Malav, S. Talukder, P. Gokulakrishnan and S. Chand, *Crit. Rev. Food Sci. Nutr.*, 2015, **55**, 1241–1245.
188. F. B. Hu, B. O. Otis and G. McCarthy, *JAMA*, 2019, **322**, 1547–1548.
189. S. Singh and A. Varma, in *Rhizobium Biology and Biotechnology*, Springer, 2017, pp. 309–330.
190. FDA, in *GRAS Notice no.737: Soy leghemoglobin preparation from a strain of Pichia pastoris*, 2018.

Subject Index

www.ingramcontent.com/pod-product-compliance
Ingram Content Group UK Ltd.
Pitfield, Milton Keynes, MK11 3LW, UK
UKHW051136291224
453001UK00030B/60